Practical Applications of Bayesian Reliability

Wiley Series in Quality & Reliability Engineering

Dr. Andre Kleyner

Series Editor

The Wiley series in Quality & Reliability Engineering aims to provide a solid educational foundation for both practitioners and researchers in Q&R field and to expand the reader's knowledge base to include the latest developments in this field. The series will provide a lasting and positive contribution to the teaching and practice of engineering.
The series coverage will contain, but is not exclusive to,

- statistical methods;
- physics of failure;
- reliability modeling;
- functional safety;
- six-sigma methods;
- lead-free electronics;
- warranty analysis/management; and
- risk and safety analysis.

Wiley Series in Quality & Reliability Engineering

Prognostics and Health Management: A Practical Approach to Improving System Reliability Using Conditioned-Based Data
by Douglas Goodman, James P. Hofmeister, Ferenc Szidarovszky
April 2019

Reliability and Service Engineering
by Tongdan Jin
February 2019

Dynamic System Reliability: Modelling and Analysis of Dynamic and Dependent Behaviors
by Liudong Xing, Gregory Levitin, Chaonan Wang
February 2019

Thermodynamic Degradation Science: Physics of Failure, Accelerated Testing, Fatigue and Reliability Applications
by Alec Feinberg
October 2016

Next Generation HALT and HASS: Robust Design of Electronics and Systems
by Kirk A. Gray, John J. Paschkewitz
May 2016

Reliability and Risk Models: Setting Reliability Requirements, 2nd Edition
by Michael Todinov
September 2015

Applied Reliability Engineering and Risk Analysis: Probabilistic Models and Statistical Inference
by Ilia B. Frenkel, Alex Karagrigoriou, Anatoly Lisnianski, Andre V. Kleyner
September 2013

Design for Reliability
by Dev G. Raheja (Editor), Louis J. Gullo (Editor)
July 2012

Effective FMEAs: Achieving Safe, Reliable, and Economical Products and Processes using Failure Mode and Effects Analysis
by Carl Carlson
April 2012

Failure Analysis: A Practical Guide for Manufacturers of Electronic Components and Systems
by Marius Bazu, Titu Bajenescu
April 2011

Reliability Technology: Principles and Practice of Failure Prevention in Electronic Systems
by Norman Pascoe
April 2011

Improving Product Reliability: Strategies and Implementation
by Mark A. Levin, Ted T. Kalal
March 2003

Test Engineering: A Concise Guide to Cost-effective Design, Development and Manufacture
by Patrick O'Connor
April 2001

Integrated Circuit Failure Analysis: A Guide to Preparation Techniques
by Friedrich Beck
January 1998

Measurement and Calibration Requirements for Quality Assurance to ISO 9000
by Alan S. Morris
October 1997

Electronic Component Reliability: Fundamentals, Modelling, Evaluation, and Assurance
by Finn Jensen
November 1995

Practical Applications of Bayesian Reliability

Yan Liu
Principal Reliability Engineer
Medtronic PLC, Minneapolis, USA

Athula I. Abeyratne
Senior Principal Statistician
Medtronic PLC, Minneapolis, USA

WILEY

This edition first published 2019

Registered Offices
John Wiley & Sons, Inc., 111 River Street, Hoboken, NJ 07030, USA
John Wiley & Sons Ltd, The Atrium, Southern Gate, Chichester, West Sussex, PO19 8SQ, UK

Editorial Office
The Atrium, Southern Gate, Chichester, West Sussex, PO19 8SQ, UK

For details of our global editorial offices, customer services, and more information about Wiley products visit us at www.wiley.com.

Wiley also publishes its books in a variety of electronic formats and by print-on-demand. Some content that appears in standard print versions of this book may not be available in other formats.

Library of Congress Cataloging-in-Publication Data applied for

ISBN: 9781119287971

Cover Design: Wiley
Cover Image: © SergeyBitos / Shutterstock

Set in 10/12pt WarnockPro by SPi Global, Chennai, India

Printed in Great Britain by TJ International Ltd, Padstow, Cornwall

10 9 8 7 6 5 4 3 2 1

Contents

Preface

Recently, groundbreaking work using Bayesian statistics for reliability analysis has emerged at various seminars and technical conferences which demonstrated great power in accurately predicting reliability and/or reducing sample size. Many engineers and scientists have expressed interest in learning Bayesian statistics. However, there is also much confusion in the learning process. This confusion comes mainly from three aspects:

1) What is Bayesian analysis exactly?
2) What are the benefits?
3) How can I apply the methods to solve my own problems?

This book is intended to provide basic knowledge and practical examples of Bayesian modeling in reliability and related science and engineering practices. We hope it will help engineers and scientists to find answers to the above common questions.

For scientists and engineers with no programming experience, coding is often considered too daunting. To help readers get started quickly, many Bayesian models using Just Another Gibbs Sampler (JAGS) are provided in this book (e.g. `3.4_Weibull.JAGS` in Section 3.4) containing fewer than ten lines of commands. Then all you need to do is to learn a few functions to run the Bayesian model and diagnose the results (discussed in Section 3.4). To help readers become familiar with R coding, this book also provides a number of short R scripts consisting of simple functions. There are some cases requiring longer R scripts. Those programs are divided into a few sections, and the function of each section is explained in detail separately.

Although some knowledge of Bayesian reliability is given to students in graduate courses of reliability engineering, the application of Bayesian reliability in industry is limited to special cases using conjugate prior distributions, due to mathematical tractability.

Thanks to the rapid development of computers in the past half century, the breakthroughs in computational algorithms and the increased computing power of personal computers have enabled complex Bayesian models to be built and solved, which has greatly promoted the progress and application of Bayesian modeling. However, most engineers and scientists may not know that these modeling and computational capabilities can help them solve more complex prediction problems, which may not have been feasible in the past using traditional statistical methods.

Bayesian models are expected to become increasingly popular among engineers and scientists. One advantage is that modern Bayesian statistics enables development of more complex reliability models for system level prediction. Some examples included in this book attempt to demonstrate this capability. These cases often require customized solutions. Most of the existing commercial statistical software provide traditional statistical methods, which are not suitable for solving complex reliability problems. In other cases, Bayesian modeling offers unique benefits to effectively utilize different sources of information to reduce sample size. This book is intended to provide readers with some examples of practical engineering applications. Hopefully readers can apply them to their own fields and get some inspiration for building new models.

The goal of this book is to help more engineers and scientists to understand Bayesian modeling capabilities, learn how to use Bayesian models to solve engineering prediction problems, and get inspiration for developing Bayesian models to solve complex problems. The main objectives of this book are

- to explain the differences and benefits of Bayesian methods compared to traditional frequentist methods
- to demonstrate how to develop models to propagate component-level reliability to the final system level and quantify reliability uncertainty
- to demonstrate how to use different sources of information to reduce sample size
- to provide model examples for complex prediction problems
- to provide R and JAGS scripts for readers to understand and to use the models
- to design Bayesian reliability and substantiation test plans.

This book is intended for industry practitioners (reliability engineers, mechanical engineers, electrical engineers, product engineers, system engineers, materials scientists, Six Sigma Master Black Belts, Black Belts, Green Belts, etc.) whose work includes predicting design or manufacturing performance. Students in science and engineering, academic scholars, and researchers can also use this book as a reference.

Prerequisite knowledge includes basic knowledge of statistics and probability theory, and calculus. The goal is to enable engineers and scientists in different fields to acquire advanced Bayesian statistics skills and apply them to their work.

Throughout this book we extensively use the Markov chain Monte Carlo (MCMC) method to solve problems using JAGS software. We made an effort to reduce the use of complex Bayesian theory in this book and therefore it is not intended for people who want to learn the theory behind MCMC simulations.

Chapter 1 introduces basic concepts of reliability engineering, including random variables, discrete and continuous probability distributions, hazard function, and censored data. The Bayesian approach to reliability inference is briefly discussed. Non-parametric estimation of survival function using the Kaplan–Meier method is introduced. The concepts of system reliability estimation, design capability prediction, and accelerated life testing are also discussed.

Basic concepts of Bayesian statistics and models are presented in Chapter 2. Basic ideas behind Bayesian reasoning, Bayesian probability theory, Bayes' theorem, selection of prior distributions, conjugate priors, Bayes' factor and its applications are discussed.

Bayesian computation, the Metropolis–Hastings algorithm, Gibbs sampling, BUGS/JAGS models for solving Bayesian problems, MCMC diagnostics and output analysis are introduced in Chapter 3. Discreate and continuous probability distributions that are frequently used in reliability analysis are discussed in detail in Chapter 4.

Applications of these distributions in solving reliability problems using the Bayesian approach are also discussed in this chapter.

Chapter 5 introduces the concept of reliability testing and demonstration. The difference between substantiation and reliability testing is discussed. Classical and Bayesian methods for developing zero-failure test plans for both substantiation and reliability testing are presented. Examples are given for developing these test plans assuming that the underlying time to failure model is Weibull.

In Chapter 6 we discuss the concepts of design capability and design for reliability. Monte Carlo simulation techniques are introduced from the Bayesian perspective for estimating design capability and reliability, with examples to demonstrate these techniques. Chapter 7 introduces Bayesian models for estimating system reliability. The theory of reliability block diagrams, fault trees, and Bayesian networks are introduced with practical examples.

Bayesian hierarchical models and their applications are discussed in Chapter 8. Chapter 9 introduces linear and logistic regression models in the Bayesian perspective. Examples and a case study are presented to show the reader how to apply Bayesian methods for solving different regression problems.

Please send comments, suggestions or any other feedback on this book to AbeyraLiu118@gmail.com.

Yan Liu

Athula I. Abeyratne

Acknowledgments

The authors would like to sincerely thank Xingfu Chen, Donald Musgrove, Alicia Thode, Paul DeGroot, Pei Li, Vladimir Nikolski, and Norman Allie for their contributions in reviewing the manuscript. The authors would also like to sincerely thank Bradley P. Carlin and Harrison Quick for helping to answer questions related to Bayesian statistics. Many thanks to Greg Peterson and Shane Sondreal for their reviews and support to make this work presentable.

Thanks to our mentor Eric Maass, who has a great passion for teaching statistical methods to engineers. Tarek Haddad and Karen Hulting also provided valuable consulting on this topic. Some examples in this book are modified from actual engineering applications. The authors want to thank many Medtronic coworkers who contributed their case studies and/or provided valuable feedback, including Roger Berg, Mun-Peng Tan, Scott Hareland, Paul Wisnewski, Patrick Zimmerman, Xiaobo Wang, Jim Haase, Anders Olmanson, and Craig Wiklund.

About the Companion Website

This book is accompanied by a companion website:

www.wiley.com/go/bayesian20

The website includes:

Computer scripts and data files

Scan this QR code to visit the companion website.

1

Basic Concepts of Reliability Engineering

This chapter reviews basic concepts and common reliability engineering practices in the manufacturing industry. In addition, we briefly introduce the history of Bayesian statistics and how it relates to advances in the field of reliability engineering.

Experienced reliability engineers who are very familiar with reliability basics and would like to start learning Bayesian statistics right away, may skip this chapter and start with Chapter 2. Bayesian statistics has unique advantages for reliability estimations and predictive analytics in complex systems. In other cases, Bayesian methods may provide flexible solutions to aggregate various sources of information to potentially reduce necessary sample sizes and therefore achieve cost effectiveness. The following chapters provide more specific discussions and case study examples to expand on these topics.

1.1 Introduction

High product quality and reliability are critical to any industry in today's competitive business environment. In addition, predictable development time, efficient manufacturing with high yields, and exemplary field reliability are all hallmarks of a successful product development process.

Some of the popular best practices in industry include Design for Reliability and Design for Six Sigma programs to improve product robustness during the design phase. One core competency in these programs is to adopt advanced predictive analytics early in the product development to ensure first-pass success, instead of over-reliance on physical testing at the end of the development phase or on field performance data after product release.

The International Organization for Standardization (ISO) defines reliability as the "ability of a structure or structural member to fulfil the specified requirements, during the working life, for which it has been designed" (ISO 2394:2015 General principles on reliability for structures, Section 2.1.8). Typically, reliability is stated in terms of probability and associated confidence level. As an example, the reliability of a light bulb can be stated as the probability that the light bulb will last 5000 hours under normal operating conditions is 0.95 with 95% confidence.

Accurate and timely reliability prediction during the product development phase provides inputs for the design strategy and boosts understanding and confidence in product reliability before products are released to the market. It is also desirable to utilize and aggregate information from different sources in an effective way for reliability predictions.

Practical Applications of Bayesian Reliability, First Edition. Yan Liu and Athula I. Abeyratne.

Companion website: www.wiley.com/go/bayesian20

Textbooks on reliability engineering nowadays are dominated by frequentist statistics approaches for reliability modeling and predictions. In a frequentist/classical framework, it is often difficult or impossible to propagate individual component level classical confidence intervals to a complex system comprising many components or subsystems. In a Bayesian framework, on the other hand, posterior distributions are true probability statements about unknown parameters, so they may be easily propagated through these system reliability models. Besides, it is often more flexible to use Bayesian models to integrate different sources of information, and update inferences when new data becomes available.

Given the benefits mentioned above, potential applications of Bayesian methods on reliability prediction are quite extensive. Historically, Bayesian methods for reliability engineering were applied on component reliability assessment where conjugate prior (will be discussed in Chapter 2) distributions were widely used due to mathematical tractability. Recent breakthroughs in computational algorithms have made it feasible to solve more complex Bayesian models, which have greatly boosted advancement and applications of Bayesian modeling. One popular algorithm is Markov chain Monte Carlo (MCMC) sampling, a method of simulating from a probability distribution based on constructing a Markov chain. MCMC methods along with rapid advancement in high-speed computing have made it possible for building and solving complex Bayesian models for system reliability.

Over the past one or two decades, Bayesian statistics books have appeared in different scientific fields. However, most existing Bayesian statistics books do not focus on reliability analysis/predictions, thus real-life practical examples on reliability modeling are often absent. This challenge prevents reliability engineers from adopting the Bayesian approach to solve real-life problems. The goal of our book is to address this gap.

A few general topics covered in this book are:

- Design for reliability
- Basic concepts of Bayesian statistics and models
- Bayesian models for component reliability estimation
- Bayesian models for system reliability estimation
- Bayesian networks
- Advanced Bayesian reliability models.

Specifically, the topics covered are:

- Design for reliability
  This topic includes reliability definition, basic probability theory and computations, statistical models, basics of component reliability prediction, basics of system reliability prediction, critical feature capability prediction, Monte Carlo simulations, and accelerated life testing (ALT), etc.
- Basic concepts of Bayesian statistics and models
  This topic includes Bayes' theorem and history, Bayesian inference vs. frequentist inference, basic statistical concepts: point estimate, confidence interval, discrete and continuous probability distributions, censored data, and selection of prior distributions (conjugate priors, non-informative priors, and informative priors), likelihood function, model selection criteria, introduction of MCMC algorithms and sampling methods, and Bayesian computation software (WinBUGS, OpenBUGS, Just Another Gibbs Sampler (JAGS), R, etc.).

- Bayesian models for component reliability estimation
  This topic includes component level reliability prediction from reliability life testing, binomial distribution, Poisson distribution, exponential distribution, Weibull distribution, normal distribution, log-normal distribution, and reliability prediction from ALT (Arrhenius model, inverse power law model, etc.).
- Bayesian models for system reliability estimation
  This topic includes reliability block diagram, series system, parallel system, mixed series and parallel system, fault tree analysis with uncertainty, process capability or design capability analysis with uncertainty, Monte Carlo simulation, and two-level nested Monte Carlo simulation and examples (strength-stress interference, tolerance stack up, etc.).
- Bayesian networks
  This topic includes basics of conditional probability, joint probability distributions, marginal probability distributions, structures of a Bayesian network, examples, and basic steps to construct a Bayesian network model.
- Advanced Bayesian reliability models
  This topic includes using hierarchical Bayesian models to predict reliability during iterative product development, to predict reliability of specific failure mechanisms, to aggregate different sources of imperfect data, to aggregate component level and system level data for system reliability prediction, and to borrow partial strength from historical product reliability information.

The first three chapters introduce commonly used reliability engineering methods and basics of Bayesian concepts and computations. The following chapters focus more on applications related to the individual topics introduced above. Readers are free to tailor their reading to specific chapters according to their interests and objectives.

1.1.1 Reliability Definition

In reliability engineering, product reliability is defined as the probability that a component or a system performs a required function under specified use conditions for a stated period of time. Note that the three key elements in the reliability definition are probability, use condition, and duration. Probability measures the likelihood of something happening. For example, when tossing a fair coin there is a 50% probability of the coin landing heads. When throwing a six-faced fair dice, the probability of observing each of the six outcomes (1, 2, 3, 4, 5, 6) is 1/6. Use conditions describe the conditions a product is operated under, e.g. temperature, humidity, pressure, voltage. Duration is usually related to the lifetime of a product. Reliability is usually estimated based on time to failure data from bench tests, accelerated life tests, or field service.

In engineering practices, it is common to define design requirements and use different types of tests, such as design verification tests or qualification tests, to ensure the product or the incoming parts meet these requirements. Here quality is measured by the probability of meeting a certain requirement, which can be thought of as reliability at time zero. Though these are quality assurance practices, the term "reliability" is sometimes used to refer to the probability of meeting a certain requirement.

Often in design verification tests, the samples are preconditioned through an equivalent lifecycle under specified stress conditions (to ensure reliable products, the stress conditions applied in the tests are usually as aggressive as or more aggressive than the actual use conditions in the field) before being tested against a requirement. In such cases, the probability of meeting the requirement can be thought of as reliability at one lifecycle. However, this may not be the case for every requirement. To ensure quality and reliability, requirements may be classified at different levels based on importance and risk. Specified confidence/reliability requirements (e.g. 95%/99%) are assigned to key product characteristics according to their risk level. A specified 95%/99% confidence/reliability requirement means that the probability of meeting a requirement shall be at least 99% at 95% confidence level. The concept of confidence interval in these reliability requirements will be elaborated on in Chapter 2. The required sample sizes needed to meet these requirements are typically stated in terms of a frequentist statistical approach. We will explore the Bayesian solution to these problems in this book.

Given these practices in the industry, in this book we use the term reliability to generally refer to both cases described above (i.e. the traditional reliability definition and the broader applications in quality assurance). To avoid confusion, in each chapter when we go through a specific topic or example, we reiterate the definition of reliability in the context of that topic or example.

1.1.2 Design for Reliability and Design for Six Sigma

One way to assess reliability is the analysis of field data to estimate the product life expectancy and the probability of failure. This approach is appropriate for estimating reliability and monitoring performance trending of released products. However, this approach is not applicable for decision making related to new design/parts, especially when there are significant changes in the design. Quality and reliability issues caught after product release can be extremely costly to both customers and manufacturers. Product recalls due to reliability issues often result in customer dissatisfaction, huge financial losses due to repair and "firefighting" expenses, and brand degradation. Some product failures can even result in safety issues depending on the risk level of the failure mode.

Another approach in traditional industry processes is that the reliability analysis generally occurs at the end of product development, after the design is complete. One challenge in this product development process is that at the early phase of design, normally there's no adequate reliability analysis to drive decisions. This leads to potential risks of over or under design. Subsequent design changes may vary component use conditions, resulting in different reliability and design margins. Failures caught later in the development process may require many iterations of design change, which are costly and time consuming. For example, design-related issues such as power density and variability over time become a concern with advanced technology in the electronics industry (Turner 2006). As a result, there is a large demand to shift responsibility for reliability assurance to designers (Turner 2006) or collaboration between reliability engineers and designers.

In industry, the Design for Reliability philosophy is often combined with a broader quality improvement initiative called Design for Six Sigma, which is a program to ensure high-quality design and manufacturing, and to minimize design iterations. Design for Six Sigma adopted many statistical techniques (including design of experiments, control charts, reliability testing, etc.) in product development processes to promote first-pass design success, to reduce manufacturing defects, to increase design robustness to environmental factors, to reduce waste, and to increase product lifetime.

Though initially the Design for Six Sigma program was invented as an initiative to improve quality, industry practices in various corporations have demonstrated that its main value is beyond quality improvement and is more on time delivery and cost savings (Hindo 2007). The cost of poor quality was estimated to be as high as 15–30% of the entire cost (Defeo 2001). Using the cost of poor quality as the driver of the project selection in the Six Sigma program, various corporations including Honeywell, General Electric, Black & Decker, and 3M reported cost savings as high as hundreds of millions of dollars or even a few billion dollars (Hindo 2007, Defeo 2001) after implementing the programs. It was estimated that corporations that have implemented Six Sigma programs spent less than 5% of revenue fixing problems, much less than the cost in other corporations who spent 25–40% of revenue fixing problems (Pyzdek 2003).

1.2 Basic Theory and Concepts of Reliability Statistics

In this chapter, some commonly used concepts and practices in reliability engineering are briefly introduced/reviewed. We will use R scripts for basic reliability analysis. R is a language and environment for statistical computing and graphics. It is a free software and can be downloaded from the website https://www.r-project.org. R is now widely used in academia and industry. R Studio is an open source software that provides a friendly user interface for R. The instructions for installing R and R Studio are provided in Appendix A. Commonly used R commands are provided in Appendix B.

1.2.1 Random Variables

A random variable (r.v.) maps an outcome of an experiment to real number. For example, if a coin is tossed the outcome is either a head or a tail. We can define a random variable, X in this experiment such that $X = 1$ if a head turns up and $X = 0$ if a tail turns up. The sample space, S of an experiment is defined as the set of all possible outcomes of the experiment. In this case the sample space is

$$S = \{H, T\}.$$

The sample space for the r.v. X is $\{0, 1\}$.

In another example, suppose we are interested in time to failure of an electronic circuit. In this case the random variable, T, is the lifetime of the electronic circuit. This is a continuous random variable and all possible outcomes consist of all non-negative real numbers. Probability distributions can be defined on the random variables to account for the uncertainty associated with the outcome of the experiment that generated the random variable. As an example, if we toss a fair coin then the probability of observing a head is 0.5. This can be stated as $P(X = 1) = 0.5$ and $P(X = 0) = 0.5$. Since X is a discrete random variable, these probabilities describe its probability mass function (PMF).

Properties of a random variable can be described by its probability density (mass for discrete r.v.) function, cumulative distribution function (CDF), reliability function, and the hazard function. The use of these functions depends on the question that we are trying to answer.

1.2.2 Discrete Probability Distributions

Discrete probability distributions are used for attribute data (binary data, e.g. good/bad, yes/no, pass/fail, etc.) or count data (the observations are non-negative integer values). Though generally continuous variable data are preferred in engineering practices, they may not always be available. In other cases, continuous data from product testing results are sometimes converted to the pass/fail type of attribute data. This type of converting is highly inefficient and results in loss of useful information, but it can still be seen in the industry in quality assurance as part of a tradition.

Commonly used discrete probability distributions include the binomial distribution and the Poisson distribution. In quality assurance, binomial distributions are often used for pass/fail data. Poisson distribution can be used for count data, e.g. to measure the distribution of the number of defects per unit area. We will revisit these distributions with examples in Chapters 2 and 4.

For a discrete random variable X with a sample space S, a PMF, $m(x)$ can be defined as

$$m(x) \geq 0, \quad x \in S,$$

and

$$\sum_{x \in S} m(x) = 1. \tag{1.1}$$

1.2.3 Continuous Probability Distributions

In reliability engineering, continuous probability distributions are often used to describe continuous data, such as time to failure, cycles to failure, etc. A Weibull distribution is often used to model time to failures. In other engineering practices, continuous probability distributions are used for dimensions, voltages, and any other continuous variables. Normal distribution is often used to model dimensions. Appendix C introduces commonly used discrete and continuous probability distributions. More details are discussed in Chapter 4.

For a continuous random variable, X, with possible values on the real line, a probability density function (PDF), $f(x)$, can be defined as

$$f(x) \geq 0, \quad -\infty < x < \infty,$$

and

$$\int_{-\infty}^{\infty} f(x)\mathrm{d}x = 1. \tag{1.2}$$

With this definition any non-negative function that integrates to 1 over the real line can be considered to be a PDF. Which PDF to use is dependent on the type of data being analyzed.

1.2.4 Properties of Discrete and Continuous Random Variables

1.2.4.1 Probability Mass Function

Probability mass function (PMF), $m(x)$, is the probability that a discrete random variable X takes the value x. As an example, the following R code and Figure 1.1 show the

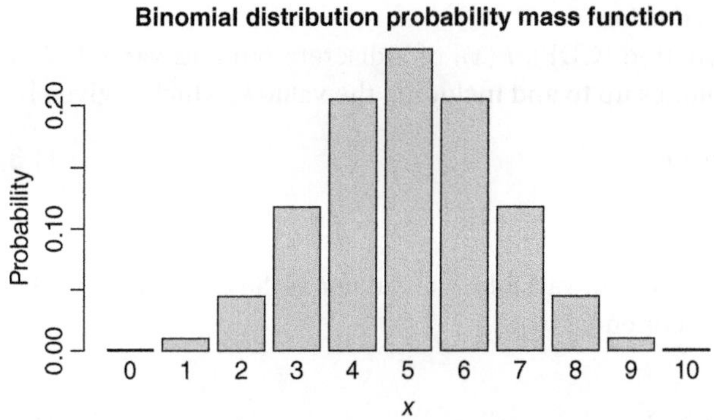

Figure 1.1 Probability mass function of a binomial distribution (size = 10, probability = 0.5).

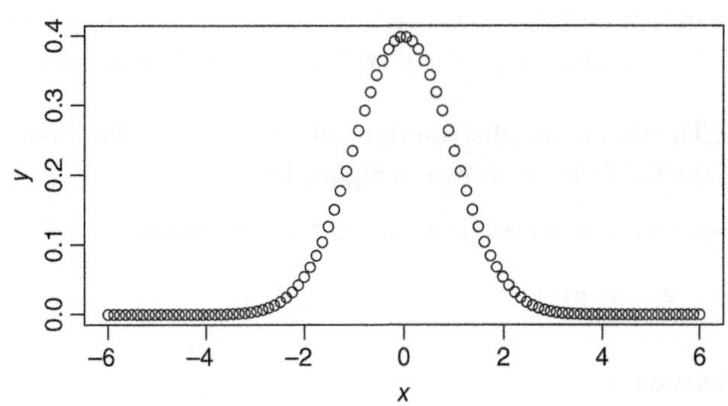

Figure 1.2 Probability density function of a normal distribution (mean = 1, sd = 1).

bar plot of a binomial distribution PMF with number of trials = 10 and probability of success = 0.5.

```
(1.2.4_Probability_Computations.R)
        ## probability mass function of a Binomial distribution

        x <- seq(0,10,by=1)
        y <- dbinom(x, size=10, prob=0.5)
        barplot(y, names=x, xlab = "x", ylab="Probability", main =
"Binomial distribution probability mass function")
```

1.2.4.2 Probability Density Function

The probability density function (PDF), $f(t)$, of a random variable T is non-negative real valued function that integrates to 1 over the range it is defined. As an example, normal distribution with mean 0 and standard deviation 1 is a PDF. In Chapter 4 we will discuss various probability distribution functions in detail.

The following R script shows how to plot the PDF of a normal distribution with mean = 0 and standard deviation = 1. Figure 1.2 shows the plotted PDF.

```
        x <- seq(-6,6,length=100)
        y <- dnorm(x,mean=0,sd=1)   # calculate the PDF of a Normal
distribution
        plot(x,y)  # generate PDF plot
```

1.2.4.3 Cumulative Distribution Function

The cumulative distribution function (CDF), $F(x)$, of a discrete random variable X at value k is the sum of all probabilities up to and including the value k, which is given by

$$F(k) = P(X \leq k) = \sum_{x \leq k} m(x), \tag{1.3}$$

where $m(x)$ is the PMF.

The CDF $F(t)$ of a continuous random variable X at value t is the cumulative probability of X having values less than or equal to t, i.e.

$$F(t) = P(x <= t) = \int_0^t f(x)\ \mathrm{d}x, \tag{1.4}$$

where $f(x)$ is the PDF. If the random variable T is the time to failure of a particular component then $F(t)$ provides the cumulative probability that the component fails on or before time t.

The R script to calculate the CDF of a normal distribution with mean = 0 and standard deviation = 1 is shown below. The CDF plot is shown in Figure 1.3.

```
        y1 <- pnorm(x,mean=0,sd=1) # calculate the CDF of a Normal
distribution
        plot(x,y1)  # generate CDF plot
```

1.2.4.4 Reliability or Survival Function

The reliability or survival function $R(t)$ is the probability of survival beyond time t. A reliability or survival function measures the percentage of products that survive a certain period of time without failures, i.e.

$$R(t) = 1 - F(t). \tag{1.5}$$

The R script to calculate the CDF of a normal distribution with mean = 0 and standard deviation = 1 is shown below. The reliability/survival curve is shown in Figure 1.4.

```
    y2 <- 1-y1  # calculate 1-CDF (reliability)
    plot(x,y2)  # generate reliability plot
```

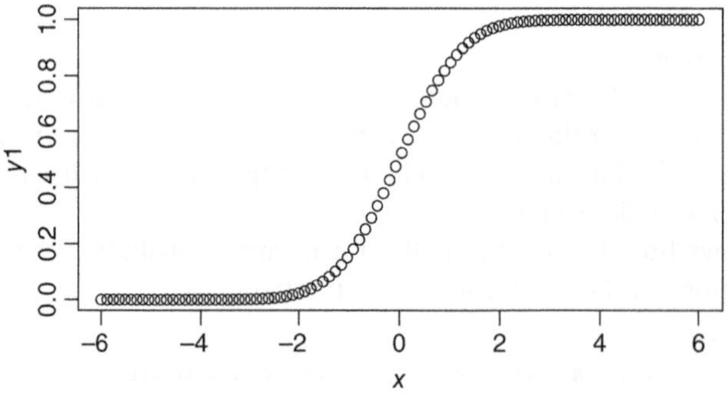

Figure 1.3 Cumulative distribution function of a normal distribution (mean = 1, sd = 1).

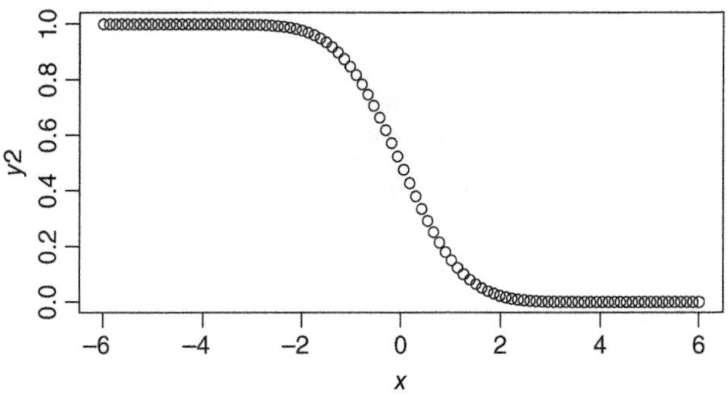

Figure 1.4 Reliability function of a normal distribution (mean = 1, sd = 1).

1.2.4.5 Hazard Rate or Instantaneous Failure Rate

The hazard rate or instantaneous failure rate, $h(t)$, for a continuous random variable, T, is the failure rate of the survivors at time t in the next moment following time t. Let $f(t)$ and $R(t)$ be the PDF and the reliability functions, respectively, then the hazard rate function is given by

$$h(t) = f(t)/R(t). \tag{1.6}$$

In this book we will use the terms "hazard rate" and "failure rate" to mean the instantaneous failure rate. Let's investigate the hazard rate through an exponential distribution. The lifetime distribution of certain electrical components such as light bulbs tends to follow exponential distributions. The PDF of an exponential distribution is given by

$$f(t) = \lambda e^{-\lambda t}, \tag{1.7}$$

where λ is the number of failures per unit time. If time is measured in hours, then it is the number of failures per hour. The reliability function of the above exponential distribution is given by

$$R(t) = P(T > t) = \int_t^{\infty} \lambda e^{-\lambda x} dx = e^{-\lambda t}. \tag{1.8}$$

Combining Eqs. (1.6)–(1.8) we get $h(t) = \lambda$. Therefore, the hazard rate for an exponential distribution is the same as the number of failures per unit time. Theoretically λ could be any positive real number. Therefore, hazard rate is not a probability. One of the defining characteristics of the exponential distribution is that it has a constant hazard rate which is the same as its parameter, λ.

There are other lifetime distributions, such as a Weibull distribution, that are more flexible than an exponential distribution to model failure rates that are varying over time. We will discuss these in Chapter 4.

A product could have different stages of failure rates. If $h(t)$ is decreasing, then the failure mechanism is called "infant mortality." A product subjected to decreasing failure rate, typically experiences more failures early on, so it is in the best interest of the manufacturer not to release such a product to the market. To get rid of the products that fail early on, they can be subjected to accelerated use conditions for a short period in the final testing. A good example of this is battery burn-in. When $h(t)$ is increasing it is an

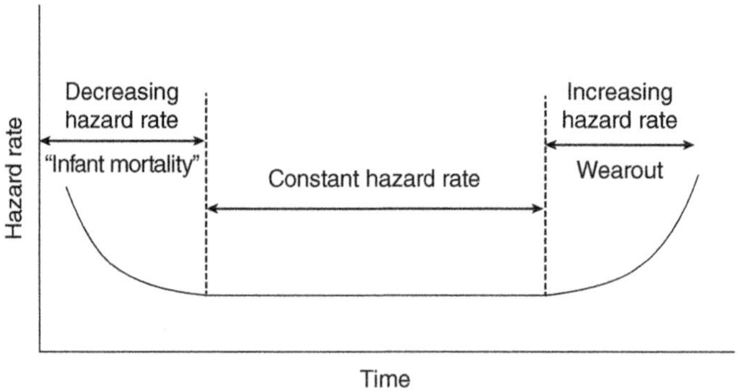

Figure 1.5 A bathtub curve showing the three regions of failure mechanism.

indication that the product has reached the wear-out phase. When $h(t)$ is a constant, the product is in useful life and the failures are considered to be random. A bathtub curve is commonly used to describe the three regions of hazard rate (Figure 1.5).

1.2.4.6 Cumulative Hazard Function

The cumulative hazard function (CHF), $H(t)$, is obtained by integrating the hazard function and is given by

$$\begin{aligned} H(t) &= \int_0^t h(x)\mathrm{d}x \\ &= \int_0^t \frac{f(x)}{R(x)}\mathrm{d}x \\ &= -\log(R(t)). \end{aligned} \tag{1.9}$$

Equation (1.9) yields an important relationship between the CHF and the reliability function, which is $R(t) = e^{-H(t)}$. For an exponential distribution, $H(t) = \int_0^t h(x)\mathrm{d}x = \int_0^t \lambda \mathrm{d}x = \lambda t$.

Therefore, the reliability function, $R(t) = e^{-\lambda t}$, which is what we have from (1.8).

1.2.4.7 The Average Failure Rate Over Time

The failure rate of a product changes over time. Therefore, the average failure rate is quoted as the typical failure rate between two time points. The average failure rate between times t_1 and t_2 is given by

$$AFR(t_1, t_2) = \frac{\int_{t_1}^{t_2} h(t)\mathrm{d}t}{(t_2 - t_1)} = \frac{H(t_2) - H(t_1)}{(t_2 - t_1)} = \frac{\log(R(t_1)) - \log(R(t_2))}{(t_2 - t_1)}. \tag{1.10}$$

1.2.4.8 Mean Time to Failure

For a continuous random variable T with a PDF $f(x)$, the mean time to failure (MTTF) is given by

$$MTTF = \int_{-\infty}^{\infty} t f(t)\mathrm{d}t. \tag{1.11}$$

1.2.4.9 Mean Number of Failures

For a discrete random variable X with a PMF $m(x)$ and sample space S, the mean number of failures (MNFs) is given by

$$\text{MNF} = \sum_{x \in S} x m(x). \tag{1.12}$$

As an example, for an exponential failure time distribution with a failure rate of λ per unit time, the MTTF is given by

$$\text{MTTF} = \int_0^\infty t \lambda e^{-\lambda t} \text{d}t = \frac{1}{\lambda}. \tag{1.13}$$

Suppose a product has an exponential failure time distribution with a failure rate of 0.025 failures a month. What is the MTTF for this product? From Eq. (1.123) we obtain MTTF = 1/0.025 = 40 hours.3

1.2.5 Censored Data

The collect time to failure data in a bench test is a common way to estimate MTTF and reliability at the specified lifetime. However, for highly reliable products, it may not be feasible to run everything to failure, due to test time constraints. For example, it might take years to run all the parts in a test to failure. A reliability engineer may have to terminate the test after six months in order to have a feasible estimation in time for a project. The engineer may observe that some of the parts have failed but some have not. In this case, the data collected is censored data. There are several different types of censored data. The concepts of different types of censored data are shown in Figure 1.6.

Right censoring means that the failure has not occurred by a certain time point (censoring time). In this case the failure would occur after the censoring time but it is not known when, for example a certain part in a reliability test survives 1000 hours and the test terminates at this point. Then the time to failure data of this part is right censored, with the censoring time being 1000 hours.

Left censoring means that the failure has occurred prior to a certain time point but it is not known exactly when. For example, a test is terminated after testing a part for

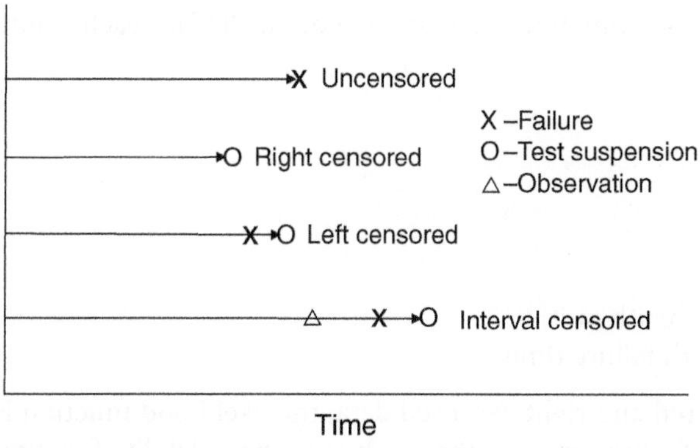

Figure 1.6 Various types of censored data.

1000 hours. An engineer only examines the part after the termination of the test and finds that the part has failed. Then the time to failure data of this part is left censored data, with the censoring time being 1000 hours.

Interval censoring means that the failure has occurred between a certain time interval but the exact time to failure is unknown. For example, a part is tested under stress conditions continuously. An engineer examines the part at 500 hours and the part has not failed. The part continues to be tested. The engineer examines the part again at 1000 hours and the part is found to have failed. The engineer thus know that the lifetime of this part is between 500 and 1000 hours, though the exact time to failure is unknown. Then the lifetime of this part is interval censored.

There are other types of censoring which are based on the cause of censoring. Type I or time censoring occurs when each unit in a fixed sample is tested for a prespecified length of time. All units not failed at the end of the test are time censored. For example, if 100 light bulbs are tested for 1000 hours each and only 60 of them have failed at the end of the test, then the 40 bulbs that have not failed are time censored.

Failure censoring or Type II censoring occurs when a test terminates after a prespecified number of failures has occurred, for example, we are test 100 light bulbs and the test terminates at the failure of the 35th light bulb. The data for the 65 light bulbs that have not failed at the test termination are called failure censored.

Type III censoring is a combination of Type I and Type II censoring. Type III censoring occurs when a test terminates if either the time criterion or failure criterion is met, for example, we test 100 light bulbs and the test terminates when 35 bulbs have failed, or each bulb is tested for 1000 hours, whichever comes first. The bulbs that have not failed at the termination of the test are called Type III censored.

Regardless of which censoring mechanism was used in generating lifetime data, how the data is handle in the analysis falls into one of the three main censoring categories mentioned: right, left, or interval censoring. The main classical approach for analyzing reliability data is called the maximum likelihood method. The likelihood function plays a very important role in estimating the unknown parameters of lifetime probability models in both classical (frequentist) and Bayesian statistics.

The contribution of each data point to the likelihood function varies by whether the data point is uncensored, right censored, left censored, or interval censored. In the following we will define the likelihood function in different censoring scenarios. Let $t_1, t_2 \ldots, t_n$ be a random sample of lifetime data obtained in a reliability test.

For uncensored data, the likelihood function is the product of the PDF for each failure time, i.e.

$$L = \prod_{i=1}^{n} f(t_i),$$

where

n is the number of data points
t_i is the observed failure time of the ith unit
$f(t_i)$ is the PDF evaluated at the ith failure time.

For a combination of uncensored and right censored data, the likelihood function is the product of the PDF for each failure time and the product of the reliability function at each right censored time, i.e.

$$L = \prod_{i=1}^{k} f(x_i) \prod_{j=k+1}^{n} R(t_j),$$

where

n is the number of data points, where the failure and censoring times are organized such that the first k data points are failures and the last $n - k$ data points are right censored data

x_i is the observed time to failure of the ith data point

$f(x_i)$ is the PDF evaluated at the ith failure time;

t_j is the recorded operating time (censoring limit) of the jth time point;

$R(t_j)$ is the reliability/survival function evaluated at the jth censoring time.

For a combination of uncensored and left censored data, the likelihood function is the product of the PDF for each failure data and the product of CDF for each left censored data, i.e.

$$L = \prod_{i=1}^{k} f(x_i) \prod_{j=k+1}^{n} F(t_j),$$

where

n is the number of data points, where the failure and censoring times are organized such that the first k data points are failures and the last $n - k$ data points are left censored data

x_i is the observed time to failure of the ith failure time

$f(x_i)$ is the PDF evaluated at the ith failure time

t_j is the censoring limit of the jth censoring time

$F(t_j)$ is the CDF evaluated at the jth censoring limit.

For a combination of uncensored data and interval censored data, the likelihood function is

$$L = \prod_{i=1}^{k} f(x_i) \prod_{j=k+1}^{n} \left[F(t_{upper\_j}) - F(t_{lower\_j})\right],$$

where

n is the number of data points, where the failure and censoring times are organized such that the first k data points are failures and the last $n - k$ data points are interval censored data

x_i is the observed time to failure of the ith failure time

$f(x_i)$ is the PDF evaluated at the ith failure time

t_{upper_j} and t_{lower_j} are the upper and lower censoring limits of the jth data point, respectively

$F(t_{upper\_j}) - F(t_{lower\_j})$ is the probability for values of time between t_{upper_j} and t_{lower_j}.

1.2.6 Parametric Models of Time to Failure Data

Parametric models use probability distributions of time to failure data to estimate reliability, CDF, and other metrics of interest. Examples are shown in Section 1.2.3. When using a distribution to model time to failure data, methods such as the maximum likelihood method is often used to estimate the unknown parameters of the distribution. Besides point estimates of the model parameters, their confidence intervals are also computed.

In addition, reliability practitioners are usually interested in point estimates and confidence intervals associated with reliability at a given time and various percentiles of the distribution. Some of the parametric reliability models we consider in this book include binomial, Poisson, exponential, gamma, beta, Weibull, normal, and log-normal. Details of these topics are further discussed in Chapters 2 and 4.

The maximum likelihood method estimates unknown parameters of a distribution by maximizing the likelihood function. An example of finding the maximum likelihood estimate of the success probability in a binomial distribution will be introduced in Section 2.5.

1.2.7 Nonparametric Estimation of Survival

Unlike parametric models, nonparametric methods don't assume any parametric probability distribution of time to failure data. These models utilize the observed number of failures (or deaths) and non-failures to determine the survivor function.

Frequently used nonparametric methods include the Kaplan–Meier (Kaplan and Meier 1958), Nelson–Aalen, and Cutler–Ederer (CE) life-table (also known as actuarial) methods. Another nonparametric method for estimating survival function is Bernard's median rank method, which is based on the rank of the ordered observations rather than the observation itself. These methods will be discussed briefly in this section using simple examples.

The main advantage of using nonparametric methods is to eliminate the need for having the data fit a specified probability distribution. The main drawback of using nonparametric methods is the inability to predict reliability beyond the time range of the data collected.

The Kaplan–Meier estimator is the most commonly used nonparametric method to estimate reliability without assuming distributions. When the data is a combination of uncensored and censored data, the Kaplan–Meier reliability estimation $\widehat{R}(t_i)$ at the failure time t_i is given by

$$\widehat{R}(t_i) = \prod_{j=1}^{i} \left(\frac{n_i - d_i}{n_i} \right), \tag{1.14}$$

where

t_i is the event time of the ith ordered failure time
n_i is the number of survivors just before the failure time t_i
d_i is the number of failures occurring at time t_i
$\widehat{R}(t_i) = \widehat{R}(t_{i-1})$, if no failures are recorded at time t_i (i.e. the event corresponds to censored data).

Example 1.1 The failure and censored times (hours) of 10 key components of a machine are 385, 450+, 475, 500, 575, 600+, 750, 750, 875, and 900+. For simplicity, the times have been ordered from low to high. A + sign next to the number indicates a right censored time. We will use the Kaplan–Meier method to estimate the survivor function of this component.

Table 1.1 Kaplan–Meier estimate of machine component survival.

Time (hours)	Number failed	Number at risk	K–M estimate
385	1	10	(10−1)/10 = 0.9
450+	0		0.9
475	1	8	0.9*(8−1)/8 = 0.787 5
500	1	7	0.787 5*(7−1)/7 = 0.675
575	1	6	0.675*(6−1)/6 = 0.562 5
600+	0		0.562 5
750	2	4	0.562 5*(4−2)/4 = 0.281 25
875	1	2	0.281 25*(2−1)/2 = 0.140 625
900+	0		0.140 625

Table 1.2 Kaplan–Meier estimates and 95% pointwise confidence intervals (generated from R-code).

Time	N.risk	N.event	Survival	Std err	Lower 95% CI	Upper 95% CI
385	10	1	0.900	0.0949	0.7320	1.000
475	8	1	0.787	0.1340	0.5641	1.000
500	7	1	0.675	0.1551	0.4303	1.000
575	6	1	0.562	0.1651	0.3165	1.000
750	4	2	0.281	0.1631	0.0903	0.876
875	2	1	0.141	0.1286	0.0234	0.844

Table 1.1 shows the Kaplan–Meier survival function estimation steps. The first column provides ordered failure and censored times. The second column from the left shows the number failed (1) or number censored (0) at each time point. The third column shows the number of units at risk just before each failure time. The fourth column shows the Kaplan–Meier estimate of the survival function using formula (1.13).

The R script below will generate the Kaplan–Meier estimates and 95% pointwise confidence intervals given in Table 1.2, and the survival plot shown in Figure 1.7 for the machine component failure time data.

(1.2.7_Kaplan_Meier_Surv_Anal_Table_1_2.R)

```
## Kaplan-Meier Survival Analysis of the data in Table 1.1

# load the package "survival"
library(survival)
# Create a time to failure data vector
TimeToEvent <- c(385,450,475,500,575,600,750,750,875,900)
# Note that censoring times are 450, 600, and 900
# Create a vector of 0s and 1s where 0 indicates censoring event
# and 1 indicates a failure event
```

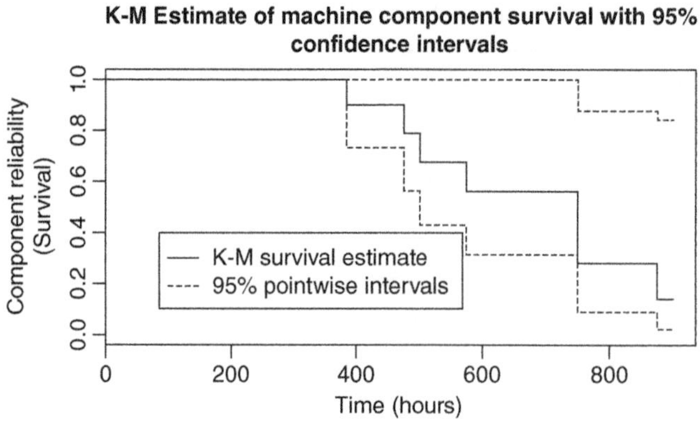

Figure 1.7 Kaplan–Meier estimate of machine component reliability (survival).

```
        Censor <- c(1,0,1,1,1,0,1,1,1,0)
        #Create a survival object
        SurvObj <- Surv(TimeToEvent, Censor)
        # Compute Kaplan-Meier survival estimates
        KMEst <- survfit(SurvObj ~ 1)
        summary(KMEst)
        cb <- confBands(y_bmt, type = "hall")
        # Generate K-M Survival plot with 95% pointwise confidence bounds
        # PDF(file="Figure1_7.pdf",width=7,height=5)
        # jpeg("Fig 1.7_Kaplan_Meier_Survival.jpeg", width = 6, height = 4,
units = 'in', res = 600)  # save the plot as jpeg format
        plot(KMEst,main = 'K-M Estimte of Machine Component Survival
with 95% Confidence Intervals',xlab="Time (Hours)",
            ylab="Component Reliability (Survival)",cex.main=0.9,
cex.lab=0.9)
        legend(100, 0.40, legend = c('K-M survival estimate','95%
pointwise intervals'), lty = 1:2)
        # dev.off()
```

Figure 1.7 shows the Kaplan–Meier survival estimates and 95% pointwise confidence interval. Notice that the survival plot is a step function and it only changes at failure times and remain constant through censored times.

1.2.8 Accelerated Life Testing

Using time to failure data to estimate product reliability from bench tests under regular use conditions can be very time consuming and costly. An alternative method, ALT, can be used instead to test products under increased level of stress conditions and thus failures can be generated in a shorter period of time (Figure 1.8).

ALT is used to address wear-out failure mechanisms. The assumption is that a product will exhibit the same failure mode and mechanism under increased stress levels in ALT as it would exhibit under actual use conditions in the field. Data generated in ALT can be used to estimate reliability or life under field use conditions. In a ALT, time to failure data at a specific stress level are fit to a life distribution. Time to failure data collected under different stress levels can be fit to a stress-life model (Figure 1.9), which is used to estimate the life distribution under the field use conditions.

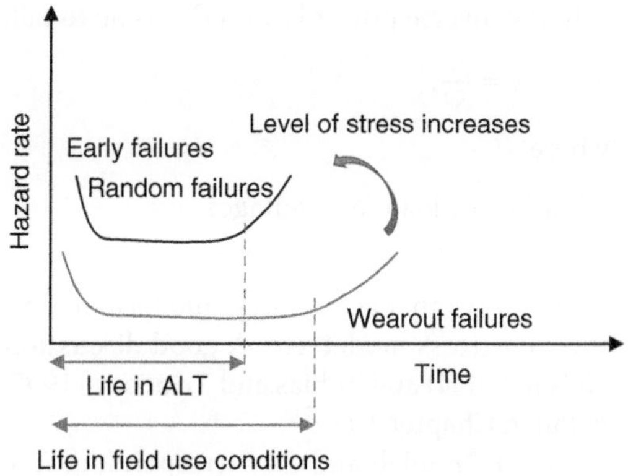

Figure 1.8 Accelerated life tests.

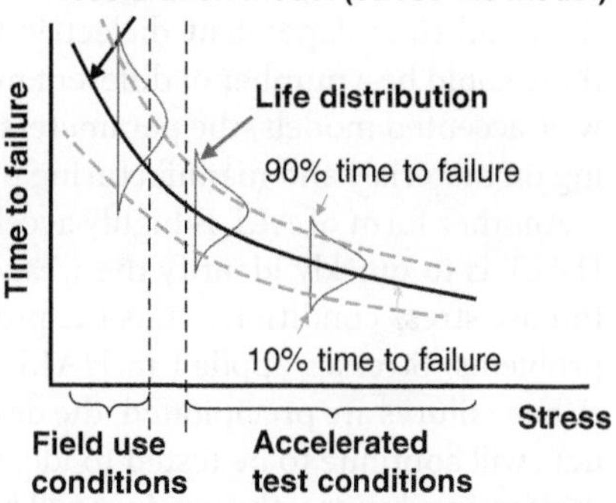

Figure 1.9 Acceleration model and life distribution.

Methods of acceleration include increasing the usage frequency, the stress levels, or both. Typical stresses in ALT include voltage, current, temperature, humidity, vibration, or a combination of various types of stresses. Here we review two commonly used acceleration models: (i) the temperature acceleration model (law of Arrhenius) and (ii) the inverse power law model.

In the temperature acceleration model, time to failure can be expressed as Eq. (1.15), which models the lifetime under a particular temperature T as

$$t = A_0 \exp(E_a/kT), \tag{1.15}$$

where

t is time to failure
A_0 is a constant
E_a is the activation energy for the failure mechanism
k is Boltzmann's constant (8.617×10^{-5} eV K^{-1})
T is the temperature measured in degrees of Kelvin (K).

In this model, the acceleration factor, AF, between two temperatures T_1 and T_2 can be obtained from Eq. (1.15) as follows

$$AF = \frac{t_1}{t_2} = \frac{A_0 e^{(E_a/kT_1)}}{A_0 e^{(E_a/kT_2)}} = e^{(E_a/k)[1/T_1 - 1/T_2]}. \tag{1.16}$$

In the inverse power law model, time to failure can be expressed as

$$t = \frac{a}{S^c}, \tag{1.17}$$

where

S is stress or load (e.g. voltage)
a and c are positive constants.

There are other acceleration models, such as the Eyring model, the degradation model, the step-stress model, etc. A good discussion of various ALT models can be found in Nelson (2004) and Tobias and Trindade (1995). We will discuss some of these models in detail in Chapter 10.

The ALT models and the corresponding acceleration stresses are specific to the underlying failure mechanism. Examples of some commonly encountered failure mechanisms in the semiconductor industry include electro-migration (Black 1974), hot carrier injection, and time-dependent dielectric breakdown. Although there is some agreement, there could be a number of different models used for some ALT failure models. Even for well-accepted models, the parameters can sometimes have a fairly wide range, depending on materials and manufacturing process, etc.

Another form of ALT is highly accelerated life testing (HALT). The main objective of HALT is to quickly identify the weakest links by testing the products to failure under intense stress conditions, thus the product quality can be improved by eliminating the problems. Stresses applied in HALT are usually well beyond normal use conditions. When failures are precipitated, the design problem is identified and fixed, and the products will continue to be tested to identify the next weakest link. With several rounds of reinforcing designs, the product will become more and more robust until it is too costly to continue fixing problems or when there are sufficient design margins. The limitation of HALT is that the test results cannot be used to predict the product reliability. In other words, HALT can be used to ensure that the product quality and reliability are better, but there is little information about how much better.

Burn-in is another type of accelerated testing intended to screen out/precipitate early failures or infant mortality type of failures usually caused by manufacturing defects. For components like integrated circuit boards, lithium/iodine batteries, and so on, it is common practice to apply burn-in tests under elevated stress conditions (voltages, current, humidity, and temperature, etc.). One debate of using burn-in is that those accelerated stress conditions should not be so high that they result in product damage.

1.3 Bayesian Approach to Reliability Inferences

1.3.1 Brief History of Bayes' Theorem and Bayesian Statistics

In the 1740s, English Reverend and mathematician Thomas Bayes developed Bayes' theorem to answer the question about the probability of a cause given an effect (called "inverse probability"). In Bayes' solution, a thought experiment is used which is described as follows: an assistant randomly throws balls on a flat square table and tells whether every new ball stops to the left or right of the first ball. With this information, how can he tell the position of the first ball? Bayes realized that this is a learning process: the more information we have on the new balls, the more we know about the first ball's position (McGrayne 2011). In brief, Bayes' theorem can be described as

initial belief(called prior)+new data/evidence=improved belief(called posterior)

Though a significant discovery, Bayes didn't publish his idea. Instead, two years after his death, in 1763, Bayes' friend Richard Price found Bayes' paper, realized the significance of this solution for "one of the most difficult problems in the doctrine of chances," and helped publish Bayes' paper entitled "An Essay towards solving a Problem in the Doctrine of Chances" in the Royal Society's *Philosophical Transactions* (McGrayne 2011).

In 1774, French mathematician Pierre-Simon Laplace independently developed and published Bayes' theorem. In the early and mid-20th century, due to the doubt about subjective probability to quantify our initial belief and the idea that science cannot allow subjectivity, Bayesian inference was attacked by statisticians including R.A. Fisher and J. Neyman, and was replaced with frequentist inference. Later, in Section 2.5, we will discuss the difference between Bayesian inference and frequentist inference. One major difference is that uncertainties of the model parameters are quantified by probability distributions in Bayesian inference and by confidence intervals in frequentist inference.

Meanwhile during early and mid-20th century, a few statisticians continued to make progress on Bayesian statistics. The subjective interpretation of probability (probability was based on personal beliefs which can be quantified) was developed by Ramsey (1927) and Finetti (1937). The objective form of Bayesian inference was further developed by Jeffreys (1946) by devising rules for selecting priors.

The revival of Bayesian statistics is related to the early development of modern computer science. During World War II, German U-boats cut off British sources of food. The German codes produced by Enigma machines were able to change rapidly and were considered unbreakable. Alan Turing, father of computer science and artificial intelligence, built a machine and used Bayesian methods to crack the Enigma code (McGrayne 2011).

However, Bayesian statistics remain dismissed in academia until in the 1990s when personal computers became popular and the MCMC was developed. MCMC is a method of simulating from a probability distribution based on constructing a Markov chain. Details of this sampling algorithm will be discussed in Chapter 3.

Historically, applications of Bayesian reliability in industry were limited to special cases where conjugate prior distributions are used, due to mathematical tractability. MCMC algorithms and the increased computational power of personal computers have made it easier to develop and solve more advanced statistical models for complex problems, which is not be feasible using the traditional statistical methods of the past.

These breakthroughs in computational algorithms have greatly boosted advancement and applications of Bayesian modeling. In the last two decades, extensive research for new Bayesian methodologies generated the practical application of complicated models in a wide range of science.

1.3.2 How Does Bayesian Statistics Relate to Other Advances in the Industry?

Two advances in the industry are observed or anticipated.

1.3.2.1 Advancement of Predictive Analytics

Predictive reliability analysis in advance provides inputs for the design strategy and boosts understanding and confidence in product reliability before products are released to the market. A Bayesian framework provides a straightforward solution for product reliability prediction with uncertainty quantified, even in complex systems, which could have been challenging using traditional reliability analysis methods (Martz and Waller 1990).

1.3.2.2 Cost Reduction

One challenge that many industries face is economic pressure to develop highly reliable products while simultaneously lowering costs. With technology maturing, the reliability of many products is increasing, and so is customer expectation. To demonstrate high reliability based on traditional methods might be an extremely expensive task with an incredibly large sample size.

In addition, Bayesian modeling enables aggregating information from different sources (e.g. historical data and bench test data) to predict reliability, which may provide unique benefits for sample size reduction. Some examples to demonstrate this benefit are included in Chapters 7 and 8. The cost associated with testing could be reduced potentially, and time to market could be improved without compromising performance and reliability.

In the following chapters, various case examples are provided to help engineers gain insights into these new capabilities, to learn how to adopt Bayesian models to solve engineering prediction problems, and to be creative in developing advanced Bayesian models to solve complex problems.

1.4 Component Reliability Estimation

Component reliability estimation often involves estimating reliability from lifetime data using the parametric or nonparametric methods introduced above. Lifetime data could be collected from the characterization test or from the field. When lifetime data are absent, component reliability can also be estimated from the data gathered in accelerated life tests. The results of these analysis are the point estimate and the specified (e.g. 95%) confidence interval (called credible intervals in Bayesian world) of the component reliability. In Chapter 4, we will go through the commonly used probability distributions one by one to discuss in detail how to estimate component reliability based on lifetime data using the Bayesian approach.

1.5 System Reliability Estimation

System reliability is the probability of the system to operate without failures for a specified period of time under specific use conditions. Since the components or subsystems in a system may experience different stress conditions, it may not be feasible to obtain lifetime data at the system level from one type of bench test during product development. Instead, reliability at the component or subsystem level is often assessed individually from different reliability tests or different sources of data.

With component/subsystem level reliability estimated individually, system level reliability is then estimated by aggregating reliability from the component/subsystem level, based on certain system reliability models. Commonly used system reliability prediction methods include the reliability block diagram, fault tree analysis, and Bayesian network, etc. which will be discussed in Chapter 7.

The results of the system reliability estimation are the point estimate and the specified (e.g. 95%) confidence interval of the system reliability. In a frequentist/classical framework, except in the simplest cases, it is often difficult or impossible to propagate classical confidence intervals though complex system models. In a Bayesian framework, on the other hand, posterior distributions are true probability statements about unknown parameters, so they may be easily propagated through these system reliability models (Hamada et al. 2008). In Chapter 7, we will discuss this topic in more detail.

1.6 Design Capability Prediction (Monte Carlo Simulations)

Monte Carlo simulation is a method that samples random values repeatedly to solve problems. Nowadays Monte Carlo simulations are widely used among engineers for reliability analysis and design capability analysis. In a design capability analysis, a design characteristic/performance is compared against the requirement to estimate the probability of the design meeting the requirement. In a Six Sigma design, a typical requirement is to demonstrate that the process capability index, $C_{pk} > 1.5$. This requirement indicates that both the lower specification limit (LSL) and the upper specification limit (USL) are at least 4.5 standard deviations from the mean. If the data is normally distributed, this implies about 7 ppm is not meeting the specification requirements. Monte Carlo simulations can also be applied to many other engineering cases to quantify the probability of an outcome/output variable given the known uncertainties of input variables and a model describing the relationship between the output variable and input variables.

Figure 1.10 shows the general flow of how Monte Carlo simulation works to estimate the probability of meeting a design requirement. Assume

- y is the output variable to be determined and to be compared against a requirement
- $x_1, x_2, \cdots, x_n$ are n known input variables
- uncertainty of each of the input variable can be described by a probability distribution
- $y = f(x_1, x_2, \cdots, x_n)$ is the transfer function that describes the relationship between the output and inputs. Note that the transfer function is a deterministic model that can be based on first principles (e.g. Newton's laws, Ohm's law, etc.) or empirical equations (e.g. linear regression, nonlinear regression, etc.).

The following steps are then performed in Monte Carlo simulations:

1) A random value is generated for each input variable x_i $(i = 1, \ldots, n)$ based on the predefined probability distribution of x_i.
2) Based on the set of values for $x_1, x_2, \cdots, x_n$ and the transfer function $y = f(x_1, x_2, \cdots, x_n)$, the y value is calculated.
3) Steps 1 and 2 are repeated k times (k is usually a large number, e.g. 100 000) to collect a set of y values.
4) The summary statistics for the set of y values (mean, standard deviation, etc.) is estimated and each of the y values compared to the requirement.

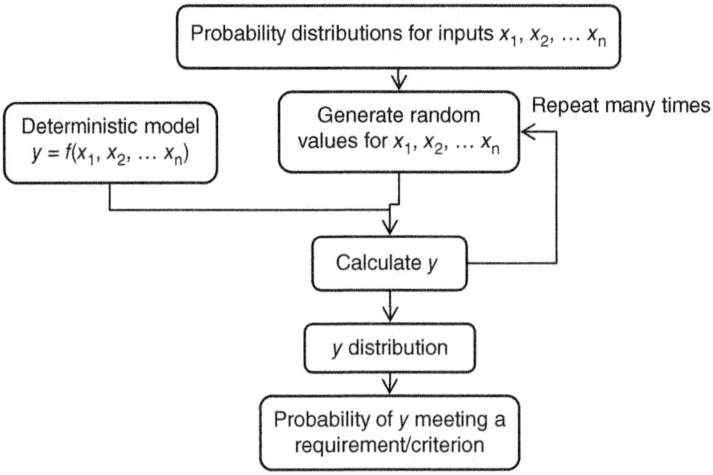

Figure 1.10 Flow chart of Monte Carlo simulations.

5) If out of all the y values collected in step 3 m out of k times the y value meet the requirement, then we say the probability of meeting the requirement based on the Monte Carlo simulation is $\frac{m}{k} \times 100\%$.

Examples and more discussions on Monte Carlo simulations to estimate design capability are included in Chapter 6.

1.7 Summary

This book introduces advanced Bayesian models to engineers and scientists with many real-life reliability case studies. In this chapter, we review basic concepts and existing commonly applied practices in reliability engineering. Good discussion of classical approach to reliability analysis can be found in Meeker and Escobar (1998) and Tobias and Trindade (1995).

We also briefly introduce the history of Bayesian inference and how Bayesian analysis relates to other advances in the industry. Classical reliability techniques were limited to reliability models where closed form solution or normal approximation was available for the parameters or statistics of interest. This limitation is no longer applied to Bayesian methods since the distributions of quantities of interest are obtained from the simulated posterior samples.

Bayesian statistics has unique advantages for reliability estimations and predictive analytics in complex systems and can allow flexible modeling solutions to reduce sample sizes and testing cost. However, historically, Bayesian modeling and analysis are limited in application due to mathematical tractability. Recent advancements in high-speed computing and breakthroughs in computational algorithms have made it feasible to solve more complex Bayesian models, which has greatly increased applications of Bayesian modeling. One popular algorithm is MCMC sampling, which will be introduced in Chapter 3. In later chapters, we will discuss these topics in detail and provide case study examples to demonstrate these benefits.

References

Black, J.R. (1974) *Physics of Electromigration.* IEEE Proceedings of the International Reliability Physics Symposium, April 2–4, 1974, Las Vegas, NV, USA.

Defeo, J.A. (2001). The tip of the iceberg – when accounting for quality, don't forget the often hidden costs of poor quality. *Quality Progress* 34 (5): 29–37.

Finetti, B. (1937, 1964). *La Prévision: ses lois logiques, ses sources subjectives.* Annales de l'Institut Henri Poincaré [*Foresight: its Logical Laws, Its Subjective Sources* (translation of the 1937 article in French)] (ed. H.E. Kyburg and H.E. Smokler) Studies in Subjective Probability. New York: Wiley.

Hamada, M.S., Wilson, A.G., Shane Reese, C., and Martz, H.F. (2008). *Bayesian Reliability.* New York: Springer.

Hindo, B. (2007) At 3M, A struggle between efficiency and creativity. *Inside Innovation – In Depth,* June 11, 2007.

ISO 2394:2015(en) (2015) General principles on reliability for structures, section 2.1.8, https://www.iso.org/obp/ui/#iso:std:iso:2394:ed-4:v1:en, accessed April 16, 2018.

Jeffreys, H. (1946). *An Invariant Form for the Prior Probability in Estimation Problems. Proceedings of the Royal Society of London. Series A, Mathematical and Physical Sciences.* 186 (1007): 453–461.

Kaplan, E.L. and Meier, P. (1958). Nonparametric estimation from incomplete observations. *Journal of America Statistics Association* 53 (282): 457–481.

Martz, H.F. and Waller, R.A. (1990). Bayesian reliability analysis of complex series/parallel systems of binomial subsystems and components. *Technometrics* 32 (4): 407–416.

McGrayne, S.B. (2011). *The Theory that Would not Die: How Bayes' Rule Cracked the Enigma Code, Hunted Down Russian Submarines, and Emerged Triumphant from Two Centuries of Controversy.* New Haven, CT: Yale University Press.

Meeker, W.Q. and Escobar, L.A. (1998). *Statistical Methods for Reliability Data.* Hoboken, NJ: Wiley-Interscience.

Nelson, W.B. (2004). *Accelerated Testing: Statistical Models, Test Plans, and Data Analysis.* Hoboken, NJ: Wiley.

Pyzdek, T. (2003). *The Six Sigma Handbook.* New York: The McGraw-Hill Companies, Inc.

Ramsey, F.P. (1927). Facts and propositions. *Aristotelian Society Supplementary* 7: 153–170.

Tobias, P.A. and Trindade, D.C. (1995). *Applied Relibility,* 2e. New York: Van Nostrand Reinhold.

Turner, T.E. (2006) *Design for reliability.* Proceedings of 13th IPFA 2006, Singapore, 257–263.

2

Basic Concepts of Bayesian Statistics and Models

This chapter introduces basic concepts of Bayesian statistics. We start with the basic idea of Bayes' theorem, followed by an introduction of Bayesian inference. Selection of prior distributions and commonly used conjugate prior distributions are briefly discussed. We also summarize the difference between Bayesian inference and frequentist inference, and how Bayesian inference works with Monte Carlo simulation methods. Basic concepts of point estimation, interval estimation, Bayes' factor, and prediction are provided. In addition, we will show how to use Bayes' factor as a model selection tool.

2.1 Basic Idea of Bayesian Reasoning

Bayesian reasoning is an approach of learning from evidence as it accumulates. To illustrate this basic idea, let's think about the following two scenarios.

Scenario 1. Tom just moved to a new city without investigating the crime rates in the area. He randomly selected a neighborhood. Over the next few months, Tom continually heard or observed crime incidents in his neighborhood. How do these observed events change Tom's impression of the community? Tom may come to the idea that he lives in a neighborhood with a relatively high crime rate.

Scenario 2. Tom decided to move to another block with a lower crime rate. This time, he learned about the crime data from the local police department. According to the crime statistics, he moved to one of the safest districts in the city. Tom lived happily in his new home for some time until one day he heard a criminal incident in his new neighborhood. This event is very rare, and such an incident has never happened in the past 50 years in that neighborhood. How did this crime incident change Tom's impression of the community? Does Tom continue to believe he lives in a safe neighborhood, or does he change his mind and deem it unsafe?

Later we will see how Bayes' theorem provides the answer to Scenario 2. In Scenario 2, which of the two perspectives is correct depends on the weight of historical crime statistics and the weight of the new event. Without complicated mathematics, the basic idea of Bayesian reasoning can be described as:

$$\text{prior belief} + \text{new data/evidence} = \text{updated belief}$$

This formula basically describes how humans learn: we learn from observing new evidence. From the new evidence, we either form an opinion about something we do not know, or change our initial opinion when the evidence contradicts it.

Practical Applications of Bayesian Reliability, First Edition. Yan Liu and Athula I. Abeyratne.

Companion website: www.wiley.com/go/bayesian20

In Scenario 1, Tom initially did not know how safe the first block was. However, when he saw enough evidence, he gradually formed the view that the community was not safe, even though he did not know any historical information. In Scenario 2, Tom reads historical crime statistics, on the basis of which he considers the second community to be fairly safe. Later, after Tom found a crime incident in the second block, two possibilities arose. Tom may ignore the new fact so that he still believes the second community is safe; Tom may also change his initial thoughts in favor of the idea that the second community is unsafe, because he thinks the new event has a greater weight in his new opinion.

How do we determine the weight of initial belief and new evidence objectively? It might be hard to calculate the weight in our mind in an intuitive way. Bayes' theorem, on the other hand, provides a mathematical method of calculating the weight of both initial belief and new evidence. Therefore, the improved/updated belief can be regarded as a weighted average of two pieces of information. The weighting is related to the uncertainty of the prior belief and the sample size of the new data, as we will discuss in more detail in Section 2.4.2.

2.2 Basic Probability Theory and Bayes' Theorem

To understand Bayes' theorem in a mathematical way, first we go over some commonly used concepts in probability theory.

The sample space of an experiment consists of all possible outcomes of that experiment. As an example, consider the experiment of drawing a card at random from a well-shuffled deck of cards. There are 52 possible outcomes of this experiment. Therefore, the sample space is a list of all 52 cards in the deck. Events can be defined on the sample space. For example, define the event A as observing a spade card. The probability of this event is computed by counting the number of outcomes belong to event by the total number (= 52) of possible outcomes in the sample space. There are 13 spade cards so, $P(A) = 13/52 = 1/4$.

We demonstrate basic probability concepts using Figures 2.1 and 2.2. Here we consider an experiment with only two possible outcomes: events A and B. The probabilities associated with these events are given by their corresponding areas in the figures.

- *Joint probability* of events A and B, $P(A, B)$ or $P(A \cap B)$, is the probability of A and B both being true. This probability is given by the area common to both events A and B, shown in light gray in Figure 2.1.
- *Conditional probability* of event A given event B, denoted by $P(A \mid B)$, is the probability of observing A is true, given that event B is true. This probability is defined as

$$P(A \mid B) = \frac{P(A \cap B)}{P(B)}.$$

Therefore, the conditional probability of $P(A \mid B)$ is the joint probability of A and B divided by the probability of event B. This implies that when two events are mutually exclusive (see Figure 2.2b), then $P(A \mid B) = P(B \mid A) = 0$.

As an example of conditional probability, consider the experiment of drawing a card at random form a well-shuffled deck of playing cards. Let A be the event of observing

Figure 2.1 Marginal and joint probabilities.

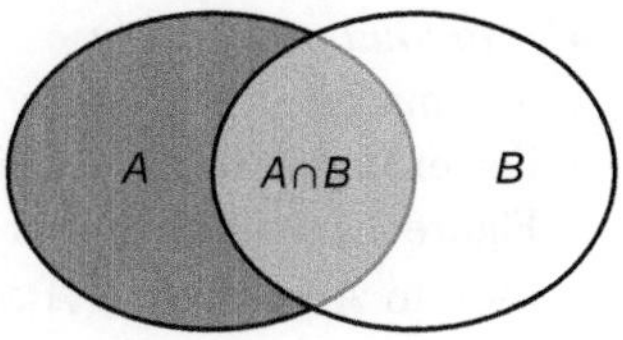

Figure 2.2 Union of two events, *A* and *B*, where (a) *A* and *B* are not mutually exclusive and (b) *A* and *B* are mutually exclusive.

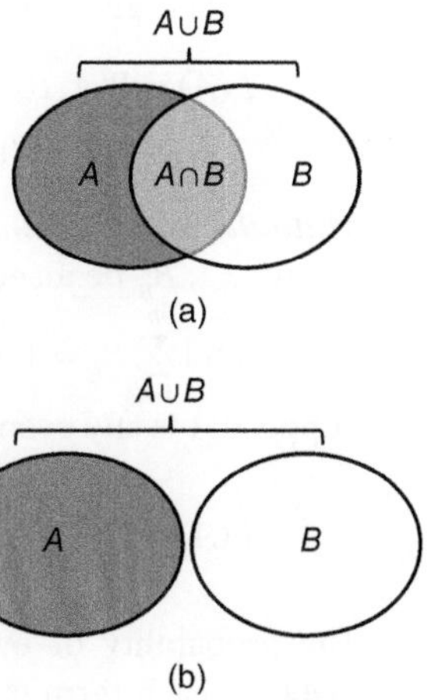

an ace card and B be the event of observing a black card. The conditional probability of A is true given that B is true is equal to observing an ace among the black cards which is 2/26 or 1/13. This probability can also be computed using the definition of conditional probability. The probability that the randomly selected card is an ace card and also black is 2/52. The probability of selecting a black card is 26/52. Therefore the ratio between the two is (2/52)/(26/52) = 2/26 = 1/13.

- *Independence*
  Events A and B are said to be *independent* if and only if
  $P(B) = P(B \mid A)$ or $P(A) = P(A \mid B)$.
  When A and B are independent,

$$P(A, B) = P(A \cap B) = P(A) \times P(B \mid A) = P(A) \times P(B).$$

- The *probability of the union* of two events A and B, is the probability that either A or B is true (see Figure 2.2a),

$$P(A \cup B) = P(A) + P(B) - P(A \cap B).$$

- When two events A and B are *mutually exclusive* (see Figure 2.2b),

$$P(A \cap B) = 0,$$
$$P(A \cup B) = P(A) + P(B).$$

- The *probability of the complement* of event A (i.e. A being false) is

$$P(\overline{A}) = 1 - P(A)$$

- *Marginal probability*
  The marginal probability of an event is given without reference to any other events. For example, the probability of event A is given by the area of the left side oval in Figure 2.1. However, this probability can also be computed by adding the areas common to A and B and A and $\overline{B}$. Since this is the union between two mutually exclusive events, $A \cap B$ and $A \cap \overline{B}$, $P(A)$ (see the sum of the dark and light gray areas in Figure 2.1) is given by

$$\begin{aligned} P(A) &= P(A, B) + P(A, \overline{B}) \\ &= P(A \mid B)P(B) + P(A \mid \overline{B})P(\overline{B}). \end{aligned}$$

- *Mutually exclusive and exhaustive events*
  Let $B_1, \ldots, B_n$ be a sequence of events in an experiment such that $P(B_i \cap B_j) = 0$ for all $i \neq j$ and $\sum_{i=1}^{n} B_i = 1$, then these events are called mutually exclusive and exhaustive. Suppose A is any event in the same experiment, $P(A)$ can be given by

$$P(A) = \sum_{i=1}^{n} P(A \mid B_i)P(B_i).$$

  The probability of event A is the weighted sum of the conditional probabilities $P(A \mid B_i)$, each term is weighted by the probability of the event that it is conditioned on. In some instances, it is easier to compute the probability of an event in this manner than direct computation.

Example 2.1 Assume there are two independent lotteries: A and B. From historical data, it is known that the probability of winning lottery A is 60% and the probability of winning lottery B is 40%. Tom decides to buy two tickets of lottery A and another two tickets of lottery B. What is the probability that Tom will win both lottery A and lottery B?

When Tom buys two tickets of lottery A, which has a 60% chance of winning, the probability that he will NOT win lottery A from either ticket is

$$(1 - 0.6) \times (1 - 0.6) = 0.16.$$

The probability that Tom will win lottery A from at least one of the two tickets is

$$1 - 0.16 = 0.84.$$

Let's call the event that Tom wins lottery A, *Win_A*, and the event that Tom does not win lottery A, *Lose_A*. The probability table of whether Tom wins lottery A is shown in Table 2.1.

Table 2.1 Probability table of whether Tom wins lottery A when he buys two tickets of lottery A.

	Win_A	*Lose_A*
Probability	0.84	0.16

Table 2.2 Probability table of whether Tom wins lottery B when he buys two tickets of lottery B.

	Win_B	*Lose_B*
Probability	0.64	0.36

Table 2.3 Probability of winning lottery A and B when buying two tickets of each lottery.

Joint probability of winning both A and B $P(Win\_A,\ Win\_B) = 0.84 \times 0.64 = 0.5376$	Joint probability of winning B, losing A $P(Lose\_A,\ Win\_B) = 0.16 \times 0.64 = 0.1024$	Marginal probability of winning B $P(Win\_B) = P(Win\_A,\ Win\_B) + P(Lose\_A,\ Win\_B) = 0.5376 + 0.1024 = 0.64$
Joint probability of winning A, losing B $P(Win\_A,\ Lose\_B) = 0.84 \times 0.36 = 0.3024$	Joint probability of losing both A and B $P(Lose\_A,\ Lose\_B) = 0.16 \times 0.36 = 0.0576$	Marginal probability of losing B $P(Lose\_B) = P(Win\_A,\ Lose\_B) + P(Lose\_A,\ Lose\_B) = 0.3024 + 0.0576 = 0.36$
Marginal probability of winning A $P(win\_A) = P(Win\_A,\ Win\_B) + P(Win\_A,\ Lose\_B) = 0.5376 + 0.3024 = 0.84$	Marginal probability of losing A $P(Lose\_A) = P(Lose\_A,\ Win\_A) + P(Lose\_A,\ Lose\_B) = 0.1024 + 0.0576 = 0.16$	

Similarly, when Tom buys two tickets of lottery B, which has a 40% chance of winning, the probability that he will NOT win lottery B from either ticket is

$$(1 - 0.4) \times (1 - 0.4) = 0.36.$$

The probability that Tom will win lottery B from at least one of the two tickets is

$$1 - 0.36 = 0.64.$$

Let's call the event that Tom wins lottery B, *Win_B*, and the event that Tom does not win lottery B, *Lose_B*. The probability table of whether Tom wins lottery B is shown in Table 2.2.

When Tom buys two tickets of lottery A and another two tickets of lottery B, the probability that Tom will win both A and B is a joint probability, i.e.

$$P(Win\_A,\ Win\_B) = 0.84 \times 0.64 = 0.5376.$$

This means that there is a 53.76% probability that Tom will win both lottery A and lottery B if he buys two tickets of each. More joint probabilities and marginal probabilities in this example are listed in Table 2.3.

What's the probability that Tom will win lottery B given that he has won lottery A? This is a conditional probability,

$$P(Win\_B \mid Win\_A) = \frac{P(Win\_B, Win\_A)}{P(Win\_A)}$$
$$= \frac{0.5376}{0.84}$$
$$= 0.64.$$

Note that this conditional probability is equivalent to the marginal probability of Tom winning lottery B, i.e.

$$P(Win\_B \mid Win\_A) = P(Win\_B) = 0.64.$$

This is not surprising, since Win_A and Win_B are independent events. In other words, the chance of winning lottery B is not dependent on whether or not Tom wins lottery A and vice versa.

Based on the basic concepts of probability theory, let us now take a look at the mathematical form of Bayes' theorem.

Let A and B be two events. The basic form of Bayes' theorem can be stated as

$$P(A \mid B) = \frac{P(B \mid A)P(A)}{P(B)}, \tag{2.1a}$$

where

- $P(A)$ is the initial belief (probability of A)
- $P(A \mid B)$ is the updated belief given the evidence B
- $P(B \mid A)$ is the probability of evidence given the initial belief A
- B is the evidence (or new data)
- $P(B)$ is the marginal probability of B.

To provide a more general definition of the Bayes' formula concerning discrete events, let's consider a mutually exclusive and exhaustive set of events $E_1, \ldots, E_n$ defined on a sample space, S. Let A and B be any two events defined on the same sample space, then a more general form of Bayes' theorem is given by

$$P(A \mid B) = \frac{P(B \mid A)P(A)}{P(B)}$$
$$= \frac{P(B \mid A)P(A)}{\sum_{i=1}^{n} P(B \mid E_i)P(E_i)}. \tag{2.1b}$$

Later in this chapter we will introduce a version of Bayes' theorem that is applicable to continuous random variables.

The basic idea of Bayesian reasoning is to learn from evidence as it accumulates and adjust prior belief accordingly. Note that prior beliefs are not necessarily beliefs based on past information. Some people tend to think that data gathered in the past should always be used as a prior belief. This may not be true in some cases. The prior belief (or probability) does not have to be conditional on anything that happened in the past. However, the posterior is always defined as the conditional probability, i.e. the probability conditional on new data.

To illustrate the basic concept, let's consider the following example.

Example 2.2 An impedance measurement is designed to monitor if there is an "open circuit" or a "short circuit" in a certain device. Specifically, abnormally high impedance indicates that there could be improper connections, broken components, wire fracture, etc. However, other causes could also result in a high impedance reading, so it doesn't always indicate an open circuit. It is known that the device failure probability is about 1%. Assume that there is a 95% chance that the impedance measurement successfully identifies a faulty device.

When a customer receives a new device and measures abnormally high impedance, is the device more likely to be defective or normal?

The prior, posterior, likelihood, and data in this example are as follows.

- *Prior probability*: Device failure probability = 1%.
- *Data*: Abnormally high impedance is measured/observed.
- *Likelihood*: The probability that impedance measurement successfully identifies a defective device is 95%.
- *Posterior probability*: Device failure probability given that abnormally high impedance is measured/observed.

Let's also introduce a few more definitions.

- *Sensitivity*: The proportion of positives that are correctly detected/identified.
- *Specificity*: The proportion of negatives that are correctly detected/identified.
- *Positive predicted value (PPV)*: The proportion of detected/identified positives that are true positives.
- *Negative predictive value (NPV)*: The proportion of detected/identified negatives that are true negatives.

In this specific example, these terms are defined as follows:

- Sensitivity = *P(identified fault|fault)*.
- Specificity = *P(identified success|success)*.
- PPV = *P(fault|identified fault)*.
- NPV = *P(success|identified success)*.

In this case, we would like to know the probability of PPV. If PPV is larger than 50%, that means the device is more likely to be defective when a customer measured abnormally high impedance.

To estimate PPV, based on Bayes' theorem,

$$
\begin{aligned}
PPV &= P(fault \mid I = identified\ \ fault) \\
&= \frac{P(I = identified\ \ fault \mid fault) \times P(fault)}{P(I = identified\ \ fault)} \\
&= \frac{P(identified\ \ fault \mid fault) \times P(fault)}{P(identified\ \ fault,\ \ true\ \ fault) + P(identified\ \ fault,\ \ true\ \ success)} \\
&= \frac{P(identified\ \ fault \mid fault) \times P(fault)}{P(identified\ \ fault \mid fault) \times P(fault) + P(identified\ \ fault \mid success) \times P(success)} \\
&= \frac{sensitivity \times P(fault)}{sensitivity \times P(fault) + (1 - specificity) \times (1 - P(fault))}.
\end{aligned}
\tag{2.2}
$$

Note that the probability in the denominator of the third expression in Eq. (2.2), i.e. *P(identified fault)*, is a marginal probability.

The prior probability is the probability of the device being defective, which is *P(fault)* = 1%.

Based on Eq. (2.2), we get that

$$\begin{aligned} PPV &= P(fault \mid identified\ fault) \\ &= \frac{sensitivity \times P(fault)}{sensitivity \times P(fault) + (1 - specificity) \times (1 - P(fault))} \\ &= \frac{0.95 \times 0.01}{0.95 \times 0.01 + 0.05 \times 0.99} \\ &= 16.1\%. \end{aligned}$$

In other words, when a customer receives a new device and sees a failure in the impedance measurement indication, the result is more likely to be a false positive rather than a true alarm. A more rational option for the customer is to ignore the measurement (a better solution would be to use other independent tests to follow the impedance measurement, see Example 7.3 for details). This may seem counter-intuitive because when we rely on the impression of intuition, human cognitive biases tend to exaggerate the diagnostic nature of the evidence (Kahneman 2013). Next, let's revisit this result using the weighted average concept introduced earlier.

Prior belief: Fault is unlikely (*P(fault)* = 1%).

Evidence/data: Fault is detected.

Note that the evidence contradicts the prior as it supports a different conclusion. Posterior belief is a weighted average of the prior belief and the data. Whether the posterior belief is closer to the prior belief or evidence depends on which one of the two is more heavily weighted.

On the other hand, *P(success|identified success)* = 99.9%, which means that when the impedance measurement indicates that the device is good, it is very likely that the device is a good one. This is not counter-intuitive, since both the evidence and the prior support the same conclusion, so the evidence strengthens the prior.

One solution to improve PPV is to increase specificity. When individual evidence is insufficient, and specificity cannot be improved, one can consider using multiple sensors (measurements from two or more independent test methods). This is widely applied in medical diagnosis. For example, a doctor uses screening tests such as ultrasound or blood tests to find out if a fetus has the risk of Down syndrome. When the screening test results indicate abnormalities, the doctor will use follow-up tests to further diagnose Down syndrome. However, if screening is normal, follow-up testing is usually not required.

2.3 Bayesian Inference (Point and Interval Estimation)

Example 2.2 shows how to apply Bayes' theorem for basic probability calculations. In that example, we assumed that the probability of a fault device, *P(fault)*, is a constant. In real-world applications, the prior probability is not always a fixed value. Let us consider Example 2.3, where the uncertainty of the prior reliability is taken into account.

Example 2.3 Historical data collected from the field helps us form a belief that the probability of a device defect has a normal distribution with a mean of 1% and a standard deviation of 0.2%. For simplicity, the same sensitivity and specificity (95%) are assumed. When a customer receives a new device and measures abnormally high impedance, what is the probability of the device being defective? We would like to know the point estimate and uncertainty of the probability of this defect.

Since the objective in Example 2.3 includes uncertainty estimation, let us first introduce the basic concepts of point estimation and interval estimation in Bayesian inference.

For a set of numbers, the *mean* (or expected value) is the average value, i.e. the total summation divided by the count of numbers. The *median* is the middle value in the set of numbers. The *mode* is the number that has the highest frequency of occurrence in the set.

In Bayesian statistics, unknown parameters are treated as random variables, with probability distribution assigned. The probability distribution represents uncertainty of this unknown parameter. The point estimate of the unknown parameter can be the mean, median, or mode of the probability distribution.

A Bayesian credible interval is usually used for interval estimation. A $(1 - \alpha) \times 100\%$ *Bayesian credible interval* is an interval for the parameter with the probability that the parameter lies in it is $(1-\alpha)$. As an example, consider that a particular parameter has a gamma probability distribution (discussed in Chapter 4) with the shape parameter 3 and the scale parameter 4. Figure 2.3 presents two different 90% credible intervals, one of which is the highest density interval and the other is the symmetric (or equal tail) credible interval. The highest density interval always contains the most probable values of the parameter. The symmetric interval is the one that equally distributes the non-coverage probability (0.10 in this case) to the left and right tails of the distribution. Equal tail intervals are the most commonly used method and may computationally be

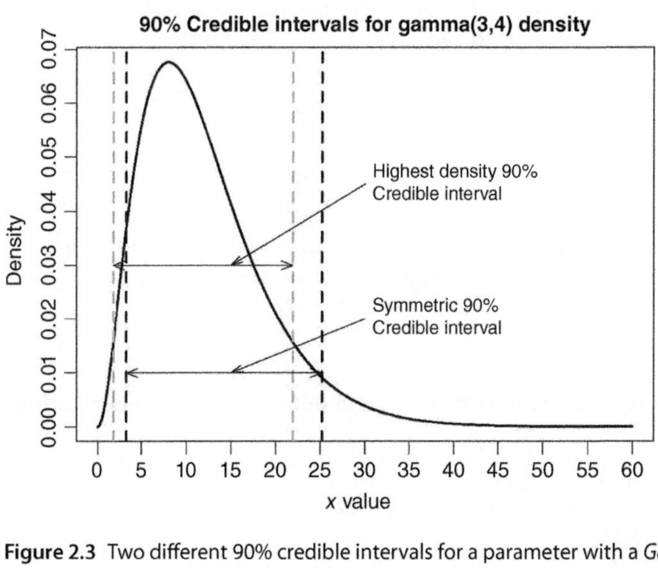

Figure 2.3 Two different 90% credible intervals for a parameter with a *Gamma*(3,4) distribution.

the simplest. In addition, it is invariant under parameter transformation. In this book, for simplicity purposes, we use equal-tail credible intervals in our examples.

Now back to Example 2.3. Given the abnormally high impedance measured, the probability of the device being defective is PPV. According to Eq. (2.2),

$$\begin{aligned} PPV &= P(fault \mid identified\ fault) \\ &= \frac{sensitivity \times P(fault)}{sensitivity \times P(fault) + (1 - specificity) \times (1 - P(fault))} \\ &= \frac{0.95 \times P(fault)}{0.95 \times P(fault) + 0.05 \times (1 - P(fault))}. \end{aligned} \tag{2.3}$$

Now let's apply Monte Carlo simulations to estimate PPV according to Eq. (2.3). The R script is shown in **`2.3_Interval_Estimation.R`**.

```
(2.3_Interval_Estimation.R)
## In order to get reproducible results, one has to set the seed at a
fixed value
set.seed(12345)

## specify_decimal is a function to show k number of decimals for
number x
specify_decimal <- function(x, k) format(round(x, k), nsmall=k)

## Number of iterations in Monte Carlo simulation
iter <- 100000

## Prior
P_fault <- rnorm(iter, mean = 0.01, sd = 0.002)

## Posterior
PPV <- 0.95*P_fault/(0.95*P_fault+0.05*(1-P_fault))

## show results of posterior mean, median, and 95% credible interval
print(paste("posterior mean is:", specify_decimal(mean(PPV),3)))
print(paste("posterior standard deviation is:", specify_decimal
(sd(PPV),3)))
print(paste("posterior median is:", specify_decimal(quantile
(PPV, 0.50),3)))
print(paste("posterior 95% credible interval is:", specify_decimal
(quantile(PPV, 0.025),3), ",", specify_decimal(quantile(PPV, 0.975),3)))

# Histogram of posterior
# jpeg("Example2.3_PPV_distribution.jpeg", width = 6, height = 4,
units = 'in', res = 1800)  # save the plot as jpeg format
hist(PPV, xlab="PPV", xlim=c(0,0.3),breaks=100)
box()
# dev.off()
```

After running the R script, the results below are obtained. The PPV distribution is shown in Figure 2.4.

```
[1] "posterior mean is: 0.160"
[1] "posterior standard deviation is: 0.027"
[1] "posterior median is: 0.161"
[1] "posterior 95% credible interval is: 0.104 , 0.212"
```

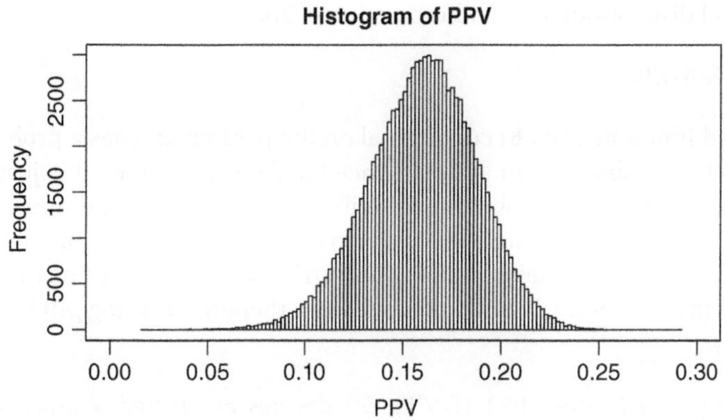

Figure 2.4 PPV distribution.

From Monte Carlo simulations, posterior distribution (PPV distribution) has a mean of 0.16 and a standard deviation of 0.03. The other statistics are:

- median = 0.16
- 2.5th and 97.5th percentiles are 0.10, and 0.21, respectively.

Since the mean and median of the posterior distribution of PPV is the same, it is a symmetric distribution. Therefore, 2.5% and 97.5% quantiles of the distribution provide a symmetric (around the mean) two-sided 95% credible interval. To summarize this information, we can state that the distribution of PPV has a mean (point estimate) of 0.16 and there is a 95% chance that it is between 0.10 and 0.21. So, when a customer receives a new device and measures abnormally high impedance, on the average there is a 16% chance that it is defective, but the defective rate could vary between 10% and 21% with probability 0.95. The range 10–21% is a 95% Bayesian credible interval (or Bayesian confidence interval) of the device defective rate.

2.4 Selection of Prior Distributions

2.4.1 Conjugate Priors

In Example 2.3, prior and posterior both have probability distributions, but the likelihood (defined as sensitivity in the example) is a constant. In Bayesian statistics, parameters of interests are always assumed to have some probability distribution.

Bayes' theorem provides a mathematical mechanism to update the prior belief (i.e. distribution) of the parameters given the observed data. Let $h(\boldsymbol{\theta})$ be the prior probability distribution of the parameter vector $\boldsymbol{\theta}$ and $f(\mathbf{y} \mid \boldsymbol{\theta})$ be the sampling distribution of data $\mathbf{y}$ given the parameters $\boldsymbol{\theta}$. The function $f(\mathbf{y} \mid \boldsymbol{\theta})$ is also known as the likelihood function. Bayes' theorem states that the posterior distribution of $\boldsymbol{\theta}$ given the data $\mathbf{y}$, $p(\boldsymbol{\theta} \mid \mathbf{y})$, can be obtained by

$$p(\boldsymbol{\theta} \mid \mathbf{y}) = \frac{f(\mathbf{y} \mid \boldsymbol{\theta})\ h(\boldsymbol{\theta})}{k(\mathbf{y})}, \tag{2.4}$$

where $k(\mathbf{y})$ is the marginal distribution of $\mathbf{y}$ and is computed from

$$k(\mathbf{y}) = \int f(\mathbf{y} \mid \boldsymbol{\theta})\, h(\boldsymbol{\theta})\, d\boldsymbol{\theta}$$

Typically, the likelihood function, $f(\mathbf{y} \mid \boldsymbol{\theta})$ conditional on the parameters, has a probability distribution prior to the observation of data. Once the data is observed it is just a function of the parameters and not a probability distribution. The following example shows how this plays out. The marginal distribution $k(\mathbf{y})$ in the denominator of Eq. 2.4 is a normalizing constant which does not depend on $\boldsymbol{\theta}$. This can be observed by noting that numerator of the right-hand side of Eq. 2.4 integrates to $k(\mathbf{y})$, therefore making $p(\boldsymbol{\theta} \mid \mathbf{y})$ a true probability density.

Example 2.4 In a design verification test (DVT), 59 devices are tested against a requirement. All 59 devices successfully passed. Suppose we do not have other information about the reliability associated with this requirement. Here we loosely define reliability as the probability that the devices meet the design verification requirements. What is the device reliability point estimate? What are the device reliability point estimate and 95% credible interval?

We apply Bayes' theorem to estimate the device reliability, θ, given that x successes out of n trials are observed. Based on Bayes' theorem,

$$p(\theta \mid x) = \frac{f(x \mid \theta)\ h(\theta)}{k(x)},$$

where

$h(\theta)$ is the prior probability density of unknown device reliability θ

$f(x \mid \theta)$ is the likelihood function of data x given reliability θ (in this case x is the number of devices that will successfully pass the test given the reliability θ)

$k(x)$ is the marginal probability density of data x obtained by integrating out all possible values of θ in the likelihood $f(x \mid \theta)$

$p(\theta \mid x)$ is the posterior probability density of reliability θ given data x. We are interested in estimating the parameter θ using its posterior distribution.

We use a beta distribution, $Beta(\alpha, \beta)$ to represent the prior belief, $h(\theta)$, of the device reliability θ. Since there is no historical information about the device reliability, we could select $Beta(1, 1)$, which is also a uniform distribution. This prior distribution indicates that the reliability of the device is between 0 and 1, and the probability of any reliability value is the same.

A beta probability density function (pdf) of device reliability θ is

$$h(\theta \mid \alpha, \beta) = \frac{\theta^{\alpha-1}(1-\theta)^{\beta-1}}{B(\alpha, \beta)}, \tag{2.5}$$

where $B(\alpha, \beta) = \frac{\Gamma(\alpha)\Gamma(\beta)}{\Gamma(\alpha+\beta)}$ and we assume that α and β are fixed with known values; $\Gamma(x)$ is the gamma function.

We use a binomial distribution as the likelihood function. A binomial distribution is commonly used for modeling x successes out of n identical and independent trials with probability of success θ. The probability mass function of a binomial distribution is

$$f(x \mid \theta) = \binom{n}{x} \theta^{x}(1-\theta)^{n-x}, \quad x = 0, \ldots, n \tag{2.6}$$

$k(x)$ is the marginal probability density of data x (prior to observing the data), i.e.

$$k(x) = \int f(x \mid \theta)h(\theta)d\theta = \int \binom{n}{x} \theta^x (1-\theta)^{n-x} \frac{\theta^{\alpha-1}(1-\theta)^{\beta-1}}{B(\alpha,\beta)} d\theta.$$

As mentioned earlier, marginal probability density evaluated at the observed data x is a normalizing constant which makes the posterior a probability density. In most cases in Bayesian statistics, the marginal probability density is hard to compute and may not have a closed-form solution. The reason we chose the beta distribution as the prior and the binomial distribution as the likelihood is that the beta distribution is the conjugate prior of the binomial likelihood function (we will talk about its definition later). Therefore, we can have a closed-form solution of the posterior distribution.

Based on Bayes' theorem, the posterior distribution of θ is

$$\begin{aligned} f(\theta \mid x) &= \frac{f(x \mid \theta)h(\theta)}{k(x)} = \frac{f(x \mid \theta)h(\theta)}{\int_0^1 f(x \mid \theta)h(\theta)\mathrm{d}\theta} \\ &= \frac{\binom{n}{x} \theta^x (1-\theta)^{n-x} \frac{\theta^{\alpha-1}(1-\theta)^{\beta-1}}{B(\alpha,\beta)}}{\int_0^1 \binom{n}{x} \theta^x (1-\theta)^{n-x} \frac{\theta^{\alpha-1}(1-\theta)^{\beta-1}}{B(\alpha,\beta)} \mathrm{d}\theta} \\ &= \frac{\frac{\binom{n}{x}}{B(\alpha,\beta)} \theta^{x+\alpha-1}(1-\theta)^{n-x+\beta-1}}{\frac{\binom{n}{x}}{B(\alpha,\beta)} \int_0^1 \theta^{x+\alpha-1}(1-\theta)^{n-x+\beta-1}\mathrm{d}\theta} = \frac{\theta^{x+\alpha-1}(1-\theta)^{n-x+\beta-1}}{\int_0^1 \theta^{x+\alpha-1}(1-\theta)^{n-x+\beta-1}\mathrm{d}\theta} \\ &= \frac{\frac{\theta^{x+\alpha-1}(1-\theta)^{n-x+\beta-1}}{B(x+\alpha,\ n-x+\beta)}}{\int_0^1 \frac{\theta^{x+\alpha-1}(1-\theta)^{n-x+\beta-1}}{B(x+\alpha,\ n-x+\beta)}\mathrm{d}\theta}. \end{aligned} \tag{2.7}$$

Note that the numerator in the equation above is a beta density, $Beta\ (x+\alpha, n-x+\beta)$. The denominator is the integral of beta density $Beta\ (x+\alpha, n-x+\beta)$, so the denominator equals 1. Thus, the posterior distribution of θ is also a beta distribution,

$$p(\theta \mid x) = \frac{\theta^{x+\alpha-1}(1-\theta)^{n-x+\beta-1}}{B(x+\alpha,\ n-x+\beta)} \sim Beta\ (x+\alpha, n-x+\beta). \tag{2.8}$$

Equations (2.7) and (2.8) show that when a prior is a beta distribution and the likelihood is a binomial distribution, the posterior has a closed-form solution, which is also a beta distribution. In this case, the beta distribution is called a conjugate prior. A conjugate prior has similar mathematical expression as the likelihood and the posterior distribution. When a conjugate prior is used in conjunction with a specific likelihood function, the posterior distribution will have similar mathematical expression as the prior after applying Bayes' theorem. In these cases, the inferences based on the posterior distributions are straightforward and simpler.

Table 2.4 Conjugate priors for commonly used discrete likelihood distributions.

Likelihood distribution (parameter)	Conjugate prior
Bernoulli (π)	Beta
Binomial (π)	Beta
Negative binomial (π) with known failure number, k	Beta
Poisson (λ)	Gamma

Table 2.5 Conjugate priors for commonly used continuous likelihood distributions.

Likelihood distribution (parameter)	Conjugate prior
Exponential (λ)	Gamma
Gamma (α *shape known, rate* β)	Gamma
Normal (μ, σ^2 *known*)	Normal
Normal (μ *known*, σ^2)	Inverse gamma
Norma (μ, σ^2)	Normal, inverse gamma
Lognormal (μ, *known precision* $\tau(={}^1/_{\sigma^2})$)	Normal
Lognormal (μ *known*, τ)	Gamma
Multivariate normal (μ, $\sum$ *known covariance matrix*)	Multivariate normal
Multivariate normal (*known mean vector* μ, $\sum$)	Inverse Wishart

Before high-speed computers were invented, statisticians used conjugate priors to compute the posterior distributions in Bayesian statistics. However, the Bayesian analysis should not be limited to priors that are mathematically tractable. With the availability of powerful computers and sophisticated sampling techniques this approach is no longer necessary.

Recent breakthroughs in computational algorithms and the application of Markov chain Monte Carlo (MCMC) sampling methods have made it feasible to solve more complex Bayesian models, or to solve problems where conjugate priors don't exist. We will introduce MCMC algorithms in Chapter 3. Nowadays, the choice of prior distribution doesn't have to be a conjugate prior to obtain a posterior distribution. However, conjugate priors still play an important role when they are flexible enough to describe the prior distributions of the parameters of interest. Tables 2.4 and 2.5 list conjugate priors for the indicated parameters of commonly used discrete and continuous likelihood distributions. We will continue to talk about other conjugate priors in Chapter 4.

2.4.2 Informative and Non-informative Priors

A prior distribution is called non-informative if it is "flat" in relation to the likelihood function. Any value of the parameter is equally likely under a flat prior distribution.

As an example, uniform (0,1) distribution is a flat prior for the device reliability θ in Example 2.4. According to this distribution, θ can be anywhere between 0 and 1 with equal probability. Sometimes non-informative priors such as flat priors could lead to an improper (non-integrable) posterior. We cannot make any predictions or statistical inferences with improper posterior distributions and therefore we must be careful when choosing non-informative priors. Choosing a non-informative prior does not imply total ignorance of the parameter of interest. Non-informative priors are often not invariant under transformation. This means a prior may be non-informative of a parameter in one parameterization, but under a certain transformation it may no longer be non-informative.

Another type of non-informative prior is called Jeffreys' prior. One of the most useful features of Jeffreys' prior is that it is invariant to parameterization (Bolstad 2007; Gelman et al. 2013).

Jeffreys' prior is proportional to the square root of the determinant of the expected Fisher information matrix of the selected model

$$p(\boldsymbol{\theta}) \propto |I(\boldsymbol{\theta})|^{1/2},$$

where $I(\boldsymbol{\theta})$ is the expected Fisher information matrix, i.e.

$$I(\boldsymbol{\theta}) = -E_{X|\theta}\left[\frac{\partial^2}{\partial\theta^2}\log f(X \mid \boldsymbol{\theta})\right],$$

where $f(\boldsymbol{X} \mid \boldsymbol{\theta})$ is the likelihood function of the data $\boldsymbol{X}$ given the parameter vector $\boldsymbol{\theta}$.

When the likelihood function is a binomial distribution $X \sim Binomial\ (n, \theta)$, the Jeffreys' prior of θ is *Beta* (0.5, 0.5). In this case Jeffreys' prior turns out to be the conjugate prior as well. However, usually this is not the case and Jeffreys' prior can even be an improper prior (i.e. distribution does not integrate to finite value). Proof of why Jeffreys' prior is invariant to reparameterization and an example of the binomial distribution can be found in Appendix D.

In Example 2.4 we used the conjugate prior (beta distribution) and the binomial likelihood distribution to estimate the posterior distribution of the reliabilityθ. Assuming that

- the prior distribution of reliability θ has a beta distribution *Beta* (α, β)
- the likelihood function is a binomial distribution with x successes out of n trials, then
- the posterior distribution of θ is also a beta distribution,

$$p(\theta \mid n, x) \sim Beta\ \ (x + \alpha, n - x + \beta) \tag{2.9}$$

- the posterior mean value of θ is given by

$$\frac{x+\alpha}{x+\alpha+n-x+\beta} = \frac{x+\alpha}{n+\alpha+\beta} = \frac{x}{n+\alpha+\beta} + \frac{\alpha}{n+\alpha+\beta}$$
$$= \frac{x/n}{1+\dfrac{\alpha+\beta}{n}} + \frac{\alpha/(\alpha+\beta)}{\dfrac{n}{\alpha+\beta}+1}.$$

Let $\omega = \dfrac{1}{1+\dfrac{\alpha+\beta}{n}}$ then $1-\omega = 1 - \dfrac{1}{1+\dfrac{\alpha+\beta}{n}} = \dfrac{(\alpha+\beta)/n}{1+\dfrac{\alpha+\beta}{n}} = \dfrac{1}{\dfrac{n}{\alpha+\beta}+1}$.

From this posterior it can be seen that the posterior mean of θ is the weighted sum of the prior mean, $\alpha/(\alpha+\beta)$, and the data mean x/n. If there is a lot of data (n is much larger than $\alpha+\beta$), then the posterior mean is dominated by the data. On the other hand, if the amount of data is small (n is much smaller than $\alpha+\beta$), the posterior mean is mostly influenced by the prior mean. The posterior mean is more sensitive to the selection of prior distributions when the amount of data is small. When the sample size is large enough, the Bayesian solution to solving a problem is fairly close to the frequentist solution, regardless of the choice of prior distribution. In the following example, we use a few non-informative and informative prior distributions of θ, and evaluate their effects on the posterior distribution.

Now we revisit Example 2.4 by applying different prior distributions. We choose the following prior distributions for reliability θ (here reliability is defined as percentage conforming to the requirement).

1) Non-informative prior *Beta* (1, 1). This is also a uniform distribution [0,1]. By using this prior, we assume there is no prior information on θ (i.e. θ can be any value between 0 and 1 with equal probability).
2) Jeffreys' prior *Beta* (0.5, 0.5). This non-informative prior puts more weight on the tails but equally between the left and right tails.
3) Informative prior *Beta* (1, 9). This prior tells us that θ has a mean of 0.1 based on previous knowledge, but we allow some uncertainty (variance = 0.0082, standard deviation =0.09) on our prior belief.
4) Informative prior *Beta* (9, 1). This prior tells us that θ has a mean of 0.9 based on previous knowledge, but we allow some uncertainty (variance = 0.0082, standard deviation = 0.09) on our prior belief.
5) Strongly informative prior *Beta* (180, 20). This prior tells us that θ has a mean of 0.9 based on previous knowledge. Assuming we are much more certain compared to our belief in case 4), here we set a stronger prior with less uncertainty (variance = 0.0004, standard deviation = 0.02).

The R script to calculate the posterior distributions of reliability θ (indicated as p in the R script) based on the different priors discussed above is shown in `2.4.2_Binomial.R`.

```
(2.4.2_Binomial.R)
#######################################################
##   Calculate and plot p based on different priors  ##
##   Prior: Beta                                      ##
##   Likelihood: Binomial                             ##
#######################################################

## Generate sequence of values for plotting the probability density
functions

p = seq(0,1,length=500)

# Flat prior
# a = 1
# b = 1

# Jeffreys prior
```

```
  # a = 0.5
  # b = 0.5

  # Informative prior beta(1,9): p mean = 0.1
  # a = 1 # Number of successes in a previous test
  # b = 9 # Number of failures in a previous test

  # Informative prior beta(9,1): p mean = 0.9
  # a = 9 # Number of successes in a previous test
  # b = 1 # Number of failures in a previous test

  # Strong Informative prior beta(180,20): p mean = 0.9
   a = 180 # Number of successes in a previous test
   b = 20 # Number of failures in a previous test

  # # Data
  d = 59 # Number of successes in current test
  g = 0 # Number of failures in current test

  # Prior density
  prior=dbeta(p,a,b)

  # Likelihood
  like=dbeta(p,d+1,g+1) #This is rescaled likelihood. Actual likelihood
= dbeta(p,d+1,g+1)/(d+g+1)

  # Posterior density
  post = dbeta(p,a+d,b+g)

  # Create plot
  jpeg("Binomial_Beta_180-20.jpeg", width = 6, height = 4, units = 'in',
res = 600)  # save the plot as jpeg format
  plot(p,like,type="l",ylab="Density",xlab="Reliability (p)",lty=1,lwd=2,
col="red",ylim=c(0,30)) # plot likelihood
  lines(p,post,lty=2,lwd=2,col="blue") # add posterior
  lines(p,prior,lty=3,lwd=2,col="purple") # add prior
  legend(.5,30,c("prior","Likelihood","Posterior"),lty=c(3,1,2),
lwd=c(2,2,2),col=c("purple","red","blue")) # add legend
  dev.off()
  #dev.copy2eps(file="Binomial_Beta_1-1.eps")

  # Posterior of p is Beta(a+d,b+g)
  # Reliability mean
  Reliability_mean = (a+d)/(a+d+b+g)
  print(paste("Mean of the reliability is:",Reliability_mean))

  # A 95% credible interval for the reliability is given by CredInt_95
= qbeta(c(.025,.975),a+d,b+g)
  CredInt_95 = qbeta(c(.025,.975),a+d,b+g)
print(paste("95% Credible interval for the reliability is given
by:",CredInt_95[1],CredInt_95[2]))
```

The five different prior distributions discussed above, along with the corresponding posterior distributions of reliability θ (p in the R script), and the rescaled likelihood are plotted in Figure 2.5. Figure 2.5a shows that since the prior is non-informative, the posterior is the same as the likelihood, which means the prior has no influence on the

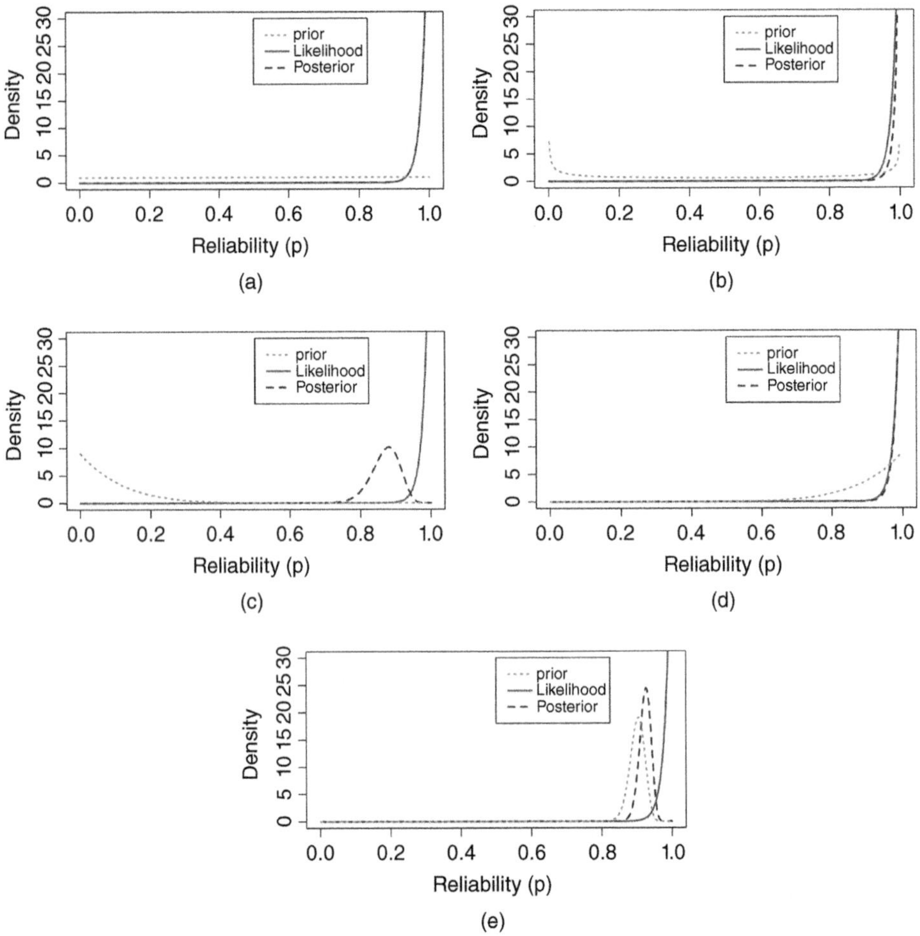

Figure 2.5 Prior, likelihood, and posterior of reliability θ (also p) with five different prior distributions: (a) *Beta*(1, 1) prior, (b) *Beta*(0.5, 0.5) prior, (c) *Beta*(1, 9) prior, (d) *Beta*(9, 1) prior, and (e) *Beta*(180, 20) prior (dotted line, prior; solid line, likelihood; dashed line, posterior).

posterior of θ. In Figure 2.5b, Jeffreys' prior is used, which is also non-informative. The posterior is very close to the likelihood, and the prior has little influence. In Figure 2.5c, the prior indicates that θ is most likely low, but the data (likelihood) shows that θ is most likely high. The prior contradicts the data, so the posterior mean is somewhere in between. This indicates that the prior data should not be used in deriving the prior distribution. In Figure 2.5d, the prior and the likelihood pretty much agree with each other (both indicate the reliability is most likely high), so the posterior is further strengthened (prior mean = 9/10 = 0.9; posterior mean = 68/69 = 0.9855). In Figure 2.5e, as the prior is very strong with less uncertainty compared to the prior in Figure 2.5d, the posterior is very close to the prior, which indicates that the prior has a much larger influence compared to the data on the posterior distribution. The 95% equal tail credible intervals of θ are given in Table 2.6. There is a substantial difference in the credible intervals of the reliability θ, depending on which prior distribution was used.

Table 2.6 Equal tail 95% credible intervals of θ.

θ prior	Mean of θ	Posterior 95% equal-tail credible interval of θ	
		Lower bound	Upper bound
Beta (1, 1)	0.9836	0.9404	0.9996
Beta (0.5, 0.5)	0.9917	0.9585	1.0000
Beta (1, 9)	0.8696	0.7813	0.9377
Beta (9, 1)	0.9855	0.9472	0.9996
Beta (180, 20)	0.9228	0.8874	0.9520

The basic question is what amount of historical data should be borrowed for estimating the parameters of interest. The amount of borrowing must depend on how much the current data agree with the historical data. We will discuss different approaches for addressing this issue in later chapters.

Selection of informative priors can sometimes result in sample size reduction. If previous data closely match the current data, Bayesian methods can be used to reduce sample size compared to traditional frequentists methods. Examples will be given in later chapters to demonstrate this advantage of Bayesian solutions.

It is also flexible to use Bayesian models to combine different sources of information (e.g. DVT data, field data, manufacturing data, domain knowledge, etc.) and update inferences when new data becomes available. When combining different sources of information, it is feasible to achieve sample size reduction using informative priors. For example, in a previous test, 22 parts passed requirement, in the following test 59 parts of the same design passed the same requirement. These data can be combined to reduce the uncertainty of the reliability estimation.

However, real-life engineering problems can be more complicated than this simple case. For example, in each of several historical tests, 22 parts passed requirement; in the following test 59 parts of a similar (but not identical) design passed the same requirement. How can we leverage the previous information? It is not fair to fully use the parts in previous tests since they do not have an identical design to the current parts. However, it is not reasonable to totally ignore the history either, since the designs are somewhat similar. In this case, we can consider Bayesian hierarchical models to leverage the partial strength of the previous data. Hierarchical models will be introduced in Chapter 8.

As another example, we may have different sources of information about the reliability of the same design, but different sources of information may have different levels of credibility. Some data sources may be less trustworthy than other sources due to potential bias or misclassification. Meanwhile, no single source of information might be sufficient to provide precise reliability estimation. In such cases, Bayesian models can be very flexible to combine all sources of information and give different weights to different data sources, thus the strength of all information can be combined. We will provide an example to illustrate this concept in Chapter 7.

Non-informative priors (similar terms include vague priors, diffuse priors, flat priors, etc.) are used when there is little prior information or knowledge about the unknown parameters. Non-informative priors are used in most of the examples in this book.

2.5 Bayesian Inference vs. Frequentist Inference

To compare Bayesian inference with frequentist inference, first let us see the maximum likelihood estimation in Example 2.4. Recall that the binomial distribution probability mass function of getting x successes out of n trials with success probability device reliability θ is

$$L = f(x \mid \theta) = \binom{n}{x} \theta^x (1-\theta)^{n-x}.$$

This is the likelihood function. The maximum likelihood estimation is to find $\hat{\theta}$ to maximize this likelihood function. We take the logarithm of the likelihood function, which is

$$\log(L) = \log \binom{n}{x} + x \log \theta + (n-x) \log(1-\theta).$$

The solution $\hat{\theta}$ to maximize the likelihood function is the same one to maximize the $\log(L)$. In order to find $\hat{\theta}$, let the derivative of $\log(L)$ be zero, which is

$$\frac{\partial \log(L)}{\partial \theta} = \frac{x}{\theta} - \frac{n-x}{1-\theta} = 0.$$

Thus, we have the maximum likelihood solution,

$$\hat{\theta} = \frac{x}{n}.$$

In Example 2.4, when there are 59 successes out of 59 trials, the reliability maximum likelihood estimation is

$$\hat{\theta} = \frac{x}{n} = \frac{59}{59} = 1.$$

Note that the Bayesian solution point estimate (using the mean, based on flat prior) is not 1 (refer to Table 2.6).

Table 2.7 shows the difference between frequentist and Bayesian inference. As shown in earlier sections in this chapter, Bayesian inference of an unknown parameter is based on the posterior distribution of that parameter.

In frequentist inference, it is assumed that parameters are fixed but unknown. In the previous example, maximum likelihood estimation is a common way to compute the point estimate of the unknown parameter. A frequentist confidence interval (e.g. 95% confidence interval) is often given to quantify the uncertainty associated with the

Table 2.7 Frequentist vs. Bayesian inference.

	Frequentist	**Bayesian**
Unknown parameters	Unknown parameters are treated as *fixed* (estimated with some confidence)	*Probability distributions* are used to summarize the uncertainty about the unknown parameters
Parameter estimation method	Methods such as maximum likelihood estimator (MLE)	Using posterior distribution computed conditional on the data based on Bayes' theorem

estimated parameter. The frequentist confidence interval is probably one of the most confusing concepts for engineers.

For example, consider a 95% confidence interval for a parameter. Before the data is observed, there is a 95% chance (or 0.95 probability) that the unknown parameter will be contained within the lower and upper bounds of the 95% confidence interval. Notice that the lower and upper confidence limits are random quantities before the data is observed. After the data is observed and the upper and lower confidence limits are computed, we cannot say that there is 0.95 probability that the unknown parameter lies between the lower and upper confidence bounds. However, many engineers believe this is the case. A frequentist confidence interval either contains the true value or does not contain the true value, i.e. the probability that a true value lies within a particular frequentist confidence interval is either 0 or 1.

In frequentist inference, a 95% confidence interval actually means the following: imagine you have the luxury to repeat the experiment again and again, and in each experiment you take a random sample and estimate the 95% confidence interval of an unknown parameter, then 95% of the estimated confidence intervals contain the true value of the unknown parameter, and the other 5% of confidence intervals do not contain the true value. Repeated sampling is a concept in frequentist inference, which means the dataset you observed from your experiment is just one random dataset out of many possible datasets, and the experiment will be repeated over and over again. However, in real engineering practice, most likely this repeated experimentation never happens. Instead, an engineer usually only takes a sample once, makes a one-time measurement, and has only one estimation based on the only dataset available. When interpreting frequentist confidence intervals, one has to imagine that the observed sample is just one of many possible samples, though the other samples are just imaginary and will probably never happen in real life. When we say 95% or 90% confidence interval, the number 95% or 90% only indicates the quality of the imaginary re-sampling process.

In Bayesian inference, once the data is observed it is fixed and then the inference will be based on the posterior probability distribution conditional on the data. In a Bayesian framework, a 95% credible interval (or Bayesian confidence interval) does mean that there is 95% chance that the unknown parameter is between the lower and higher confidence bounds, which is more intuitive compared to the meaning of the frequentist confidence interval. There is no need to talk about repeated sampling to describe the Bayesian credible interval. Additional samples (if any) from the same process may help get a better estimate of the posterior distribution of the parameter. This does not mean that the sampling variability in the data will not have any impact on the posterior distribution of the parameter. With a small sample, the posterior distribution of the parameter is susceptible to more uncertainty than with a large sample. We will show examples of how sample size affects the Bayesian credible interval in later chapters. When there is large amount of data, the frequentist confidence interval and the Bayesian credible interval may produce similar results.

We use the following example to further illustrate the interpretation of the frequentist confidence interval to help readers understand the difference between frequentist confidence interval and Bayesian credible/confidence interval.

Example 2.5 It is known that in a container there are 10 000 balls in total, out of which there are 2000 red balls to represent defective parts and the rest are white balls to represent non-defective parts. So the true non-defective rate, θ, is 1–2000/10000 = 80%. Each time the reliability engineer randomly takes 59 balls out of the container and calculates the 95% confidence interval of the non-defective rate. He puts all the balls back in the container and repeats this process 100 times. This method of sampling is called sampling with replacement. Now let's examine what percentage of the 100 calculated 95% confidence intervals for the non-defective rate actually contain the true rate of 0.80. Because we are constructing a 95% confidence interval, the expectation is that approximately 95% of the intervals would contain the true rate.

From the previous analysis, a point estimate of the non-defective rate, θ, can be obtained by the maximum likelihood estimation method and is given by

$$\hat{\theta} = \frac{x}{n},$$

where x is the number of successes and n is the sample size (59 in this case).

A frequentist confidence interval for θ can be obtained using the fact that the sampling distribution of $\hat{\theta}$ is asymptotically normally distributed with a mean value of θ and an approximate standard deviation of $\sqrt{\hat{\theta}(1-\hat{\theta})/n}$. When both $n\hat{\theta}$ and $n(1-\hat{\theta})$ are at least 5, an approximate $(1-\alpha) \times 100\%$ two-sided symmetric confidence interval for θ is given by $(\hat{\theta} + z_{\alpha/2}\hat{\sigma},\ \hat{\theta} - z_{\alpha/2}\hat{\sigma})$, where $\hat{\sigma} = \sqrt{\hat{\theta}(1-\hat{\theta})}$, $(1-\alpha) \times 100\%$ is the confidence level, and $z_{\alpha/2}$ is the $\alpha/2$ quantile of the standard normal distribution. In our example, $(1-\alpha) = 0.95$ and therefore $\alpha = 0.05$. This leads to $z_{\alpha/2} = -1.96$.

When the sample size is small (typically, $n < 30$) or the number of either failures or successes in the sample is less than five then the normal approximation to the sampling distribution of $\hat{\theta}$ is not appropriate and therefore the resulting confidence interval may not have proper coverage. In these cases, one can choose from other alternative confidence interval calculation methods that are more appropriate than the normal approximation. One of the most commonly used exact methods is the Clopper–Pearson method (see Agresti 2002 for details).

The following R script (`2.5_One_Proportion_CI.R`) shows a simulation approach to obtain an estimate of the coverage probability of the frequentist 95% confidence interval.

```
(2.5_One_Proportion_CI.R)
##########################################
#   Calculate confidence interval        #
#   of 1 proportion sampling             #
##########################################

set.seed(123)# To be able to reproduce the results choose a fixed seed

# create a vector containing 2000 bad parts. A bad part is indicated by
number 0.
bad <- rep(0, 2000)

# create a vector containing 8000 good parts. A good part is indicated
by number 1.
good <- rep(1, 8000)
```

```
  # Now let's select 59 parts at random each time from these 10000 parts
(combining good
  # and bad parts) with replacement and calculate the frequentist 95%
confidence interval.
  # If each confidence interval contains the true non-defective rate,
we count it as 1.
  # Repeat the process 100 times and
  # compute the percentage of 1.  This is an estimate of the coverage
probability of the
  # frequentist 95% confidence interval

  sample_NDrate <- numeric(100)
  sample_NDrate_LCB <- numeric(100)
  sample_NDrate_UCB <- numeric(100)

  for (i in 1:100) {
    sample_data <- sample(c(good, bad), 59)
    sample_NDrate[i] <- sum(sample_data)/59

    # confidence interval calculated using Normal approximation
    sample_NDrate_LCB[i] <- sample_NDrate[i] - 1.96*sqrt
(sample_NDrate[i]*(1-sample_NDrate[i])/59)
    sample_NDrate_UCB[i] <- sample_NDrate[i] + 1.96*sqrt
(sample_NDrate[i]*(1-sample_NDrate[i])/59)
  }

  #  hist(sample_NDrate)

  require(plotrix)
  # jpeg("Example2.5_Confidence_interval.jpeg", width = 8, height = 4,
units = 'in', res = 1800)  # save the plot as jpeg format
  plotCI(1:100, sample_NDrate, ui=sample_NDrate_UCB, li=sample_NDrate_LCB)
  abline(h = 0.8, lwd = 1)
  box()
  # dev.off()
```

The 100 estimated confidence intervals are plotted in Figure 2.6 and compared with the true non-defective rate, 0.8 (indicated by the horizontal solid line). Note that out of the 100 confidence intervals, there are six confidence intervals that do not contain the true non-defective rate of 0.8. These six confidence intervals are highlighted by the ovals. So, the coverage probability of the frequentist confidence intervals in this example is 94%, pretty close to the expected coverage of 95%. If we had repeated the simulation 1000 times or 10 000 times we would have been able to get a better estimate of the coverage probability.

This example demonstrates what frequentist confidence interval really means. A 95% confidence interval means that, if repeating the experiment over and over again, out of all the calculated confidence intervals using the same method, about 95% of the confidence intervals contain the true value. But in real life, people usually only sample once, and calculate the confidence interval once. The calculated confidence interval either contains or does not contain the true non-defective rate.

Suppose we are dealing with a system that comprises many subsystems and each subsystem has many components. Suppose we are interested in the system level reliability. We can use frequentist methods to estimate component level reliability estimates and

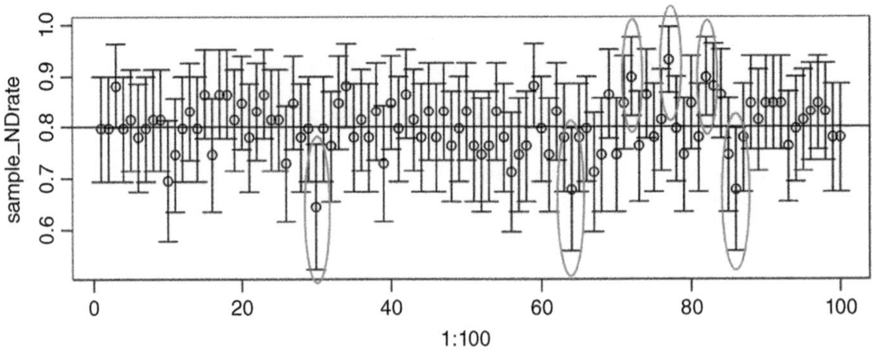

Figure 2.6 100 confidence intervals compared to the true non-defective rate of 0.8 (indicated by the horizontal solid line).

the corresponding uncertainties of these estimates as confidence intervals. However, it may not be easy to use frequentist methods to combine component level reliabilities to estimate the subsystem or system level reliability. As an alternative, Bayesian methods can easily be used (with MCMC methods; discussed in Chapters 3, 6, and 7) for solving this type of complex problems. In the Bayesian approach, each component reliability and the corresponding uncertainty is given by probability distributions and they can be combined through an appropriate model(s) to estimate the reliability of the subsystems and hence the entire system. This is the main advantage of the Bayesian approach compared to frequentist methods (Hamada et al. 2008).

2.6 How Bayesian Inference Works with Monte Carlo Simulations

Note that since the Bayesian posterior is a probability distribution, it can easily work in a system reliability estimation framework via Monte Carlo simulations. An example is illustrated below. More examples are provided in Chapters 6 and 7.

Example 2.6 An engineer needs to estimate the reliability of a system consisting of five components connected in series. This means the system fails whenever any of the five components fails. Each component has only pass/fail data from assurance testing. The test data is summarized in Table 2.8. The objective is to estimate system reliability and its 95% credible interval.

To assess system level reliability, first we need to evaluate the reliability of individual components. In this example, the test data for each of the five components are identical. According to the discussion in Section 2.4, if each component prior reliability has a flat distribution *Beta* (1, 1) (assuming no prior information on component level reliability), and if the data is 299 successes out of 299 trials, the posterior distribution of the reliability of each component is *Beta* (300, 1).

Let R_1, R_2, … R_5 be the probabilities that components 1 through 5, respectively, pass the quality assurance test. The system will pass the quality assurance test if and only if all five components pass their individual tests. Assuming that components are independent

Table 2.8 Testing data of a series system.

Component #	Sample size	Number of successes
1	299	299
2	299	299
3	299	299
4	299	299
5	299	299

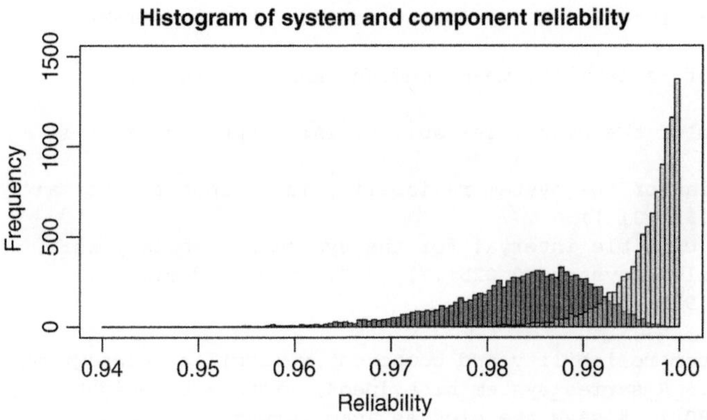

Figure 2.7 Histograms of simulated system reliability (dark gray) and component reliability (light gray).

from one another, the system reliability, R_{system} (i.e. the probability that the system will pass the test) is given by

$$R_{\text{system}} = \prod_{i=1}^{5} R_i. \tag{2.10}$$

The system reliability and its 95% credible interval can be computed using Monte Carlo simulation as shown in the following R script (`2.6_Simple_Series_System.R`). Figure 2.7 shows histograms of simulated system reliability and component reliability. More discussions on system level reliability estimation are given in Chapter 7.

```
(2.6_Simple_Series_System.R)
##########################################
#   Estimate system reliability;         #
#   Series system                        #
##########################################

# Assume the prior reliability of each component has a flat
distribution Beta(1,1),
# reliability posterior is Beta(x+1, n-x+1) - see conjugate prior
distributions
```

```
     # Sample 10000 iterations from each reliability posterior
distribution
     R1 <- rbeta(10000,300,1)
     R2 <- rbeta(10000,300,1)
     R3 <- rbeta(10000,300,1)
     R4 <- rbeta(10000,300,1)
     R5 <- rbeta(10000,300,1)

     # Calculate the system reliability based on series system
reliability block diagram
     R_system <- R1*R2*R3*R4*R5

     ## specify_decimal is a function to show k number of decimals
for number x
     specify_decimal <- function(x, k) format(round(x, k), nsmall=k)

     ## show results of reliability mean, median, and 95% credible
interval
     print(paste("mean of the system reliability is:", specify_decimal
(mean(R_system),3)))
     print(paste("median of the system reliability is:", specify_decimal
(quantile(R_system, 0.50),3) ))
     print(paste("95% credible interval for the system reliability is:",
specify_decimal(quantile(R_system, 0.025),3), ",", specify_decimal
(quantile(R_system, 0.975),3)))

     # Overlap of system reliability and component reliability histograms
     # jpeg("Example2.6_R_series_system_hist.jpeg", width = 6, height = 4,
units = 'in', res = 1800)  # save the plot as jpeg format
     hist(R_system, col=rgb(0.1,0.1,0.1,0.5), main = " Histogram of
system and component reliability", xlab="Reliability", xlim=range(0.94,1),
ylim=range(0,1500),breaks=100 )
     hist(R1, col=rgb(0.8,0.8,0.8,0.5), breaks=100, add=T)
     box()
# dev.off()
```

The simulation results are shown below:

```
[1] "mean of the system reliability is: 0.984"
[1] "median of the system reliability is: 0.985"
[1] "95% credible interval for the system reliability is: 0.967, 0.995"
```

In this problem, a frequentist solution of the system reliability is 1. The frequentist's uncertainty associated with each component reliability can be computed as a confidence interval, however these confidence intervals cannot be easily combined to get a confidence interval for the system reliability. One can try the Monte Carlo simulation approach, but the challenge is how to assign an appropriate probability density to quantify the uncertainty of each component reliability estimate.

2.7 Bayes Factor and its Applications

Inferences concerning the parameters of a Bayesian analyses are always based on their posterior distributions. When Bayes' theorem is used to compute the posterior distribution of the parameters, the marginal distribution of the data is used as a normalizing parameter (see Eqs. 2.2 and 2.3) to make a posterior a true probability density. The marginal distribution of the data depends on the theoretical probability distribution (model) selected to describe the data. Therefore, it can be used as a tool for model selection.

Suppose we want to compare two competing models, M_1 and M_2. Let $h_1(\boldsymbol{\theta}_1 \mid M_1)$ and $h_2(\boldsymbol{\theta}_2 \mid M_2)$ be the prior distributions and $f_1(\mathbf{y} \mid \boldsymbol{\theta}_1, M_1)$ and $f_2(\mathbf{y} \mid \boldsymbol{\theta}_2, M_2)$ the likelihood functions of the data under the models M_1 and M_2, respectively. Let $P(M_1) = p_1$ and $P(M_2) = 1 - p_1$ be the prior probabilities of selecting models M_1 and M_2, respectively. Let $P(M_1 \mid \mathbf{y})$ and $P(M_2 \mid \mathbf{y})$ be the posterior probabilities of models M_1 and M_2 being true given the data, $\mathbf{y}$, respectively. Notice that $P(M_2 \mid \mathbf{y}) = 1 - P(M_1 \mid \mathbf{y})$. The posterior odds that model M_1 is true are given by

$$\begin{aligned}\frac{P(M_1|\mathbf{y})}{P(M_2 \mid \mathbf{y})} &= \frac{P(\mathbf{y} \mid M_1)\,P(M_1)}{P(\mathbf{y} \mid M_2)\,P(M_2)}\\ &= \frac{P(M_1)\int_{\boldsymbol{\theta}_1} f_1(\mathbf{y} \mid \boldsymbol{\theta}_1, M_1)\,h_1(\boldsymbol{\theta}_1 \mid M_1)\,d\boldsymbol{\theta}_1}{P(M_2)\int_{\boldsymbol{\theta}_2} f_2(\mathbf{y} \mid \boldsymbol{\theta}_2, M_2)\,h_2(\boldsymbol{\theta}_2 \mid M_2)\,d\boldsymbol{\theta}_2}\\ &= \frac{P(M_1)}{P(M_2)} \times \frac{k_1(\mathbf{y} \mid M_1)}{k_2(\mathbf{y} \mid M_2)},\end{aligned} \tag{2.11}$$

where $k_1(\mathbf{y} \mid M_1)$ and $k_2(\mathbf{y} \mid M_2)$ are the marginal densities of the data under the models M_1 and M_2, respectively.

Therefore, we have

$$\text{posterior odds} = \text{prior odds} \times \text{Bayes' factor}.$$

The Bayes' factor is defined as the ratio of the marginal densities of the data $\mathbf{y}$ under the models M_1 and M_2. When there are more than two competing models, we can consider one pair at a time and compute the odds ratios and thus determine which model is most favorable given the data. The following example demonstrates this situation.

Let's revisit Example 2.4 and calculate the Bayes' factor in order to compare the three models considered:

$$M_1 = Beta\ (1, 1)$$
$$M_2 = Beta\ (1, 9)$$
$$M_3 = Beta\ (9, 1)$$

Recall that we used the beta prior on the product conformance rate, θ. For any generic model, M(a beta prior distribution with parameters α and β) we can compute the marginal probability of observing 59 successes out of 59 samples as follows

$$\begin{aligned}P(x = 59 \mid n = 59,\ M) &= \int_0^1 f(x \mid \theta) f(\theta) d\theta\\ &= \int_0^1 \binom{n}{x} \theta^x (1-\theta)^{n-x} \frac{\theta^{\alpha-1}(1-\theta)^{\beta-1}}{B(\alpha, \beta)} d\theta\\ &= \int_0^1 \binom{59}{59} \theta^{59} (1-\theta)^{59-59} \frac{\theta^{\alpha-1}(1-\theta)^{\beta-1}}{B(\alpha, \beta)} d\theta\\ &= \int_0^1 \frac{\theta^{\alpha+59-1}(1-\theta)^{\beta-1}}{B(\alpha, \beta)} d\theta\end{aligned}$$

When $M_1 = Beta\ (1, 1)$,

$$\begin{aligned}
P(x=59 \mid n=59,\ M_1) &= \int_0^1 \frac{\theta^{60-1}(1-\theta)^{1-1}}{B(1,\ 1)} d\theta \\
&= \int_0^1 \theta^{60-1}(1-\theta)^{1-1} d\theta \\
&= B(60,\ 1) \int_0^1 \frac{\theta^{60-1}(1-\theta)^{1-1}}{B(60,\ 1)} d\theta \\
&= B(60,\ 1) = \frac{\Gamma(60)\Gamma(1)}{\Gamma(60+1)} = \frac{59!}{60!} = \frac{1}{60} = 0.016667.
\end{aligned}$$

When $M_2 = Beta\ (1, 9)$,

$$\begin{aligned}
P(x=59 \mid n=59,\ M_2) &= \int_0^1 \frac{\theta^{60-1}(1-\theta)^{9-1}}{B(1,\ 9)} d\theta \\
&= \frac{B(60,\ 9)}{B(1,\ 9)} \int_0^1 \frac{\theta^{60-1}(1-\theta)^{9-1}}{B(60,\ 9)} d\theta \\
&= \frac{B(60,\ 9)}{B(1,\ 9)} = \frac{\Gamma(60)\Gamma(9)}{\Gamma(60+9)} \frac{\Gamma(1+9)}{\Gamma(1)\Gamma(9)} = \frac{59!8!}{68!} \frac{9!}{8!} \\
&= 2.0292 \times 10^{-11}.
\end{aligned}$$

The Bayes' factor of model M_1 compared to model M_2 is

$$0.016667/(2.0292 \times 10^{-11}) = 8.2133 \times 10^{8}.$$

This implies that with the observed data, model M_1 is highly favored compared to model M_2 as the appropriate prior distribution.

When $M_3 = Beta\ (9, 1)$,

$$\begin{aligned}
P(x=59 \mid n=59,\ M_3) &= \int_0^1 \frac{\theta^{68-1}(1-\theta)^{1-1}}{B(9,\ 1)} d\theta \\
&= \frac{B(68,\ 1)}{B(9,\ 1)} \int_0^1 \frac{p^{68-1}(1-p)^{1-1}}{B(68,\ 1)} d\theta \\
&= \frac{B(68,\ 1)}{B(9,\ 1)} = \frac{\Gamma(68)\Gamma(1)}{\Gamma(68+1)} \frac{\Gamma(9+1)}{\Gamma(9)\Gamma(1)} = \frac{67!}{68!} \frac{9!}{8!} \\
&= \frac{9}{68} = 0.13235.
\end{aligned}$$

The Bayes' factor of model M_3 compared to model M_1 is

$$0.13235/0.016667 = 7.94.$$

If we assigned equal probabilities a priori to the three models (i.e. $P(M_1) = P(M_2) = P(M_3) = 1/3$) then we can compute the probabilities that each model is true given the data.

$$P(M_1 \mid x = 59) = \frac{P(x = 59 \mid M_1)P(M_1)}{P(x = 59 \mid M_1)P(M_1) + P(x = 59 \mid M_2)P(M_2) + P(x = 59 \mid M_3)P(M_3)}$$
$$= \frac{0.016667}{1/60 + 2.0292 \times 10^{-11} + 0.13235}$$
$$= 0.11844$$

Similarly, we can compute $P(M_2 \mid x = 59) = 1.36\text{E} - 10$ and $P(M_3 \mid x = 59) = 0.888156$. The model that is most consistent with the observed data is $M_3 - Beta(9, 1)$.

2.8 Predictive Distribution

In Bayesian analysis, all inferences about parameters are based on their posterior distribution. There may be situations in which the prediction of the outcome of a future sample is important. We could use posterior distributions for this purpose as well.

In most product development operations in industry, the design characterization test (DCT) is typically performed prior to formal DVT, to ensure a first-pass success. Though generally continuous data are preferred to be collected during DVT, sometimes these tests are based on attribute sampling plans where a number of parts, e.g. 59 parts, are sampled and if all pass the required specifications then the test is passed. With the frequentist approach, if the test is passed this sampling plan guarantees with 95% confidence that 95% of the untested parts will meet the specification requirements. Suppose that a DCT is passed and that another 59 parts will be tested for the DVT. We would like to know the probability that the DVT will be passed and more broadly the probability of passing a specified number of parts. In the frequentist approach there is no natural way of incorporating the results of the DCT to predict the outcomes of the DVT. However, this is very easy to do in the Bayesian setting, as demonstrated in the next example.

Example 2.7 A random sample of 59 parts were tested for a DCT and all parts met the product specification requirements. In the following DVT, 59 parts will be tested against the same requirement. We would like to predict the number of parts that will pass the DVT.

This example reveals a common misconception in engineering practices: if in the previous DCT 59 out of 59 parts passed, engineers tend to be confident that they will observe success in the upcoming DVT (i.e. all 59 parts in the DVT will pass)!

However, we don't know how confident one should be when observing 59 parts passing in the DCT, as the chance that another 59 parts pass the DVT obviously is not 100%. We will go through the math in this example to provide quantitative results.

As mentioned before, the 95% one-sided lower confidence bound for the event rate (the probability that a randomly selected part meets the specification requirements) is 0.95. This can be shown using the sampling distribution (likelihood) of a random sample of 59 parts taken from a population with event rate θ. The probability of observing x parts meeting the requirements in a random sample of n parts is given by Eq. (2.12). To determine the 95% lower confidence bound of the event rate θ, we need to find the smallest value of θ (say θ_0) for which the probability of passing the DCT is 0.05 (= 1 – 0.95). For

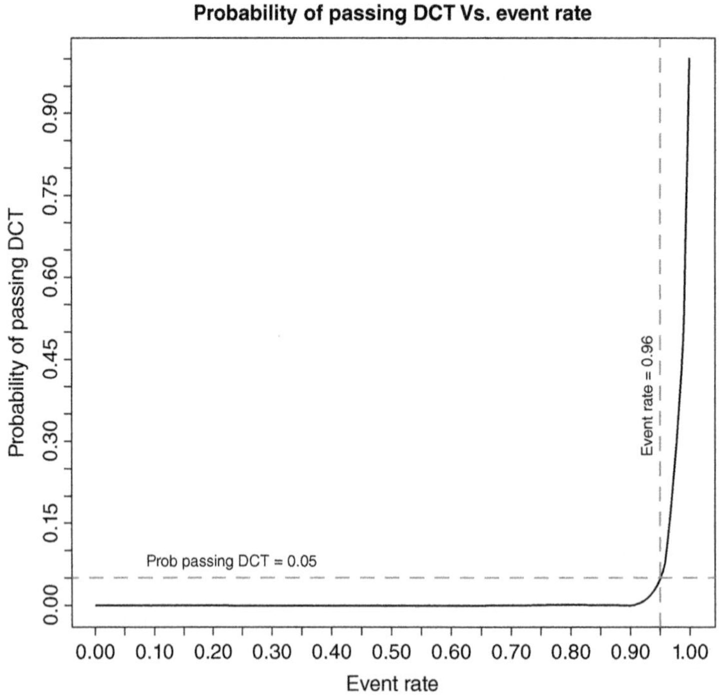

Figure 2.8 Probability of passing the DCT.

any value $\theta < \theta_0$, the probability of passing the DCT would be less than 5%. Therefore, the interval $(\theta_0, 1)$ is considered to be the 95% confidence bound for θ. This is achieved when $\theta = 0.95$, as shown in the calculations below:

$$f(x \mid \theta, n = 59) = \binom{n}{x} \theta^x (1-\theta)^{n-x} \tag{2.12}$$

For a binomial distribution with number of trials $n = 59$ and event probability $\theta = 0.95$, the probability of getting $x = 59$ parts meeting the requirements (i.e. passing the DCT) is given by

$$f(x = 59 \mid \theta = .95, n = 59) = \binom{59}{59} (0.95)^{59}(1-0.95)^{59-59}$$
$$= (0.95)^{59} = 0.04849 \simeq 0.05.$$

Figure 2.8 shows the probability of passing the DCT (59 of 59 parts tested meeting the requirements) for different event rates. When the event rate is 0.95 or more, then the probability of passing the DCT is 5% or more. Thus the one-sided 95% lower confidence bound for the event rate θ is 0.95.

If we were to use the frequentist approach to determine the outcome of the DVT involving another 59 parts, we have to choose a value for the event rate, θ. Should we use a single value or a range of values? Since we do not know the exact value, we could use a range of values and Monte Carlo simulation to estimate the DVT success rate. However, the challenge is to assign a probability distribution to this range of values. All these difficulties can be overcome naturally if we use the Bayesian approach to solve this problem. The Bayesian solution to this problem essentially incorporates the uncertainty associated with the estimation of the event rate, θ, through its posterior distribution. In Example 2.7, if we assume the non-informative uniform distribution (*Beta* (1, 1)) as the prior distribution of the event rate, θ, after observing 59 parts passing the DCT, its posterior distribution is given by

$$g(\theta) = f(\theta \mid x = 59, n = 59) \sim Beta\ (60, 1).$$

For a binomial distribution with number of trials $n = 59$ and event probability θ, the probability of getting $x = 59$ observations out of $n = 59$ trials is shown in Eq. (2.13).

$$\begin{aligned} f(x = 59 \mid \theta, n = 59) &= \binom{n}{x} \theta^x (1-\theta)^{n-x} \\ &= \binom{59}{59} \frac{59!}{59!(59-59)!} \theta^{59} (1-\theta)^{59-59} \\ &= \theta^{59}. \end{aligned} \tag{2.13}$$

Equation (2.13) provides the probability of observing 59 parts passing the DVT out of 59 parts tested given the event rate of θ. Having seen the results of the DCT, the uncertainty about the event rate is now model by the *Beta*(60, 1) distribution. To compute the expected probability of passing the DVT we need to integrate out the uncertainty associated with θ as follows:

$$\begin{aligned} f(x) &= \int_0^1 \theta^{59} \frac{\theta^{(60-1)}(1-\theta)^{(1-1)}}{B(60,1)} d\theta \\ &= \frac{\theta^{119}}{119} \times \frac{\Gamma(61)}{\Gamma(60)\Gamma(1)} \Big|_0^1 \\ &= \frac{60}{119} = 0.5042. \end{aligned} \tag{2.14}$$

An approximation to this analytical solution and other related quantities of interest such as the expected number of parts passing the DVT and a 95% credible interval for the number of parts passing the DVT can be easily obtained through Monte Carlo simulation. The R-script **2.8_Prediction_of_DVT.R** demonstrates the Monte Carlo simulation approach.

```
    (2.8_Prediction_of_DVT.R)
    #####################################################################
    ## To estimate the probability of the No. of successes in the DVT  ##
    ## (59 trials)                                                     ##
    ## Posterior: Beta(60,1)                                           ##
    ## Likelihood: Binomial                                            ##
    #####################################################################

    set.seed(123)

    ## Reliability posterior: Beta(60,1)
    ## Sample 100000 iterations from the reliability posterior
distribution
    iter <- 100000
    p <- rbeta(iter,60,1)
```

```
    ## For a Binomial distribution with number of trials = 59
    ## and event probability p,
    ## use Monte Carlo simulation to calculate the probability of getting
x_predict observations (No. of successes)
    x_predict <- rbinom(iter, 59, p)

    ## Show summary statistics of the number of successes
    print(paste("mean of x_predict is:", mean(x_predict)))
    print(paste("median of x_predict is:", quantile(x_predict, 0.50)))
    print(paste("95% credible interval for the x_predict is (two sided):",
quantile(x_predict, 0.025), ",", quantile(x_predict, 0.975)))
    print(paste("95% credible interval lower bound for the x_predict is
(one sided):", quantile(x_predict, 0.05) ))

    # Probability of DVT success
    P_DVT_success <- length(which(x_predict > 58)) / iter
    print(paste("probability of DVT success is:", P_DVT_success))

    # Barplot showing the probability of the number of successes in the DVT
    barplot(table(x_predict)/iter, xlab="No. of successes (x)",
ylab="Probability")
    box()
```

The simulation results are shown below. Note that the simulated probability (0.5054) of passing the DVT is very close to its analytical result (0.5042). The probability of getting x successes in the DVT is shown in Figure 2.9. Note that the probability of getting 59 successes out of 59 trials in this DVT is only approximately 50%, just like tossing a coin! This result is somewhat counter institutive. One would have expected much higher DVT success probability having seen the DCT success. This is one of the main disadvantages of dealing with attributes (in this case, pass/fail) data.

```
    [1] "mean of x_predict is: 58.03439"
    [1] "median of x_predict is: 59"
    [1] "95% credible interval for the x_predict is (two sided): 54 , 59"
    [1] "95% credible interval lower bound for the x_predict is
(one sided): 55"
        [1] "probability of DVT success is: 0.50539"
```

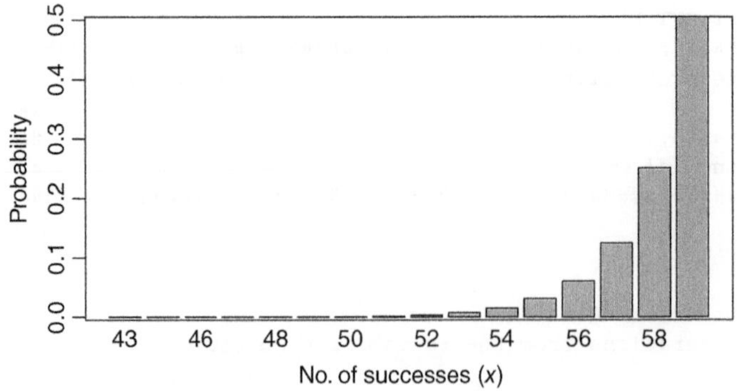

Figure 2.9 Probability of getting x successes in a DVT.

2.9 Summary

Key messages in this chapter are summarized as follows.

- Bayesian reasoning is an approach of learning from evidence as it accumulates.
- In Bayesian statistics, unknown parameters are treated as random variables, with probability distribution assigned.
- It is flexible to use Bayesian models to combine different sources of information and update inferences when new data become available.
- Simple Monte Carlo solutions can provide good approximations to analytical solutions when they exist.
- It is feasible to achieve sample size reduction using informative priors.
- Bayesian models are especially useful for complex system reliability estimation, where traditional statistics may not have closed-form solution or the frequentist solution may not exist at all.

References

Agresti, A. (2002). *Categorical Data Analyis*, 2e. Hoboken, NJ: Wiley.

Bolstad, W.M. (2007). *Introduction to Bayesian Statiscs*. Hoboken, NJ: Wiley.

Gelman, A., Carlin, J.B., Stern, H.S. et al. (2013). *Bayesian Data Analysis*, 3e. Boca Raton, FL: CRC Press.

Hamada, M.S., Wilson, A.G., Shane Reese, C., and Martz, H.F. (2008). *Bayesian Reliability*. New York: Springer.

Kahneman, K. (2013). *Thinking, Fast and Slow*. New York: Farrar, Straus and Giroux.

3

Bayesian Computation

This chapter introduces commonly used Bayesian computation methods. For distributions where conjugate priors are not available and for complex Bayesian models, posterior distributions may not be analytically tractable. Nowadays, with computers and Markov chain Monte Carlo (MCMC) algorithms, solving posteriors with various types of prior distributions and building complex models has become feasible. In this chapter, MCMC algorithms, including the Metropolis algorithm and Gibbs sampling, are briefly introduced.

This chapter also introduces a widely used Bayesian computation software, Just Another Gibbs Sampler (JAGS), to provide MCMC sampling of Bayesian posteriors. We will walk through R and JAGS example codes to explain the function of various portions of the codes, including creating and running the model, summarizing posterior samples, and different methods of MCMC chain convergence diagnostics. Lastly, methods for model comparisons are introduced.

3.1 Introduction

In Chapter 2 we introduced conjugate prior distributions, in which cases the corresponding posterior distributions have closed forms. For some distributions, conjugate prior distributions are not available. For example, we would like to estimate both the shape and scale parameters of a two-parameter Weibull distribution, a popular distribution in reliability applications. Closed-form solutions for this case are not available.

The Weibull distribution is one of the most commonly used lifetime distributions in reliability engineering. It is used to describe the life of certain electronic components, capacitors, dielectrics, and ball bearings. It is also often used to model material properties such as elongation, electrical resistance, mechanical/electrical strength, metal fatigue, etc.

The Weibull distribution is very flexible in that it can model all three regions of a hazard rate bathtub curve (i.e. decreasing hazard rate, constant hazard rate, and increasing hazard rate).

The probability density function (PDF) of a two-parameter Weibull distribution is

$$f(t \mid \beta, \eta) = \frac{\beta}{\eta}\left(\frac{t}{\eta}\right)^{\beta-1} e^{-(t/\eta)^{\beta}}, \tag{3.1}$$

Practical Applications of Bayesian Reliability, First Edition. Yan Liu and Athula I. Abeyratne.

Companion website: www.wiley.com/go/bayesian20

where

t is the time to failure ($t \geq 0$)
β is the shape parameter ($\beta > 0$)
η is the scale parameter ($\eta > 0$).

The hazard function of the Weibull distribution is

$$h(t \mid \beta, \eta) = \frac{\beta}{\eta}\left(\frac{t}{\eta}\right)^{\beta-1}. \tag{3.2}$$

When $\beta > 1$ there is increasing hazard rate or wear-out failures, for $\beta < 1$ there is decreasing hazard rate or early failures, and $\beta = 1$ indicates constant hazard rate or random failures. For most products and materials, β is in the range of 0.5–5. The exponential distribution is a special case of a Weibull distribution with a shape parameter $\beta = 1$ and the characteristic life $\eta = 1/\lambda = MTTF$ (mean time to failure).

The reliability at time t is

$$R(t \mid \beta, \eta) = e^{-(t/\eta)^{\beta}}. \tag{3.3}$$

With the above parameterization the mean is $\eta\Gamma(1+1/\beta)$ and the variance is $\eta^2[\Gamma(1+2/\beta) - (\Gamma(1+1/\beta))^2]$, where $\Gamma(x)$ is the gamma function given by

$$\Gamma(k) = \int_0^{\infty} x^{k-1}e^{-x}dx.$$

The scale parameter η is also known as the characteristic life of the Weibull distribution because when $t = \eta$, 63.2% ($=1 - e^{-1}$) of the units will have failed. This result does not depend on the value of the shape parameter, β.

Example 3.1 Collected cycles to failure data from a fatigue test are shown in Table 3.1. Our goal is to fit the data to a Weibull distribution and estimate the Weibull shape and scale parameters.

First, we apply the maximum likelihood method using Minitab® to estimate the shape and scale parameters of the Weibull distribution. Later, we will compare these results with the corresponding Bayesian estimators. The maximum likelihood estimators (MLEs) are shown below:

```
                       Standard   95.0% Normal CI
Parameter  Estimate       Error     Lower     Upper
Shape       1.83198    0.140108   1.57696   2.12823
Scale       100.834     5.80687   90.0715   112.882

Log-Likelihood = -526.257
```

To find the Bayesian estimates of the shape and scale parameters in this case, we use a discretization method to approximate Bayesian posterior distributions for these two parameters

3.2 Discretization

Discretization is a process to transform continuous distributions into discrete quantities. When discretizing a continuous distribution, the larger the number of discrete

Table 3.1 Cycles to failure data.

75	60	51	14	41
28	89	99	290	209
52	130	30	151	20
67	79	64	105	123
78	58	114	87	56
5	102	159	41	99
46	79	95	137	93
132	63	134	95	86
169	72	193	155	70
97	57	26	47	67
102	90	158	145	184
189	75	88	85	116
150	29	80	79	121
118	95	38	173	80
42	119	27	147	54
144	201	67	48	81
61	26	48	105	100
152	25	59	65	132
49	17	34	24	39
155	67	55	51	66

categories, the closer the result will be to the true continuous distribution. When the prior distribution is discretized into a finite number of categories, calculation of the posterior distribution can be simplified and approximated. This is one method of Bayesian computation when the prior distribution is not a conjugate prior and the closed-form posterior distribution cannot be solved. In this section, we will use the discretization method to approximate the scale and shape posterior distributions of the Weibull distribution in Example 3.1.

Assume the prior distributions of the scale parameter η and the shape parameter β are $g_1(\eta)$ and $g_2(\beta)$, respectively, and the likelihood is a Weibull distribution $f(\boldsymbol{t} \mid \beta, \eta)$. Then, the joint posterior distribution is

$$p(\beta, \eta \mid \boldsymbol{t}) = \frac{g(\beta, \eta) \times f(\boldsymbol{t} \mid \beta, \eta)}{k(\boldsymbol{t})},$$

where $g(\beta, \eta)$ is the prior joint density function of β and η, and $k(\boldsymbol{t})$ is the normalizing constant, which depends only on the data $\boldsymbol{t}$ (vector of failure times). Since the scale parameter η and the shape parameter β are *a priori* independent,

$$\begin{aligned} p(\beta, \eta \mid \boldsymbol{t}) &= \frac{g_1(\eta) \times g_2(\beta) \times f(\boldsymbol{t} \mid \beta, \eta)}{k(\boldsymbol{t})} \\ &\approx \frac{g_1(\eta) \times g_2(\beta) \times f(\boldsymbol{t} \mid \beta, \eta)}{\sum_{\eta} \sum_{\beta} g_1(\eta) \times g_2(\beta) \times f(\boldsymbol{t} \mid \beta, \eta)}. \end{aligned}$$

Since the denominator is a constant, the joint posterior distribution is proportional to the numerator,

$$p(\beta, \eta \mid \boldsymbol{t}) \propto g_1(\eta) \times g_2(\beta) \times f(\boldsymbol{t} \mid \beta, \eta). \tag{3.4}$$

The marginal posterior distributions of the scale parameter η and the shape parameter β are calculated as

$$h_1(\eta \mid \boldsymbol{t}) = \sum_{\beta} p(\beta, \eta \mid \boldsymbol{t}), \tag{3.5}$$

$$h_2(\beta \mid \boldsymbol{t}) = \sum_{\eta} p(\beta, \eta \mid \boldsymbol{t}). \tag{3.6}$$

The R code using the discretization method to approximate Bayesian posterior distributions is shown in **`3.1_Weibull_Parameters_Estimation.R`**. The code requires the package rgl to run some of the functions. To install the package, type the following command.

```
> install.packages("rgl")
```

First, we will use diffuse priors on the shape and scale parameters, so that the resulting posterior estimates can be easily compared with the MLE estimates. We choose the following uniform distributions that include the MLE estimates:

Shape $\sim$ Uniform (1, 3)
Scale $\sim$ Uniform (60, 130).

There is no specific reason to select these uniform priors. One can choose different uniform prior distributions that include the MLE estimates of the parameters. Later, when we introduce more advanced Bayesian computation algorithms, we will use other diffuse prior distributions and assess the sensitivity of Bayesian solutions to the choice of prior distributions.

(3.1_Weibull_Parameters_Estimation.R)

```
##########################################################################
## Bayesian analysis for Weibull distribution parameter inference using ##
## discretization                                                       ##
##########################################################################

No_of_steps <- 100
iteration <- No_of_steps+1
Shape_step <- (3.0-1) / No_of_steps
Scale_step <- (130-60) / No_of_steps

shape_prior <- rep(1/(No_of_steps+1), No_of_steps+1) # Shape prior is
assumed flat/Uniform
scale_prior <- rep(1/(No_of_steps+1), No_of_steps+1) # Scale prior is
assumed flat/Uniform
shape <- seq(1.0, 3.0, Shape_step)  # a vector lists the states of shape
parameter
scale <- seq(60, 130, Scale_step) # a vector lists the states of scale
parameter

# Read data from a file
dataTable <- read.table("Example3.1Data.txt",header=F,sep="")
```

```
data <- dataTable[,1]

# dev.off()  # for new graphics with default settings

L_times_prior <- matrix(0, nrow = No_of_steps+1, ncol = No_of_steps+1)

for (i in 1:iteration){

    for (j in 1:iteration){

        likelihood <- 1

        for (k in 1:length(data)){

            likelihood <- likelihood * dweibull(data[k], shape[i],
scale[j], log = FALSE)
        }

        #log_L_times_prior[i,j] <- log(shape_prior[i]*scale_prior[j]*
likelihood)
        L_times_prior[i,j] <- shape_prior[i]*scale_prior[j]*likelihood*
1E235
    }
}

require(rgl)
zlim <- range(L_times_prior,na.rm=T)
zlen <- (zlim[2] - zlim[1]) + 1
#color.range <- rev(rainbow(zlen))       # height color lookup table
color.range <- rev(terrain.colors(zlen))
colors  <- color.range[(L_times_prior-zlim[1])+1] # assign colors to
heights for each point
persp3d(shape, scale, L_times_prior, zlab="Scaled Joint Posterior Dist",
col=colors)

# Estimate the marginal distributions of shape and scale posteriors
shape_posterior <- rep(NA, No_of_steps+1)
scale_posterior <- rep(NA, No_of_steps+1)

for (i in 1:iteration){
  temp <- 0
  for (j in 1:iteration){
    shape_posterior[i] <- temp + L_times_prior[i,j]
  }
}

shape_posterior <- shape_posterior/sum(shape_posterior)  # Normalize shape
posterior

plot(shape, shape_posterior)

for (j in 1:iteration){
  temp <- 0
  for (i in 1:iteration){
    scale_posterior[i] <- temp + L_times_prior[i,j]
  }
```

```
}

scale_posterior <- scale_posterior/sum(scale_posterior)  # Normalize scale
posterior

plot(scale, scale_posterior)

# Estimate the mean and sd of the posterior shape and scale distributions
shape_mean <- sum(shape*shape_posterior)
scale_mean <- sum(scale*scale_posterior)
shape_sd <- sum((shape - shape_mean)^2*shape_posterior)^0.5
scale_sd <- sum((scale - scale_mean)^2*scale_posterior)^0.5

print(paste("shape posterior mean is:", signif(shape_mean,5)))
print(paste("shape posterior standard deviation is:", signif(shape_sd,5)))
print(paste("scale posterior mean is:", signif(scale_mean,5)))
print(paste("scale posterior standard deviation is:", signif(scale_sd,5)))
```

For the two unknown parameters, the joint posterior distribution is displayed in a three-dimensional space, where the shape and scale parameters are on the two horizontal axes and the scaled joint probability density is on the vertical axis. In the R code, the joint posterior distribution is calculated as the product of the likelihood, joint prior distribution, and a large constant 1E235 (such a large value is used to increase the range for plotting purposes). The calculated joint posterior distribution is shown in Figure 3.1. Note that the calculated joint posterior distribution is not a true posterior distribution as it is not normalized – it is proportional to the true joint posterior distribution, per Eq. (3.4). The idea here is to show the relative frequency of the posterior parameter values. The relevant normalizing constant can be computed using a similar discretization approach. When divided by this normalizing constant, Figure 3.1 can be transformed to a true joint posterior distribution.

The shape and scale marginal posterior distributions are shown in Figure 3.2, calculated based on Eqs. (3.5) and (3.6). The y axes in the respective plots are the corresponding marginal densities.

Use the following R code to check if the shape and scale posteriors are normalized. Since the summation of the shape or the scale posterior is 1, we can tell that the shape and scale posteriors are both normalized.

```
> sum(shape_posterior)
[1] 1
> sum(scale_posterior)
[1] 1
```

The mean and standard deviation of both shape and scale parameters are shown below. The mean shape parameter is close to the MLE result while the mean of the scale parameter is lower than the MLE result. However, we expected both parameter estimates to be closer to their MLE values since we used vague priors.

```
[1] "shape posterior mean is: 1.8298"
[1] "shape posterior standard deviation is: 0.15154"
[1] "scale posterior mean is: 89.041"
[1] "scale posterior standard deviation is: 5.3040"
```

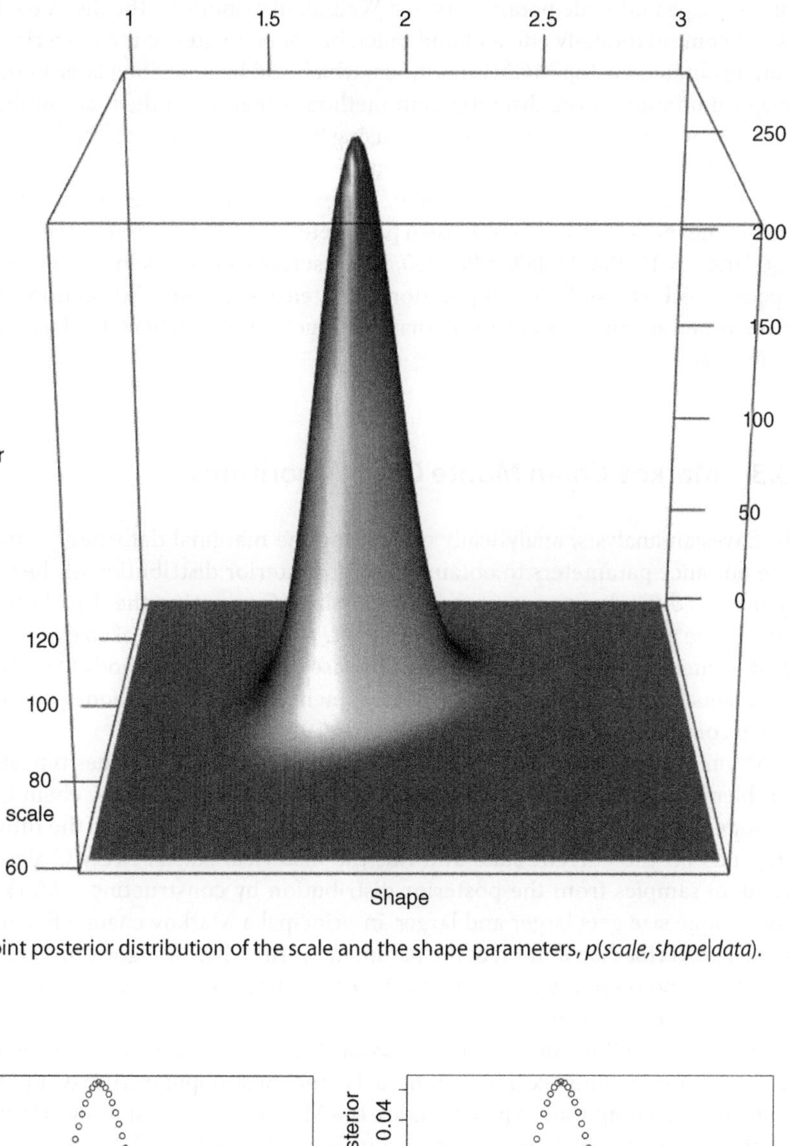

Figure 3.1 Joint posterior distribution of the scale and the shape parameters, *p(scale, shape|data)*.

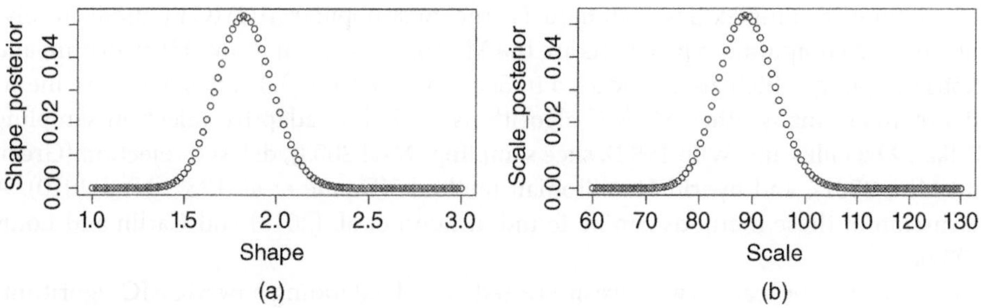

Figure 3.2 Marginal posterior distributions of shape and scale.

In this section, we used the discretization method to approximate Bayesian estimates of the shape and scale parameters in a Weibull distribution. The discretization method is not computationally efficient and much better estimates of the posterior distribution can be obtained using MCMC methods, which will be described later in this chapter. A major limitation of the discretization method is that when there are multiple parameters, the number of discrete categories using the discretization method increases rapidly, which makes the method computationally intensive or even infeasible to implement due to increased computational time. For example, if there are six unknown parameters in the model, and each unknown parameter has 1000 discrete categories, then there are $1000^6 = 1000, 000, 000, 000, 000, 000$ discrete categories in the multi-dimensional space, which certainly is computationally intensive. In the next section, we will introduce other advanced algorithms that are much more efficient for Bayesian posterior computation.

3.3 Markov Chain Monte Carlo Algorithms

In Bayesian analysis, analytically computing the marginal densities by integrating out the nuisance parameters to obtain the joint posterior distribution of the parameters in general is difficult or in most cases impossible. Computing the distribution of a function of the parameters is even more difficult. These difficulties of analytically computing high-dimensional integrals prevented the use of Bayesian methods for solving complex problems for a long time until the development of MCMC methods and the rise of personal computers in late the 1980s and early 1990s.

Monte Carlo simulation is a method that samples random values repeatedly to solve problems. MCMC is one type of Monte Carlo method. A Markov chain is a collection of states where the transition probability from the current state to the future state only depends on the current state and not the historical states. MCMC algorithms draw random samples from the posterior distribution by constructing a Markov chain. As the sample size gets larger and larger, in principal a Markov chain will converge to the actual posterior distribution of the parameters. The MCMC algorithms are easily implementable in computer code and as the computing power of computers grows, more complex problems can be solved.

Nowadays MCMC methods have become the standard approach for solving posterior distributions in complex Bayesian models. The most popular MCMC methods in solving Bayesian computation problems are the Metropolis–Hastings (M-H) algorithm and Gibbs sampling, which are introduced in Sections 3.3.2 and 3.3.3. Besides these methods, there are many other MCMC algorithms, including adaptive rejection sampling (Gilks 1992; Gilks and Wild 1992), slice sampling (Neal 2003), delayed rejection (Green and Mira 2001), and hybrid/Hamiltonian methods (Duane et al. 1987; Neal 2010). A discussion of these methods can be found in Lunn et al. (2013) and Carlin and Louis (2009).

Section 3.3 is for readers who are interested in understanding how MCMC algorithms work. Practitioners don't actually need to worry about writing MCMC algorithms themselves, as these algorithms have been integrated into MCMC-based software packages, which will be discussed in Section 3.4.

3.3.1 Markov Chains

A sequence of random variables X_1, X_2,...,X_k, X_{K+1},... is called a Markov chain if for all k, the distribution of the $(k+1)$th variable depends only on the value of the kth variable, X_k, not on the preceding values of X_1, X_2,..., X_{k-1}. Mathematically this can be expressed as

$$P(X_{k+1} = x \mid X_1 = x_1, X_2 = x_2, \ldots, X_{k-1} = x_{k-1}, X_k = x_k)$$
$$= P(X_{k+1} = x \mid X_k = x_k) \text{ for all } k. \tag{3.7}$$

The right-hand term of Eq. (3.7) gives the conditional probability distribution of moving to any value x in the next step given the current value of the chain. This probability distribution is called the transition distribution. The data produced by a Markov chain are correlated. Under fairly general regulatory conditions, as $k \to \infty$, the marginal density of X_{K+1} will converge to a unique stationary distribution. Regardless of where the chain started, eventually it will converge to the same distribution. When we exclude a certain number of initial values (called burn-in) the remaining values of the chain can be considered as a sample generated from the target distribution. This property of the Markov chains helps us develop algorithms for estimating the posterior probability distributions of a set of parameters in the Bayesian analysis. Here we consider two commonly used algorithms in Bayesian analysis: Metropolis–Hastings and the Gibbs sampler.

3.3.1.1 Monte Carlo Error

By the nature of a Markov chain, the samples generated by it are positively correlated. In addition, due to its random nature, the results obtained vary every time it is run. A model parameter estimated from a Markov chain is subjected to these uncertainties even after the chain has converged. If we have a random sample from the target population, we could use the sample standard deviation to quantify the sampling error. However, this approach is inadequate for a Markov chain and it would underestimate the actual uncertainty due to the positive correlations of MCMC samples. The JAGS "summary" results provide several types of Monte Carlo error estimates. We shall discuss these options next.

Suppose $\theta_1, \ldots, \theta_N$ is a Markov chain of length N obtained after post burn-in for the parameter θ. That is, the chain is considered to be a correlated sample from the distribution of the parameter θ. The posterior mean, $\hat{\theta}$, which is an estimate of the mean of θ, is given by

$$\hat{\theta} = \frac{1}{N}\sum_{i=1}^{N} \theta_i \tag{3.8}$$

If we assume that the sample is independent and identically distributed, then an estimate of the variance of $\hat{\theta}$ is given by

$$\widehat{Var}_{\text{iid}}(\hat{\theta}) = \frac{1}{N}\left(\sum_{i=1}^{N} \frac{(\theta_i - \hat{\theta})^2}{(N-1)}\right) = \frac{S^2}{N} \tag{3.9}$$

The term $S^2 = (\theta - \hat{\theta})^2/(N-1)$ inside the parentheses in (3.9) is called the sample variance, which is an unbiased estimate of the variance of θ given that the sample is independent. In the JAGS "summary" output, the square root of S^2 is shown as "SD." Since the MCMC sample is positively correlated, the estimate given in (3.9) underestimates the true variance of $\hat{\theta}$. This variance is called the naïve variance of the posterior mean estimate, $\hat{\theta}$, and its square root is called the "Naïve SE" (naïve standard error) in the JAGS "summary" output. One way to reduce the effect of correlation between samples on the variance estimate is to use thinning, which keeps only certain sampled values, say the mth. The value of m should be selected such that the *lag m* autocorrelation is negligible (say, <0.1). However, this approach causes some information in the samples to be thrown away. Another simpler but somewhat naïve way to estimate the variance of $\hat{\theta}$ is to use the batch mean estimator, which is described in detail by Carlin and Louis (2000, p. 172). According to this method, the MCMC (post burn-in) chain is divided into m batches each of size k (i.e. $N = mk$). Suppose that the batch means are denoted by $\overline{B}_1, \ldots, \overline{B}_m$, then their average is the same as $\hat{\theta}$ (i.e. $\hat{\theta} = \overline{B} = \left(\sum_{i=1}^{m} \overline{B}_i\right)/m$). Assuming that the batch means are fairly close to being independent, an estimate of the variance of $\hat{\theta}$ can be given by

$$\widehat{Var}_{\text{batch}}(\hat{\theta}) = \frac{1}{m(m-1)} \sum_{i=1}^{m} (\overline{B}_i - \hat{\theta})^2 \tag{3.10}$$

The square root of this variance estimate is designated as "Time-series SE" in the JAGS "summary" output. When making this estimate, to obtain an approximately independent sample of batch means, it is important to make sure that the batch size is large enough so that the *lag* 1 autocorrelation between $\overline{B}_i$ values is negligible (say, <0.1). If this is not the case, then it is necessary to increase the batch size, k, and in turn the length of the chain, N.

3.3.2 Metropolis–Hastings Algorithm

Metropolis et al. (1953) originally developed the Metropolis algorithm. Subsequently, Hastings (1970) generalized the original algorithm to what is known as the Metropolis–Hastings (M-H) algorithm. This latter version of the algorithm is considered to be the foundation of all MCMC methods. A more generalized version of the algorithm that allows sampling from parameter spaces of different dimensions was developed by Green (1995). This modern algorithm is called the reversible jump M-H algorithm.

The M-H algorithm provides a simple prescription for obtaining samples (draws) from the posterior distribution of a set of parameters. Suppose $\boldsymbol{\theta} = (\theta_1, \theta_2 \cdots \theta_k)$ is a k-dimensional parameter vector and we would like to estimate its joint posterior distribution given the "data." Suppose $h(\boldsymbol{\theta}^* \mid \boldsymbol{\theta}^{(i-1)})$ is the proposal density of generating $\boldsymbol{\theta}^*$ from $\boldsymbol{\theta}^{(i-1)}$, where $\boldsymbol{\theta}^{(i-1)} = (\theta_1^{(i-1)}, \theta_2^{(i-1)}, \cdots \theta_k^{(i-1)})$ is the $(i-1)$th simulated value of the chain. In theory any density can be used as a proposal density if it satisfies three conditions. First, the proposal density must allow movements between any subsets of the parameter space in a finite number of steps. Second, it cannot be periodic. Informally, this means moving to any part of the parameter space is possible at any time. Finally, the proposal density must satisfy the following rule

$$0 < \frac{h(\theta^* \mid \theta^{(i-1)})}{h(\theta^{(i-1)} \mid \theta^*)} < \infty,$$

for all values of θ^* and $\theta^{(i-1)}$. To determine whether the generated candidate value θ^* is accepted as the next simulated value of the sequence, a quantity called acceptance probability, a, is computed as follows

$$a = \min\left(1, \frac{f(\theta \mid y)}{f(\theta^{(i-1)} \mid y)} * \frac{h(\theta^{(i-1)} \mid \theta^*)}{h(\theta^* \mid \theta^{(i-1)})}\right), \tag{3.11}$$

where $f(x \mid y)$ denotes the posterior density evaluated at x and $y = (y_1, y_2, \cdots y_n)$ represents the observed data.

If the proposal density is symmetric (i.e. $h(\theta^{(i-1)} \mid \theta^*) = h(\theta^* \mid \theta^{(i-1)})$) then the second ratio can be dropped from the (3.11) while calculating acceptance probability. After computing the acceptance probability, a, generate a *Uniform* (0, 1) random number, u. Accept the candidate value θ^* as the next simulated value of the sequence if $u \le a$ and set $\theta^i = \theta^*$. If $u > a$ then reject the candidate value θ^* and keep the current value and set $\theta^i = \theta^{(i-1)}$. This process is repeated a large number (N) of times. As $N \to \infty$, the MCMC sequence will converge to the joint posterior distribution, $f(\theta \mid y)$, of the parameter vector, θ. In this version of the M-H algorithm, all components of the parameter vector θ are updated at each step of the MCMC sequence. An alternative approach is to update the individual components of the parameter vector sequentially and this is called the single-component M-H algorithm. In this method, in each step, each individual component of the parameter vector $\theta = (\theta_1, \theta_2, \cdots \theta_k)$ is updated sequentially. The single-component M-H algorithm can be described as follows:

Let $\theta = (\theta_1, \theta_2, \cdots \theta_k)$ be the parameter vector of interest and $h_m(\theta^*_m \mid \theta^{(i-1)})$ is the proposal density for generating the mth component of the parameter vector θ in the ith step of the MCMC sequence, given the current value, $\theta^{(i-1)}$, of the parameter vector. The steps in the single-component M-H algorithm are summarized in the following steps.

1) Select the initial values of the parameter vector, $\theta^{(0)} = (\theta_1^{(0)}, \theta_2^{(0)}, \cdots \theta_k^{(0)})$.
2) To generate the ith draw of the sequence, $\theta^{(i)} = (\theta_1^{(i)}, \theta_2^{(i)}, \cdots \theta_k^{(i)})$, set
   a) $\theta = \theta^{(i-1)} = (\theta_1^{(i-1)}, \theta_2^{(i-1)}, \cdots \theta_k^{(i-1)})$
   b) For $m = 1, \cdots, k$
      I. Generate the candidate value θ^*_m for the mth component of the parameter vector $\theta^{(*)} = (\theta_1^{(*)}, \theta_2^{(*)}, \cdots \theta_k^{(*)})$ from the proposal density $h_m(\theta^*_m \mid \theta)$.
      II. Compute the acceptance probability,

$$\begin{aligned} a &= \min\left(1, \frac{f(\theta^*_m \mid \theta_{-m}, y)\, h_m(\theta_m \mid \theta^*_m, \theta_{-m})}{f(\theta_m \mid \theta_{-m}, y)\, h_m(\theta^*_m \mid \theta_m, \theta_{-m})}\right) \\ &= \min\left(1, \frac{f(y \mid \theta_{-m}, \theta^*_m)\, g(\theta^*_m, \theta_{-m})\, h_m(\theta_m \mid \theta^*_m, \theta_{-m})}{f(y \mid \theta_{-m}, \theta_m)\, g(\theta_m, \theta_{-m})\, h_m(\theta^*_m \mid \theta_m, \theta_{-m})}\right), \end{aligned} \tag{3.12}$$

where $\theta_{-m} = (\theta_1, \theta_2, \cdots, \theta_{m-1}, \theta_{m+1}, \cdots \theta_k)$ is the parameter vector θ excluding its mth component, $f(\theta^*_m \mid \theta_{-m}, y)$ is the conditional posterior density evaluated at the candidate value θ^*_m given the parameter vector

$\boldsymbol{\theta}_{-m}$, and the data, $\boldsymbol{y}$, $g(\theta_m^*, \boldsymbol{\theta}_{-m})$ is the prior joint density of $\boldsymbol{\theta}$ evaluated at $\boldsymbol{\theta} = (\theta_1^{(i-1)}, \theta_2^{(i-1)}, \cdots, \theta_{m-1}^{(i-1)}, \theta_m^*, \theta_m^{(i-1)} \cdots \theta_k^{(i-1)})$ and so on.

III. Update $\theta_m = \theta_m^*$ with probability a.

In a given step, use already updated values in the current step and yet un-updated values from the previous step for evaluating the conditional densities for updating any given component. Example 3.1 shows how to use this approach. Once all components are updated, set $\boldsymbol{\theta}^{(i)} = \boldsymbol{\theta}^*$. The order of the updating sequence can be chosen randomly. In general, the updating sequence of the parameter components does not have an impact on the convergence of the MCMC sequence.

Special cases of the M-H algorithm include the random-walk Metropolis, the independence sampler, and the Gibbs sampler. In the original Metropolis algorithm, only symmetric proposal distributions where $h(\boldsymbol{\theta} \mid \boldsymbol{\theta}^*) = h(\boldsymbol{\theta}^* \mid \boldsymbol{\theta})$ were considered. In the random-walk Metropolis sampler, proposal distributions were restricted to the type with $h(\boldsymbol{\theta} \mid \boldsymbol{\theta}^*) = h(|\theta - \theta^*|)$. In both of these cases, the acceptance probability in (3.11) reduces to

$$a = \min\left(1, \frac{f(\boldsymbol{\theta}^* \mid \boldsymbol{y})}{f(\boldsymbol{\theta}^{(i-1)} \mid \boldsymbol{y})}\right) = \min\left(1, \frac{f(y \mid \boldsymbol{\theta}^*)g(\boldsymbol{\theta}^*)}{f(y \mid \boldsymbol{\theta}^{(i-1)})g(\boldsymbol{\theta}^{(i-1)})}\right) \tag{3.13}$$

In the independence sampler M-H algorithm, proposal distributions that do not depend on the parameters are used. For example, consider the binomial likelihood of x successes in n identical and independent trials with success probability θ. Assume that the prior distribution of θ is $Uniform(a, b)$ with $0 \le a < b \le 1$. The posterior distribution of θ given the data x is given by

$$f(\theta \mid x) \propto \theta^x (1-\theta)^{(n-x)}, a \le \theta \le b$$

If we consider the proposal distribution to be the same as the prior distribution which does not depend on θ, then it is called an independent sampler.

The key elements of the M-H algorithm are:

- target distribution (posterior distribution), which the Markov chain will converge to
- proposal distribution, where the Metropolis algorithm draws candidate values from
- initial values of unknown parameters for each Markov chain
- acceptance probability.

The target distribution is the posterior distribution. Note that the target distribution does not have to be normalized, so the target distribution can be defined as prior × likelihood instead, which is proportional to the posterior distribution. After a certain number of iterations, sampled values gradually converge to a stable distribution, which is the target distribution or the posterior distribution that we are trying to estimate. Before the MCMC converges, the earlier iterations are called the burn-in period. Samples from the burn-in period will be discarded. We will only collect samples after the convergence is achieved, and any further inferences are based on these samples.

Selecting an appropriate proposal distribution can be tricky. The proposal distribution cannot be too narrow (less variability). If it is too narrow, the step size is really small and the random walk can be really slow to cover the entire range of the target distribution, thus it may take a long time to converge and the resulting sample may have high autocorrelation. In this case, the acceptance rate would be high. On the other hand, the proposal distribution cannot be too wide either (higher variability). If it is too wide, the step size can be very large, the candidate values have a low acceptance rate, and therefore the chain may stick around the same place for a long time. This will also result in high autocorrelation and therefore take a long time to converge. The ideal proposal distribution is one that is closest to the posterior distribution. A proposal distribution with an acceptance rate of 20–40% is recommended according to Spiegelhalter et al. (2003). A more detailed discussion about various aspects of the acceptance rates of different types of proposal distributions can be found in Ntzoufras (2009). An implementation of the component-wise M-H algorithm is given in the example below.

Example 3.2 The heights of 30 men selected at random from the electronic health records of a medical clinic are shown in Table 3.2. Historically, it is known that height measurements can be modeled by a normal distribution. We will show the step-by-step process for using the component-wise M-H algorithm to estimate the mean and standard deviation of the underlying normal distribution. The R code used for estimating these parameters is also provided.

The density function of a normal distribution is given by

$$f(x \mid \mu, \sigma^2) = \frac{1}{\sqrt{2\pi\sigma^2}} e^{-\frac{(x-\mu)^2}{2\sigma^2}}$$

Consider a relatively vague (i.e. covers a reasonable range of values) prior distribution for the mean μ that is normally distributed with the mean $\mu_0 = 10$ and the standard deviation $\sigma_0 = 100$. Also, consider a relatively vague prior distribution for the variance σ^2 that has an inverse-gamma distribution with shape $\alpha = 3$ and scale $\beta = 0.025$. This will result in the mean and the variance of σ^2 being 20 $(=1/(\beta(\alpha-1)))$ and 400 $(=1/(\beta^2(\alpha-1)^2(\alpha-2)))$, respectively. Let's assume the proposal distribution for the parameter μ to be normal with mean $\mu^{(i-1)}$ and standard deviation $\sigma_1 = 4$, and the proposal distribution of σ^2 to be a lognormal with mean $\log(\sigma^2)^{(i-1)}$ and standard deviation $\sigma_2 = 1$. This implies $\log(\sigma^2) \sim N(\log((\sigma^2)^{(i-1)}), \sigma_2 = 1)$. The selection of the proposal distributions in this manner were done for mathematical tractability. Their parameter selections were made by running the following code with different choices and then selecting the values that produce a reasonable (20–40%) acceptance rate.

Table 3.2 Height (inches) measurements of 30 men.

72	72	74
66	68	79
67	63	73
74	67	70
70	65	72
70	67	74
64	70	69
65	71	66
66	65	76
71	69	77

In addition, the proposal densities were selected to cover the possible range of values of the respective posterior distributions of the parameters μ and σ^2.

The joint posterior distribution of the parameters μ and σ^2 is given by

$$f(\mu, \sigma^2 \mid \mathbf{y}) = \frac{g(\mathbf{y} \mid \mu, \sigma^2) h_1(\mu) h_2(\sigma^2)}{k(\mathbf{y})}, \tag{3.14}$$

where $g(\mathbf{y} \mid \mu, \sigma^2)$ is the likelihood function of the data $\mathbf{y}$ given the parameters μ and σ^2, $h_1(\mu)$ and $h_2(\sigma^2)$ are the prior densities of the parameters μ and σ^2, respectively, and $k(\mathbf{y})$ is the marginal distribution of $\mathbf{y}$.

With the above assumptions, (3.14) can be written as

$$f(\mu, \sigma^2 \mid \mathbf{y}) = \prod_{j=1}^{n} \frac{e^{-\frac{(y_j-\mu)^2}{2\sigma^2}}}{\sqrt{2\pi\sigma^2}} \frac{e^{-\frac{(\mu-\mu_0)^2}{2\sigma_0^2}}}{\sqrt{2\pi\sigma_0^2}} \frac{1}{\Gamma(\alpha)} \frac{(\sigma^2)^{-(\alpha+1)} e^{-\left(\frac{1}{\beta\sigma^2}\right)}}{\beta^{\alpha}}$$

$$\propto (\sigma^2)^{-(\alpha+1+n/2)} e^{-\left[\sum_{j=1}^{n} \frac{(y_j-\mu)^2}{2\sigma^2} + \frac{(\mu-\mu_0)^2}{2\sigma_0^2} + \frac{1}{\beta\sigma^2}\right]} \tag{3.15}$$

The following steps are implemented for generating MCMC samples of the posterior distribution of the parameters.

Steps:

1. Select the initial values $i = 0$, $\mu^0 = 20$ and $(\sigma^2)^{(0)} = 5$. Selection of these values is arbitrary.
2. In the ith step, draw μ^* from the $N(\mu^{(i-1)}, \sigma_1 = 4)$ distribution.
3. Compute the acceptance probability a_1 as follows:

$$a_1 = \min\left\{1, \frac{f(\mu^*, (\sigma^2)^{(i-1)} \mid \mathbf{y})}{f(\mu^{(i-1)}, (\sigma^2)^{(i-1)} \mid \mathbf{y})} \frac{q_1(\mu^{(i-1)} \mid \mu^*)}{q_1(\mu^* \mid \mu^{(i-1)})}\right\},$$

where $q_1(\mu^{(i-1)} \mid \mu^*)$ is the proposal density $N(\mu^*, \sigma_1 = 4)$ evaluated at $\mu^{(i-1)}$ and $q_1(\mu^* \mid \mu^{(i-1)})$ is $N(\mu^{(i-1)}, \sigma_1 = 4)$ evaluated at μ^*. Since the normal density is symmetric, the ratio $q_1(\mu^{(i-1)} \mid \mu^*/q_1(\mu^* \mid \mu^{(i-1)}) = 1$.

Thus,

$$a_1 = \min\left\{1, \frac{\exp\left(-\left[\sum_{j=1}^{n}(y_j-\mu^*)^2/2(\sigma^2)^{(i-1)} + (\mu^*-\mu_0)^2/2\sigma_0^2 + 1/\beta(\sigma^2)^{(i-1)}\right]\right)}{\exp\left(-\left[\sum_{j=1}^{n}(y_j-\mu^{(i-1)})^2/2(\sigma^2)^{(i-1)} + (\mu^{(i-1)}-\mu_0)^2/2\sigma_0^2 + 1/\beta(\sigma^2)^{(i-1)}\right]\right)}\right\}$$

$$= \min\left\{1, \frac{\exp\left(-\left[\sum_{j=1}^{n}(y_j-\mu^*)^2/2(\sigma^2)^{(i-1)} + (\mu^*-\mu_0)^2/2\sigma_0^2\right]\right)}{\exp\left(-\left[\sum_{j=1}^{n}(y_j-\mu^{(i-1)})^2/2(\sigma^2)^{(i-1)} + (\mu^{(i-1)}-\mu_0)^2/2\sigma_0^2\right]\right)}\right\}$$

4. Generate the *Uniform*(0, 1) random variable u.
5. If $u \le a_1$ then set $\mu^{(i)} = \mu^*$, otherwise set $\mu^{(i)} = \mu^{(i-1)}$.
6. Draw $(\sigma^2)^*$ from its proposal distribution, $\log(\sigma^2) \sim N(\log((\sigma^2)^{(i-1)}), \sigma_2 = 1)$.

7. Compute the acceptance probability a_2 as follows:

$$a_2 = \min\left\{1, \frac{f(\mu^{(i)}, (\sigma^2)^* \mid \boldsymbol{y})}{f(\mu^{(i)}, (\sigma^2)^{(i-1)} \mid \boldsymbol{y})} \frac{q_2((\sigma^2)^{(i-1)} \mid (\sigma^2)^*)}{q_2((\sigma^2)^* \mid (\sigma^2)^{(i-1)})}\right\},$$

where $q_2((\sigma^2)^{(i-1)} \mid (\sigma^2)^*)$ is the proposal density $LN(\log((\sigma^2)^*), \sigma_1 = 1)$ evaluated at $(\sigma^2)^{(i-1)}$ and $q_2((\sigma^2)^* \mid (\sigma^2)^{(i-1)})$ is $LN((\sigma^2)^{(i-1)}, \sigma_2 = 1)$ evaluated at $(\sigma^2)^*$. Therefore, we get

$$\frac{q_2((\sigma^2)^{(i-1)} \mid (\sigma^2)^*)}{q_2((\sigma^2)^* \mid (\sigma^2)^{(i-1)})} = \frac{\frac{1}{(\sigma^2)^{(i-1)}} \frac{1}{\sqrt{2\pi\sigma_2^2}} \exp\left(-\left[\frac{(\log((\sigma^2)^{(i-1)}) - \log((\sigma^2)^*)^2}{2\sigma_2^2}\right]\right)}{\frac{1}{(\sigma^2)^*} \frac{1}{\sqrt{2\pi\sigma_2^2}} \exp\left(-\left[\frac{(\log((\sigma^2)^*) - \log((\sigma^2)^{(i-1)})^2}{2\sigma_2^2}\right]\right)}$$

$$= \frac{(\sigma^2)^*}{(\sigma^2)^{(i-1)}}$$

Thus,

$$a_2 = \min\left\{1, \frac{((\sigma^2)^*)^{(-\alpha-1-n/2)}}{((\sigma^2)^{(i-1)})^{(-\alpha-1-n/2)}} \times \frac{\exp\left(-\left[\sum_{j=1}^{n}(y_j - \mu^{(i)})^2/2(\sigma^2)^* + (\mu^{(i)} - \mu_0)^2/2\sigma_0^2 + 1/\beta(\sigma^2)^*\right]\right)}{\exp\left(-\left[\sum_{j=1}^{n}(y_j - \mu^{(i)})^2/2(\sigma^2)^{(i-1)} + (\mu^{(i)} - \mu_0)^2/2\sigma_0^2 + 1/\beta(\sigma^2)^{(i-1)}\right]\right)} \frac{(\sigma^2)^*}{(\sigma^2)^{(i-1)}}\right\}$$

$$= \min\left\{1, \frac{((\sigma^2)^*)^{-(\alpha+n/2)}}{((\sigma^2)^{(i-1)})^{-(\alpha+n/2)}} \frac{\exp\left(-\left[\sum_{j=1}^{n}(y_j - \mu^{(i)})^2/2(\sigma^2)^* + 1/\beta(\sigma^2)^*\right]\right)}{\exp\left(-\left[\sum_{j=1}^{n}(y_j - \mu^{(i)})^2/2(\sigma^2)^{(i-1)} + 1/\beta(\sigma^2)^{(i-1)}\right]\right)}\right\}$$

8. Generate the *Uniform*(0, 1) random variable u.
9. If $u \le a_2$ then set $(\sigma^2)^{(i)} = (\sigma^2)^*$, otherwise set $(\sigma^2)^{(i)} = (\sigma^2)^{(i-1)}$.
10. Increment i by 1 and repeat steps 2–9.

In the above steps, the order of updating the sequence between the mean μ and the variance σ^2 can be selected at random, however it will not have much impact on the convergence of the Markov chains. The following R script generates MCMC samples based on the M-H algorithm described in the above steps. This R script also generates Figures 3.3 and 3.4. The starting value of the random seed of the code was fixed so that the results of the code can be reproduced.

(3.3.2_Metropolis_Hastings_Example.R)

```
    # The following code performs Markov Chain Monte Carlo Simula-
tion using
Metropolis
    # -Hastings componentwise random-walk algorithm. The data is from
Example 3.2
    # There are two parameters of interest; the mean, mu, and the
variance, SigSqr.
    #
    library(forecast) # package to access autocorrelation plot
function, "Acf"
```

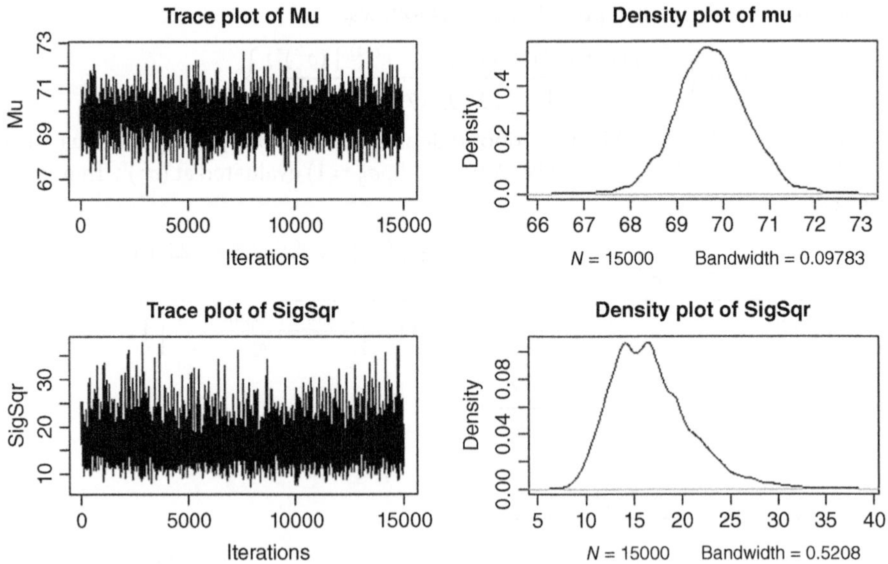

Figure 3.3 Trace and density plots of "mu" and "SigSqr" obtained from M-H posterior samples.

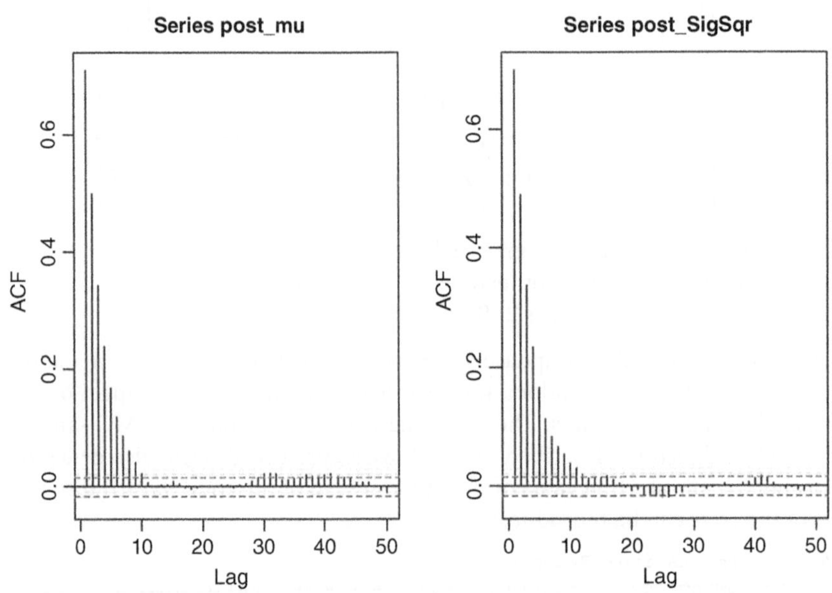

Figure 3.4 Autocorrelation plots of different lags for "mu" and "SigSqr" obtained from M-H posterior samples.

```
    library(mcmcse) # package to compute effective sample size of a Markov
Chain
    #library(invgamma) # package to sample from the inverse gamma
distribution
```

```
    # Create a function to compute the acceptance probability for the
mean, mu
    # Do the calculation in log-scale to avoid numeric overflow issues

    AccProb_mu <- function(y,mu0,sig0,muNew,muOld,SigSqr) {
      logA1 <- sum((y-muOld)**2)/(2*SigSqr)+((muOld-mu0)**2)/(2*sig0**2)
- (sum((y-muNew)**2)/(2*SigSqr)+((muNew-mu0)**2)/(2*sig0**2))
      A1 <- min(1,exp(logA1))
      return(A1)
    }
    # Create a function to compute the acceptance probability for the
variance,
    # SigSqr
    # Do the calculation in log-scale to avoid numeric overflow issues
    AccProb_Sq <- function(y,SigSqrOld,SigSqrNew,Alpha,Beta,mu) {
      n <- length(y)
      logA2 <- -(Alpha+n/2)*(log(SigSqrNew/SigSqrOld)) +  sum((y-mu)**2)
/(2*SigSqrOld) + 1/(SigSqrOld*Beta) - (sum((y-mu)**2)/(2*SigSqrNew)
+ 1/(SigSqrNew*Beta))
      A2 <- min(1,exp(logA2))
      return(A2)
    }

    # Get number of simulations
    NS <- 20000
    # Get number of burn-in simulations
    NB <- 5000
    # Get the parameters of the prior normal distribution of "mu" to cover
    # reasonable range of values
    mu0 <- 10
    sig0 <- 100

    # The following parameters of the prior inverse-Gamma
distribution of "SigSqr"
    # is selected to cover reasoanble range of values;
mean=1/(0.025*(3-1))=20,
    # variance=1/(0.025**2*(3-1)**2*(3-2))=400
    Alpha <- 3
    Beta <- 0.025

    # Get Standard deviations of the proposal distributions of "mu" and
SigSqr
    # These two parameters have been adjusted to get a reasonable
acceptance rate
    # for "mu" and "SigSqr"
    Sig1 <- 4
    Sig2 <- 1
    #
    # Get initial values of mu and SigSqr
    muIn <- 20
    SigSqrIn <- 5
    # Initialize the vector to store sampled "mu" values
    muVec <- numeric(NS)
    # Initialize the vector to store sampled "SigSqr" values
    SigSqrVec <- numeric(NS)
```

```
    # read the data
    y <- c(72,66,67,74,70,70,64,65,66,71,72,68,63,67,65,67,70,71,65,69,74,
79,73,70,72,74,69,66,76,77)

    # Initiallize the counters to monitor acceptance rates
    set.seed(1234)
    a1cnt <- 0
    a2cnt <- 0
    for (i in 1:NS) {
      if(i==1){
        SigSqrOld <- SigSqrIn
        muOld <- muIn
      }
      muNew <- rnorm(1,mean=muOld,sd=Sig1) # Draw the new value of "mu"
      SigSqr <- SigSqrOld
      a1 <- AccProb_mu(y,mu0,sig0,muNew,muOld,SigSqr) # Compute the
acceptance
    # probability of "muNew"
      u <- runif(1,0,1)
      if(u <= a1) (a1cnt <- a1cnt+1)
      if (u > a1) muNew=muOld #accept the drwan muNew with
probability, a1, and
    # reject it and keep the old value with prob (1-a1).
      SigSqrNew<- rlnorm(1, meanlog=log(SigSqrOld), sdlog=Sig2) # Draw the
new
    #  value of "SigSqr"

      mu <- muNew
      a2 <- AccProb_Sq(y,SigSqrOld,SigSqrNew,Alpha,Beta,mu) # Compute the
    #   acceptance probability of "muNew"
      u <- runif(1,0,1)
      if(u <= a2) (a2cnt <- a2cnt+1)
      if (u > a2) SigSqrNew <- SigSqrOld #accept the drwan SigSqrNew with
    #  probability, a2, and reject it and keep the old value with prob
(1-a2).
      muVec[i] <- muNew # store sampled values of "mu"
      SigSqrVec[i] <- SigSqrNew # store sampled values of "SigSqr"
      muOld <- muNew # Reset the values of "muOld" and "SigSqrOld" for the
next
    # run of simulations
      SigSqrOld <- SigSqrNew
    }

    print(paste("The acceptance rate of mu:",round(100*a1cnt/NS,2),"%"))
    print(paste("The acceptance rate of SigSqr:",round(100*a2cnt/NS,2),
"%"))

    #summary(muVec)
    Post_mu <- muVec[-seq(1:NB)]
    #summary(SigSqrVec)
    Post_SigSqr <- SigSqrVec[-seq(1:NB)]

    #length(Post_mu)
    par(mfrow=c(1,2))
    Acf(Post_mu, lag.max = 50,type = "correlation", plot = TRUE)
```

```
      Acf(Post_SigSqr, lag.max = 50,type = "correlation", plot = TRUE)
      #acf(Post_mu, lag.max = 50,type = "correlation", plot = TRUE)
      #acf(Post_SigSqr, lag.max = 50,type = "correlation", plot = TRUE)

      ## Get the summary statistics of "mu" using the samples after burn-in
      summary(Post_mu)
      ## Get various percentiles of "mu" using the samples after burn-in
      print("Various Percentiles of mu:", quote=FALSE)
      quantile(Post_mu, probs=c(0.025, 0.05, 0.10, 0.50, 0.90, 0.95, 0.975))

      ## Get the summary statistics of "SigSqr" using the samples after burn
      in summary(Post_SigSqr)
      ## Get various percentiles of "SigSqr" using the samples after burn-in
      print("Various Percentiles of SigSqr:", quote=FALSE)
      quantile(Post_SigSqr, probs=c(0.025, 0.05, 0.10, 0.50, 0.90, 0.95,
0.975))

      par(mfrow=c(2,2))
      XVal <- seq(1:length(Post_mu))
      # Create trace plot of "mu"
      #jpeg("Trace_plot_Metropolis_Hastings_mu.jpeg", width = 6, height = 4,
units = 'in', res = 600)  # save the plot as jpeg format
      plot(XVal,Post_mu,type="l",ylab="Mu",xlab="Iterations",main="Trace
plot of Mu")
      #dev.off()

      # Create a Kernel Density Plot using the Posterior samples for "mu"
      d_mu <- density(Post_mu)
      plot(d_mu, main="Density Plot of mu")


      #jpeg("Trace_plot_Metropolis_Hastings_SiSqr.jpeg", width = 6,
height = 4, units = 'in', res = 600)  # save the plot as jpeg format
      plot(XVal,Post_SigSqr,type="l",ylab="SigSqr",xlab="Iterations",
main="Trace plot of SigSqr")
      #dev.off()


      # Create a Kernel Density Plot using the Posterior samples for
"SigSqr"
      d_SigSqr <- density(Post_SigSqr)
      plot(d_SigSqr, main="Density Plot of SigSqr")
```

The output resulting from running the above code is shown below. The acceptance rates for μ and σ^2 are satisfactory. The posterior means of μ and σ^2 are 69.7 and 16.7, respectively. Their respective 95% credible intervals are (68.3, 71.3) and (10.4, 26.6), respectively. Figure 3.3 provides the trace and density plots of μ and σ^2. The trace plots for both parameters show no pattern and the MCMC chains appear to converge. The density plots are not very smooth. This behavior was expected because the M-H algorithm almost always produces correlated data. The autocorrelation plots of simulated samples of μ and σ^2 are shown in Figure 3.4. According to these plots, for both parameters the autocorrelation is greater than 0.2 for the first four lags. For subsequent lags, the autocorrelation decreases rapidly.

```
> print(paste("The acceptance rate of mu:",round(100*a1cnt/NS,2),"%"))
[1] "The acceptance rate of mu: 22.48 %"
> print(paste("The acceptance rate of SigSqr:",round(100*a2cnt/NS,2),
"%"))
[1] "The acceptance rate of SigSqr: 28.2 %"

> summary(Post_mu)
   Min. 1st Qu.  Median    Mean 3rd Qu.    Max.
  66.37   69.25   69.73   69.75   70.25   72.79

[1] Various Percentiles of mu:
> quantile(Post_mu, probs=c(0.025, 0.05, 0.10, 0.50, 0.90, 0.95,
0.975))
     2.5%       5%      10%      50%      90%      95%    97.5%
68.28345 68.50246 68.82909 69.73374 70.72109 71.01945 71.25764
>
> summary(Post_SigSqr)
   Min. 1st Qu.  Median    Mean 3rd Qu.    Max.
  7.031  13.770  16.190  16.720  19.070  37.610

 [1] Various Percentiles of SigSqr:
> quantile(Post_SigSqr, probs=c(0.025, 0.05, 0.10, 0.50, 0.90, 0.95,
0.975))
     2.5%       5%      10%      50%      90%      95%    97.5%
10.37612 11.10002 12.05624 16.19006 22.22020 24.21308 26.63937
>
```

The function "ess" in the R package "mcmcse" can be used to determine the effective sample size (ESS) from the MCMC samples generated for μ and σ^2. ESS is defined as the size of an identically and independently distributed (iid) sample with the same variance as the current sample. The method for computing ESS is given by

$$ESS = N(S^2/\delta^2),$$

where N is the length of the Markov chain, S^2 is the sample variance, and δ^2 is the true variance of the target population. Since the true variance is unknown, it must be estimated. One way to estimate δ^2 is to use the batch mean estimator given in (3.10). The package "mcmcse" has batch means method as one of the options for computing the ESS. After loading the package "mcmcse," the following code can be run to estimate the ESSs for μ and σ^2.

```
EffSS1 <- ess(Post_mu, g=NULL) # Compute effective sample size for
"mu" in the posterior samples
print(EffSS1)
n1 <- ceiling(EffSS1)

EffSS2 <- ess(Post_SigSqr, g=NULL) # Compute effective sample size for
"mu" in the posterior samples
print(EffSS2)
n2 <- ceiling(EffSS2)
```

The following results were generated by running the above code. The ESS for μ is 2689 while for σ^2 the ESS is 2494. These numbers indicate that there is a considerable amount of correlated data in the simulated samples ($N = 15\,000$). Since most of the data values are correlated, we might be able to estimate the two parameters using random samples of size equal to the ESS, taken from their respective MCMC sequences:

```
> print(EffSS1)
[1] 2688.506
> print(EffSS2)
[1] 2493.961
```

We use the ESSs to obtain random samples from the Markov chains of μ and σ^2. For the reader to be able to reproduce the results, the random seeds are fixed as shown in the code below:

```
set.seed(2345)
RS_mu_post <- sample(Post_mu,n1,replace = FALSE)
## Get the summary statistics of "mu" from a random sample of size,
n1 (=effective
## sample size) from its posterior samples
summary(RS_mu_post)
## Get various percentiles of "mu" from a random sample of size,
n1 (=effective
## sample size)from its posterior samples
quantile(RS_mu_post, probs=c(0.025, 0.05, 0.10, 0.50, 0.90, 0.95,
0.975))
set.seed(2384)
RS_SigSqr_post <- sample(Post_SigSqr,n1,replace = FALSE)
## Get the summary statistics of "SigSqr" from a random sample of
size, n2 (=
## effective sample size) from its posterior samples
summary(RS_SigSqr_post)
## Get various percentiles of "SigSqr" from a random sample of size,
n2 (=effective
## sample size) from its posterior samples
quantile(RS_SigSqr_post, probs=c(0.025, 0.05, 0.10, 0.50, 0.90, 0.95,
0.975))
```

The following results were obtained by running the above code. The summary statistics and percentile estimates based on the random samples are very close to those obtained from the full posterior samples. For example, the estimated posterior mean of μ for the full sample is 69.75 while for the random sample it is 69.73. The estimated posterior mean of σ^2 is 16.72 for the full sample and 16.71 for the random sample. These estimates are very close. The differences are little bit wider in the tails of the distributions, but are within 2% for all the estimated percentiles for both μ and σ^2.

```
> summary(RS_mu_post)
   Min. 1st Qu.  Median    Mean 3rd Qu.    Max.
  66.88   69.23   69.72   69.73   70.22   72.48

> quantile(RS_mu_post, probs=c(0.025, 0.05, 0.10, 0.50, 0.90, 0.95,
0.975))
     2.5%       5%      10%      50%      90%      95%    97.5%
68.25365 68.47386 68.79504 69.72179 70.70825 70.98914 71.24812
>
> summary(RS_SigSqr_post)
   Min. 1st Qu.  Median    Mean 3rd Qu.    Max.
  7.835  13.710  16.140  16.710  18.990  36.760
```

```
> quantile(RS_SigSqr_post, probs=c(0.025, 0.05, 0.10, 0.50, 0.90, 0.95,
0.975))
      2.5%       5%      10%      50%      90%      95%    97.5%
10.42009 11.09284 12.02181 16.13749 22.38771 24.45265 27.07860
```

3.3.3 Gibbs Sampling

The M-H algorithm allows sampling from general posterior distributions with little or no restriction on the type of distributions allowed for both the likelihood and the priors of the parameters. However, the success of obtaining good posterior samples depends on the use of appropriate proposal densities, which in some cases can be a difficult task. The drawbacks of selecting poor proposal densities were discussed in the previous section. A better approach for drawing samples from the posterior distribution of a parameter is to use its full conditional posterior density. The algorithm that implements this approach for generating MCMC samples is called the Gibbs sampler. The full conditional distributions can be easily obtained when working with conjugate priors of the parameters.

To provide a formal description of the Gibbs sampler, suppose $\boldsymbol{\theta} = (\theta_1, \cdots \theta_k)$ is a k-dimensional parameter vector of interest. Suppose $f(\theta_j \mid \theta_1, \cdots \theta_{j-1}, \theta_{j+1}, \cdots \theta_k, \boldsymbol{y})$, $i = 1, \cdots k$ denotes the full conditional distribution of the component θ_j conditional on the remaining parameters $(\theta_1, \cdots \theta_{j-1}, \theta_{j+1}, \cdots \theta_k)$ and the data $\boldsymbol{y}$. The Gibbs sampler can be described as follows:

1) Select an arbitrary set of initial values of the parameters, $\boldsymbol{\theta}^{(0)} = (\theta_1^{(0)}, \theta_2^{(0)}, \cdots \theta_k^{(0)})$.
2) To generate the ith draw of the parameter vector of the MCMC sequence:

   Step 1: Draw $\theta_1^{(i)}$ from $f(\theta_1 \mid \theta_2^{(i-1)}, \cdots \theta_k^{(i-1)}, \boldsymbol{y})$.
   Step 2: Draw $\theta_2^{(i)}$ from $f(\theta_2 \mid \theta_1^{(i)}, \theta_3^{(i-1)}, \cdots \theta_k^{(i-1)}, \boldsymbol{y})$.
   ⋮
   Step k: Draw $\theta_k^{(i)}$ from $f(\theta_k \mid \theta_1^{(i)}, \theta_2^{(i)}, \cdots \theta_{k-1}^{(i)}, \boldsymbol{y})$.

   After updating all components of the parameter vector, set $\boldsymbol{\theta}^{(i)} = (\theta_1^{(i)}, \theta_2^{(i)}, \cdots \theta_k^{(i)})$.

Repeat steps 1–k for each $i = 1, \cdots N$. For sufficiently large N, under certain regulatory conditions, the Markov chain will converge to the true posterior density $f(\boldsymbol{\theta} \mid \boldsymbol{y}) = f(\theta_1, \theta_2, \cdots \theta_k \mid \boldsymbol{y})$. As in the M-H algorithm, we discard the initial samples of the sequence, the burn-in. All inferences concerning the parameters $\boldsymbol{\theta} = (\theta_1, \cdots \theta_k)$ can be made using the post burn-in samples and they are simply called posterior samples.

As an example of the application of the Gibbs sampler, consider the data in Example 3.2. Again, we can use the same assumptions of Example 3.2 concerning the likelihood and the prior distributions of the parameters μ and σ^2. With these assumptions, full conditional distributions of μ and σ^2 are given as follows.

Recall that from (3.12) the joint posterior distribution of the parameters is given by

$$f(\mu, \sigma^2 \mid y) = \prod_{j=1}^{n} \frac{e^{-\frac{(y_j-\mu)^2}{2\sigma^2}}}{\sqrt{2\pi\sigma^2}} \frac{e^{-\frac{(\mu-\mu_0)^2}{2\sigma_0^2}}}{\sqrt{2\pi\sigma_0^2}} \frac{1}{\Gamma(\alpha)} \frac{(\sigma^2)^{-(\alpha+1)} e^{-\left(\frac{1}{\beta\sigma^2}\right)}}{\beta^\alpha}$$

$$\propto (\sigma^2)^{-(\alpha+1+n/2)} e^{-\left[\sum_{j=1}^{n} \frac{(y_j-\mu)^2}{2\sigma^2} + \frac{(\mu-\mu_0)^2}{2\sigma_0^2} + \frac{1}{\beta\sigma^2}\right]}$$

Assuming that the prior distribution of μ is $Normal(\mu_0, \sigma_0^2)$ and σ^2 is held fixed, then it can be shown (see Section 4.3.4 for derivation) that the full conditional posterior distribution of μ is given by

$$f_1(\mu \mid \sigma^2, y) = Normal\left(\frac{\left(\sigma_0{}^2 \sum_{j=1}^{n} y_j + \sigma^2 \mu_0\right)}{(n\sigma_0{}^2 + \sigma^2)}, \frac{\sigma^2 \sigma_0{}^2}{(n\sigma_0{}^2 + \sigma^2)}\right) \tag{3.16}$$

Assuming that the prior distribution of σ^2 is $InverseGamma(shape = \alpha, scale = \beta)$ and μ is held fixed, then the full conditional posterior distribution of σ^2 can be derived as follows:

$$f_2(\sigma^2 \mid \mu, y) \propto (\sigma^2)^{-(\alpha+1+n/2)} e^{-\left[\sum_{j=1}^{n} \frac{(y_j-\mu)^2}{2\sigma^2} + \frac{1}{\beta\sigma^2}\right]}$$

$$= (\sigma^2)^{-(\alpha+n/2+1)} e^{-\left[\frac{1}{\sigma^2}\left(\sum_{j=1}^{n} \frac{(y_j-\mu)^2}{2} + \frac{1}{\beta}\right)\right]}.$$

Therefore, the full conditional posterior distribution of σ^2 is an inverse gamma distribution with

$$shape = (\alpha + n/2)$$

$$scale = \frac{1}{\left(\sum_{j=1}^{n} (y_j - \mu)^2/2 + 1/\beta\right)} \tag{3.17}$$

To implement the Gibbs sampler for this problem, we will sample from the normal distribution given in (3.16) for μ and the inverse-gamma distribution given by (3.17) for σ^2. The following R code ("3.3.3_Implementation_Gibbs_Sampler.R") uses the Gibbs sampler to generate MCMC samples for μ and σ^2. The first 2000 samples are burn-in samples and therefore not considered as posterior samples of the parameters. The 5000 samples post burn-in will be used for estimating the posterior distributions of the two parameters. In the next section, we will discuss various software that are available for MCMC simulations. In this book we shall use JAGS software for solving most of the problems. In the next section, a step-by-step approach for setting up the model, prior distributions, initial values, and then running MCMC simulations and convergence diagnosis using JAGS software will be presented.

(3.3.3_Implementation_Gibbs_Sampler.R)

```
    # The following code perform Markov Chain Monte Carlo Simulation using
Gibbs sampling
    # algorithm
    # using the data given Example 3.2. There are two parameters of
interest; the mean,
    # mu, and the variance, SigSqr.
    #
    library(forecast) # package to access autocorrelation plot
function, "Acf"
    library(mcmcse) # package to compute the effective sample size of a
Markov Chain
```

```
    library(invgamma) # package to sample from the inverse gamma
distribution

    # Get number of simulations
    NS <- 7000
    # Get number of burn-in simulations
    NB <- 2000
    # Get the parameters of the vague nomral prior distribution of "mu"
    mu0 <- 10
    sig0 <- 100
    #mu0 <- 50
    #sig0 <- 10

    # The following parameters of the prior inverse-Gamma
distribution of "SigSqr"
    # is selected to cover reasoanble range of values;
mean=1/(0.025*(3-1))=20,
    # variance=1/(0.025**2*(3-1)**2*(3-2))=400
    Alpha <- 3
    Beta <- 0.025

    # Get initial values of mu and SigSqr
    muIn <- 20
    SigSqrIn <- 5
    # Initialize the vector to store sampled "mu" values
    muVec <- numeric(NS)
    # Initialize the vector to store sampled "SigSqr" values
    SigSqrVec <- numeric(NS)


    # read the data
    Y <- (72,66,67,74,70,70,64,65,66,71,72,68,63,67,65,67,70,71,65,69,74,
79,73,70,72, 74,69,66,76,77)
    n <- length(y)

    # Initiallize the counters to monior acceptance rates
    set.seed(1234) #Fix the random seed to be able to reprodcue the
results

    for ( i in 1:NS) {
      if(i==1){
        muOld=muIn
        SigSqrOld=SigSqrIn
      }
      # Determine the mean and the standard deviation of the Normal
distribution which is
      # the full conditional
      # posterior distribution of "mu"
      Norm_Mu = (sig0**2*sum(y)+mu0*SigSqrOld)/(n*sig0**2+SigSqrOld)
      Norm_SigSqr = SigSqrOld*sig0**2/(n*sig0**2+SigSqrOld)
      Norm_Sig = sqrt(Norm_SigSqr)
      muNew <- rnorm(1,mean = Norm_Mu, sd=Norm_Sig)
      muOld <- muNew # Set the value of "muOld" the next draw

      #Determine the shape and the scale parameters of the full
conditional distribution
```

```
        #of "SigSqr" which has
        #a inverse-gamma distribution
        IG_shape <- (Alpha+n/2)
        IG_Scale <- 1/(sum((y-muOld)**2)/2 + 1/Beta)
        SigSqrNew <- rinvgamma(1,shape=IG_shape, scale=IG_Scale)
        SigSqrOld <- SigSqrNew # Set the value of "SigSqrOld" for the next
draw

        muVec[i] <- muNew # store sampled values of "mu"
        SigSqrVec[i] <- SigSqrNew # store sampled values of "SigSqr"
      }

      #summary(muVec)
      PostSamp_mu <- muVec[-seq(1:NB)]
      #summary(SigSqrVec)
      PostSamp_SigSqr <- SigSqrVec[-seq(1:NB)]

      #length(PostSamp_mu)
      par(mfrow=c(1,2))
      Acf(PostSamp_mu, lag.max = 50,type = "correlation", plot = TRUE)
      Acf(PostSamp_SigSqr, lag.max = 50,type = "correlation", plot = TRUE)
      #acf(PostSamp_mu, lag.max = 50,type = "correlation", plot = TRUE)
      #acf(PostSamp_SigSqr, lag.max = 50,type = "correlation", plot = TRUE)

      ## Get the summary statistics of "mu" using the samples after burn in
      summary(PostSamp_mu)
      ## Get various percentiles of "mu" using the samples after burn-in
      print("Various Percentiles of mu:", quote=FALSE)
      quantile(PostSamp_mu, probs=c(0.025, 0.05, 0.10, 0.50, 0.90, 0.95,
0.975))

      ## Get the summary statistics of "SigSqr" using the samples after burn
in
      summary(PostSamp_SigSqr)
      ## Get various percentiles of "SigSqr" using the samples after burn-in
      print("Various Percentiles of SigSqr:", quote=FALSE)
      quantile(PostSamp_SigSqr, probs=c(0.025, 0.05, 0.10, 0.50, 0.90, 0.95,
0.975))

      par(mfrow=c(2,2))
      XVal <- seq(1:length(PostSamp_mu))
      # Create trace plot of "mu"
      #jpeg("Trace_plot_Metropolis_Hastings_mu.jpeg", width = 6, height = 4,
units = 'in', res = 600) # save the plot as jpeg format
      plot(XVal,PostSamp_mu,type="l",ylab="Mu",xlab="Iterations",main="Trace
plot of Mu")
      #dev.off()

      # Create a Kernel Density Plot using the Posterior samples for "mu"
      d_mu <- density(PostSamp_mu)
      plot(d_mu, main="Density Plot of mu")

      #jpeg("Trace_plot_Metropolis_Hastings_SiSqr.jpeg", width = 6,
height = 4, units = 'in', res = 600) # save the plot as jpeg format
```

```
    plot(XVal,PostSamp_SigSqr,type="l",ylab="SigSqr",xlab="Iterations",
main="Trace plot of SigSqr")
    #dev.off()


    # Create a Kernel Density Plot using the Posterior samples for
"SigSqr"
    d_SigSqr <- density(PostSamp_SigSqr)
    plot(d_SigSqr, main="Density Plot of SigSqr")
```

The trace and density plots shown in Figure 3.5, the autocorrelation plots shown in Figure 3.6, and the summary results shown below were generated by running the above code. The summary results generated by the Gibbs sampler match closely with those produced by the M-H sampler in the previous example. For example, the estimated posterior means of μ and σ^2 obtained from the Gibbs sampler are 69.72 and 16.69, respectively. The estimates obtained from the M-H sampler in 3.2 are 69.75 and 16.72, respectively. The 95% credible intervals of μ and σ^2 obtained from the Gibbs sampler are (68.26, 71.24) and (10.47, 26.38), respectively. The intervals obtained from the M-H sampler are (68.28, 71.26) and (10.37, 26.64), respectively. The results for the two samplers agree pretty closely. Comparison of Figures 3.3 and 3.5 shows that Gibbs sampler converges faster than the M-H sampler. In general, the Gibbs sampler is more efficient than the M-H sampler. The comparison of the autocorrelation plots in Figures 3.4 with 3.6 shows that Gibbs samples have much less autocorrelation than those generated by M-H samples.

```
> summary(PostSamp_mu)
   Min. 1st Qu.  Median    Mean 3rd Qu.    Max.
  66.95   69.21   69.72   69.72   70.22   73.43
 [1] Various Percentiles of mu:
```

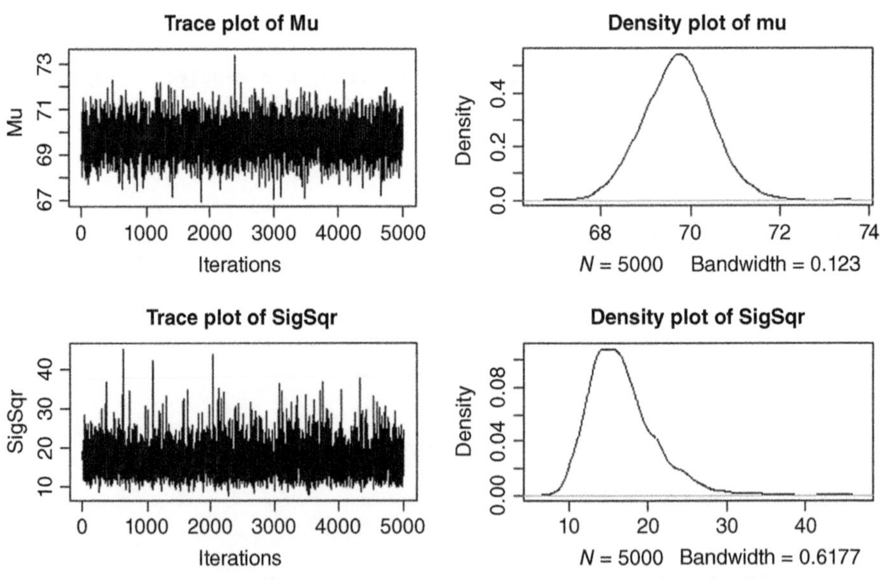

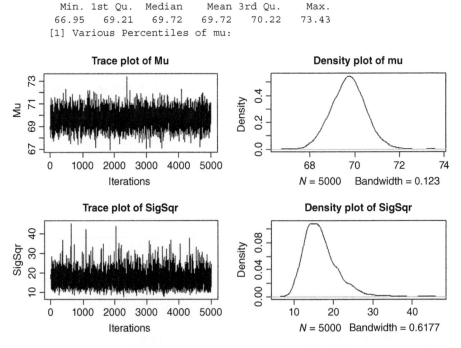

Figure 3.5 Trace and density plots of "mu" and "SigSqr" obtained from Gibbs posterior samples.

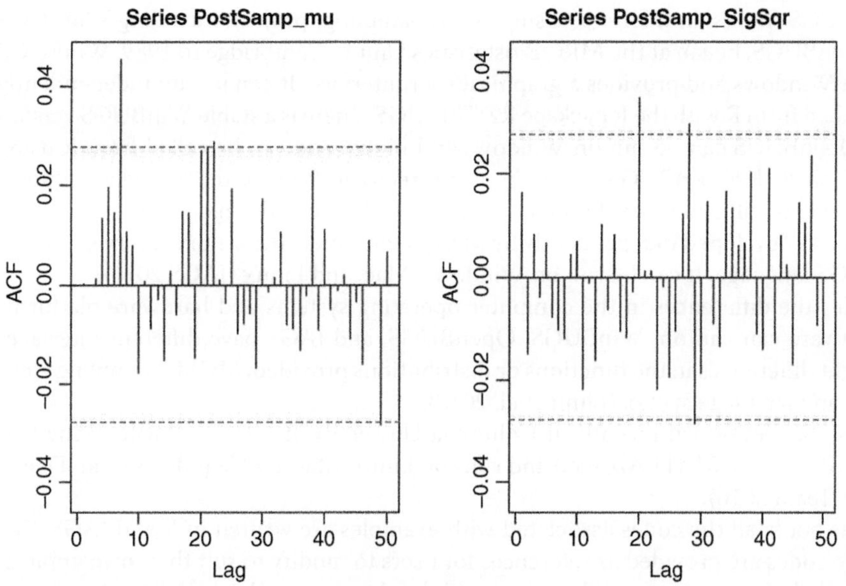

Figure 3.6 Autocorrelation plots of different lags for "mu" and "SigSqr" obtained from Gibbs posterior samples.

```
> quantile(PostSamp_mu, probs=c(0.025, 0.05, 0.10, 0.50, 0.90, 0.95,
0.975))
     2.5%       5%      10%      50%      90%      95%    97.5%
68.25938 68.48347 68.74203 69.72453 70.66449 70.96017 71.23695

> summary(PostSamp_SigSqr)
  Min. 1st Qu.  Median    Mean 3rd Qu.    Max.
 7.851  13.760  16.040  16.690  18.820  45.300
> ## Get various percentiles of "SigSqr" using the samples after
burn-in
> print("Various Percentiles of SigSqr:", quote=FALSE)
[1] Various Percentiles of SigSqr:
> quantile(PostSamp_SigSqr, probs=c(0.025, 0.05, 0.10, 0.50, 0.90,
0.95, 0.975))
     2.5%       5%      10%      50%      90%      95%    97.5%
10.47443 11.13712 12.11973 16.04428 22.09690 24.46349 26.37533
```

3.4 Using BUGS/JAGS

If you think contents in Section 3.3 may seem a little daunting, don't worry! The good news is that there are various MCMC-based non-commercial software packages available, including WinBUGS, OpenBUGS, JAGS, and STAN. These software packages provide convenient ways to carry out Bayesian analysis using MCMC methods. With the help of this software, practitioners do not need to write their own MCMC algorithms and can focus on defining the logical structure of Bayesian models. You do not even need to select a particular MCMC algorithm when using these software packages to run a script. The software automatically chooses the appropriate algorithm for you!

The BUGS (Bayesian Inference Using Gibbs Sampling) project, including WinBUGS and OpenBUGS, began at the MRC Biostatistics Unit in Cambridge in 1989. WinBUGS runs on Windows and provides a graphical user interface. It can be run independently or be called from R with the R package R2WinBUGS. There is a stable WinBUGS version 1.4.3. OpenBUGS can be run on Windows and Linux, and can be called from R using the R package R2OpenBUGS or can be run directly in R using the R package BRugs. OpenBUGS is still under development (University of Cambridge n.d.; Lunn et al. 2009).

JAGS was developed independently by Martyn Plummer. The syntax is very similar to the BUGS language. It can be run on Windows, Mac, and Linux (JAGS 2016).

Besides the differences in the computer operating systems and hardware platforms the software can run on, WinBUGS, OpenBUGS, and JAGS have different interfaces and slight differences in the functions or distributions provided, MCMC sampling algorithms, and language syntax (Lunn et al. 2013).

STAN was developed recently at Columbia University. It interfaces with many languages (R, Python, MATLAB, etc.) and runs on Linux, Mac, and Windows (Stan Development Team 2016).

In this book, all the codes associated with examples are written in R and JAGS. The example codes are provided as references for users to modify to suit their own applications. In order to use the example codes in this book, users need to install the R software and the following packages. Installation instructions can be found in Appendix A.

- *R*: for reading and processing data
- *JAGS*: for Bayesian inference using MCMC algorithms
- *R package "rjags"*: for calling JAGS from within R.

There are other packages used in solving some of the problems presented in this book and they are referred to within the R code for those problems.

Note that there is some syntax difference between R and JAGS. For example, in R the mean and standard deviation are used to define a normal distribution, while in JAGS the mean and precision (1/variance) are used to define a normal distribution. More details can be found in the JAGS user manual (Plummer 2015).

Before running the R scripts in this book, please save all the data files (.txt files), .JAGS files, and .R files in the working directory. Use command `setwd` to set a working directory in R, for example

```
> setwd("C:/Bayesian")
```

The working directory is a default location for files to be read from or written to. Use R command `getwd()` to find your current working directory. Other commonly used R commands can be found in Appendix B.

3.4.1 Define a JAGS Model

We will revisit Example 3.1 to estimate the Weibull scale and shape posterior distributions using R and JAGS. The first thing one needs to do is to define a JAGS model. The JAGS model description for Example 3.1 is shown in **3.4_Weibull.JAGS.** It is a text file that starts with `model {` and ends with `}`.

(3.4_Weibull.JAGS)

```
# Estimate Weibull scale and shape parameters
```

```
# Bayesian model in JAGS language

model {
# Likelihood:
lambda <- 1/pow(scale, shape)    # power function pow(x,z) = x^z

for( i in 1:n) {
ttf[i] ~ dweib(shape, lambda)
}

# Uniform distributions as shape and scale prior
shape ~ dunif(1,3)
scale ~ dunif(60,130)
}
```

This JAGS model defines an equivalent graphical model shown in Figure 3.7, known as a directed acyclic graph. A directed acyclic graph is a graphical model with no cycles, consisting of a set of nodes connected by edges which indicate directions or direct dependencies. More discussions on directed acyclic graphs (Bayesian network) can be found in Chapter 7.

In the directed acyclic graph shown in Figure 3.7a, node *lambda* directly depends on nodes *shape* and *scale*, as indicated by the directions of the edges connecting them. *lambda* is introduced in the model as an additional parameter, since the parameterization of the Weibull distribution is different in JAGS language compared to the parameterization in Eq. (3.1). Nodes *shape* and *scale* are called the parent nodes of node *lambda*, and *lambda* is the child node of *shape* and *scale*. Their relation is described by the following command in **3.4_Weibull.JAGS**,

```
lambda <- 1/pow(scale, shape)
```

with the child node on the left and the parent nodes on the right side of the symbol <- . Symbol <- indicates that node *lambda* is deterministic.

Similarly, each of the n nodes $ttf[i]$ $(i = 1,\ldots,n$; n is the number of observations) directly depends on nodes *shape* and *lambda*. Nodes *shape* and *lambda* are parent

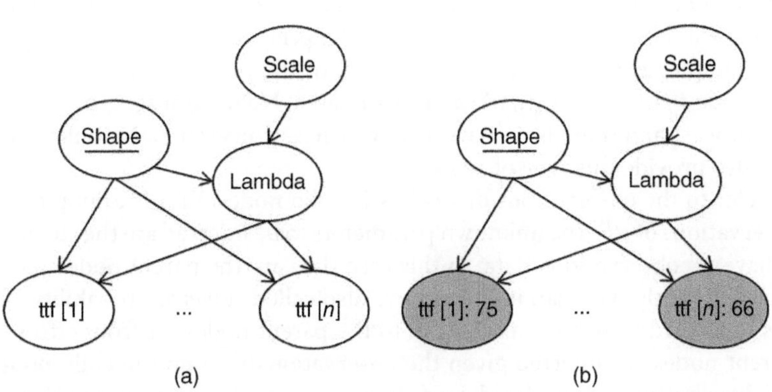

Figure 3.7 (a) Equivalent directed acyclic graph of the JAGS model shown in **3.4_Weibull.JAGS**; (b) JAGS model with *ttf* data loaded (the gray nodes).

nodes of each $ttf[i]$ node. Their relation is described by the following command in **`3.4_Weibull.JAGS`**,

```
ttf[i] ~ dweib(shape, lambda)
```

with the child node on the left and the parent nodes on the right side of the symbol $\sim$. Symbol $\sim$ indicates that node $ttf[i] \quad (i = 1,\ldots, n)$ is stochastic. Basically, this relation describes that the likelihood function is a Weibull distribution given by Eq. (3.1). In the JAGS model, a `for` loop is used to define the likelihood for each of the n data points in the vector ttf, since every time to failure data point has an individual likelihood function.

The two nodes *shape* and *scale* have no parent nodes. These are the two unknown parameters. Uniform distributions are used as priors here. To compare with the results from discretization method, the same uniform distributions used in Section 0 are defined as prior distributions for the *shape* and *scale* parameters.

There are two points of confusion that are common among engineers and scientists. We will clarify them here.

1) *The order of commands in a JAGS model does **not** matter*
   The JAGS model only describes the dependencies (relations) among variables/nodes and the prior distributions of unknown parameters. The order of commands in a JAGS model does not matter. In other words, the prior distributions can be described first, or the likelihood functions can be specified first; changing the order of these command lines does not affect the results. These commands will be compiled together. On the other hand, the order of R commands is important. R commands are executed sequentially.
2) *Bayesian inference requires defining data*
   Data need to be defined for Bayesian inference. Prior distributions are *not* the data we discussed here. Prior distributions are described in the directed acyclic graph (e.g. in JAGS model **`3.4_Weibull.JAGS`**) for top-level nodes that have no parent nodes. Data are not defined in **`3.4_Weibull.JAGS`**, but are covered by the `jags.model` function, which will be covered in Section 3.4.2.

A JAGS model can be run without data being loaded. In such cases, the distribution of each child node will be estimated conditional on the prior distributions of the parent nodes. For example, given the prior distributions of the parent nodes *shape* and *scale*, each child node $ttf[i] \quad (i = 1,\ldots, n)$ can be estimated. No data/observations were provided in this process and there is no Bayesian inference. Conventional Monte Carlo simulations can also provide this type of answer.

Data usually refer to the observations of stochastic child nodes. In this example, the data are the observations of ttf. The unknown parameters to be inferred are the stochastic nodes that have no observations/data; in this case they are the parent nodes *shape* and *scale*. This is why early Bayesian inference was also called "inverse probability," as the inference is backward from the child nodes to the parent nodes, or from effect to causes. The parent nodes are inferred given the observations/data on the child nodes and their prior distributions are updated according to Bayes' theorem. In addition to the observations of stochastic nodes, the values of constants can also be included in the data list.

3.4.2 Create, Compile, and Run the JAGS Model

Next, we will use some functions to create a JAGS model object, to run burn-in iterations, and to collect posterior samples from MCMC simulations after burn-in. There are three functions to learn here (in the "rjags" package):

- `jags.model`: creates a JAGS model object (compilation, initialization, and adaptation)
- `update`: runs a number of iterations for burn-in purposes
- `coda.samples`: collects posterior samples.

These functions are included in R script **`3.4_Weibull_Parameters_Bayesian.R`**. We will explain these functions one by one.

(3.4_Weibull_Parameters_Bayesian.R)

```
    ####################################################################
    ## Estimate shape and scale parameters for a Weibull distribution ##
    ####################################################################
    ### Load package rjags (to connect to JAGS from R for Bayesian
analysis)
    library(rjags)
```

The R command `library` loads R package `rjags`, since this code requires JAGS for Bayesian inference. Note that rjags automatically loads the "coda" package for MCMC diagnosis. We will discuss diagnostics in Section 3.4.3.

First, the time to failure data are read from a .txt file into a vector named `ttf`:

```
ttf <- scan(file = "Example3.1Data.txt",sep="")
```

Then a data list, `BayesianData`, is created which contains two vectors, `ttf` and `n` (the length of `ttf`). This is the data list to be loaded for the JAGS model, to infer posterior distributions of unknown parameters.

```
    # Data
    BayesianData <- list(ttf = ttf,
                         n = length(ttf)
    )
```

Next, we will create a JAGS model object using the `jags.model` function. The Bayesian model file in JAGS language and data are loaded in this step via arguments `file` and `data`, respectively.

```
    # Create a JAGS model object
    ModelObject <- jags.model(file = "3.4_Weibull.JAGS",
                              data=BayesianData,
                              n.chains = 2, n.adapt = 1000
    )
```

In this example we defined two MCMC chains, using command `n.chains = 2`. The purpose of using multiple chains is to diagnose if convergence is achieved. `n.adapt` is an argument to specify the number of adaptation iterations (the JAGS program uses this initial sampling phase to adapt behavior for efficiency purposes). `Inits` is an optional argument to specify initial values. In this case, the initial values for each of the chains are not specified, so they will be randomly generated by the software. Most of the cases in this book are simple Bayesian models, so it is fine to let the software specify initial values for you.

Alternatively, readers can specify identical or different initial values for multiple chains. Example code for adding identical initial values for different chains is shown below, followed by function `jags.model`.

```
InitialValues <- list(shape=2,scale=100)

ModelObject <- jags.model(file = "3.4_Weibull.JAGS",
                          data=BayesianData, inits=InitialValues,
                          n.chains = 2
)
```

Note that results from the above functions are not reproducible since random number generators (RNGs) are involved. To generate reproducible results, users also need to specify the type and the starting states of the RNGs. There are four pseudo-RNGs supplied by JAGS, namely,

```
"base::Wichmann-Hill"
"base::Marsaglia-Multicarry"
"base::Super-Duper"
"base::Mersenne-Twister"
```

Example codes are shown below.

```
    # Define different initial values for multiple chains
    # Use .RNG.name and .RNG.seed to make results reproducible
    Initial1 <- list(.RNG.name="base::Super-Duper", .RNG.seed=3453,
shape=1.5, scale=100)
    Initial2 <- list(.RNG.name="base::Super-Duper", .RNG.seed=3453,
shape=2.5, scale=120)
    InitialValues <- list(Initial1, Initial2)

    # Create a JAGS model object
    ModelObject <- jags.model(file = "3.4_Weibull.JAGS",
                              data=BayesianData, inits=InitialValues,
                              n.chains = 2, n.adapt = 1000
```

After running the above commands, the following results are shown, indicating that a JAGS model object has been created.

```
Compiling model graph
   Resolving undeclared variables
   Allocating nodes
Graph information:
   Observed stochastic nodes: 100
   Unobserved stochastic nodes: 2
   Total graph size: 110

Initializing model

  |++++++++++++++++++++++++++++++++++++++++++++++++++++| 100%
```

If the model is syntactically correct and successfully compiled, the JAGS model can be run now using the `update` command. Here we define 2000 iterations as the burn-in iterations (each of the two chains will run 2000 iterations). In the burn-in process, MCMC chains are in the process of getting converged, so data generated during burn-in will be discarded. When the burn-in iterations are completed and the MCMC chains are converged, we can start to collect posterior samples for analysis. In Section 3.4.3 we will introduce various ways to diagnose MCMC convergence.

```
# Burn-in stage
update(ModelObject, n.iter=2000)
```

After burn-in iterations, we run MCMC simulations again to collect selected posterior samples. The posterior samples are collected in coda format via command `coda.samples`. Note that only the specified parameters will have posterior samples monitored. Specifically, `variable.names` specify for which parameters we need to collect posterior samples. The number of iterations to run MCMC simulations for each MCMC chain is specified by `n.iter`. At the end of the code below, we also saved the *shape* and *scale* posterior samples as individual vectors for analysis purposes.

```
    # Run MCMC and collect posterior samples in coda format for selected
variables
    codaList <- coda.samples(model=ModelObject, variable.names =
c("shape", "scale"), n.iter = 30000, thin = 1)

    codaMatrix <- as.matrix( codaList )

    shape_samples <- codaMatrix[,"shape"]
    scale_samples <- codaMatrix[,"scale"]
```

3.4.3 MCMC Diagnostics and Output Analysis

When the MCMC simulations are completed, further output analysis can be used to summarize posterior samples, to diagnose convergence, to check autocorrelation for the posterior samples of one parameter, or to check correlation between different parameters.

The R package "coda" is used for output analysis (including generating the output summary statistics and plots) and MCMC diagnostics. Users do not need to load the "coda" package separately, since it is automatically loaded when loading the R package "rjags." MCMC diagnostics with the "coda" package include trace plots, Gelman–Rubin diagnostic, etc.

3.4.3.1 Summary Statistics

Summary statistics can be shown via the `summary` command for saved posterior samples from MCMC chains.

```
    # Show summary statistics for collected posterior samples.
    # Quantiles of the sample distribution can be modified in the
quantiles argument.
    summary(codaList, quantiles = c(0.025, 0.05, 0.5, 0.95, 0.975))
```

After running the command, the summary statistics results shown below are obtained. This example shows the number of burn-in iterations, the thinning interval, the number of chains, the number of iterations in each chain, and the mean, standard deviation, standard errors, and quantiles of the saved posterior sample for each specified parameter.

```
> summary(codaList, quantiles = c(0.025, 0.05, 0.5, 0.95, 0.975))

Iterations = 3001:33000
Thinning interval = 1
Number of chains = 2
Sample size per chain = 30000

1. Empirical mean and standard deviation for each variable,
   plus standard error of the mean:

         Mean     SD  Naive SE Time-series SE
scale 101.583 5.8809 0.0240088      0.0336359
shape   1.831 0.1409 0.0005753      0.0008103

2. Quantiles for each variable:

        2.5%     5%     50%     95%   97.5%
scale 90.320 92.155 101.445 111.492 113.530
shape  1.564  1.605   1.829   2.068   2.115
```

The collected posterior samples of the shape and scale parameters can be plotted as histograms, as shown in Figure 3.8. Note that these results are close to the results of the maximum likelihood method presented in Section 3.1.

```
# Plot histograms of shape and scale posterior samples
par(mfrow=c(1,2))
hist(shape_samples,sub=1)
hist(scale_samples,sub=2)
```

3.4.3.2 Trace Plots

Use the command `traceplot(codaList)` to show the trace plots of all saved parameter posterior samples. In this case, the codes to plot the trace plots are shown below. The trace plots are shown in Figure 3.9. In each of the plots in Figure 3.9 two traces are shown, since there are two MCMC chains in the simulations. If two chains are mixed well (not separated), and there are no patterns observed, then we can assume that convergence is achieved.

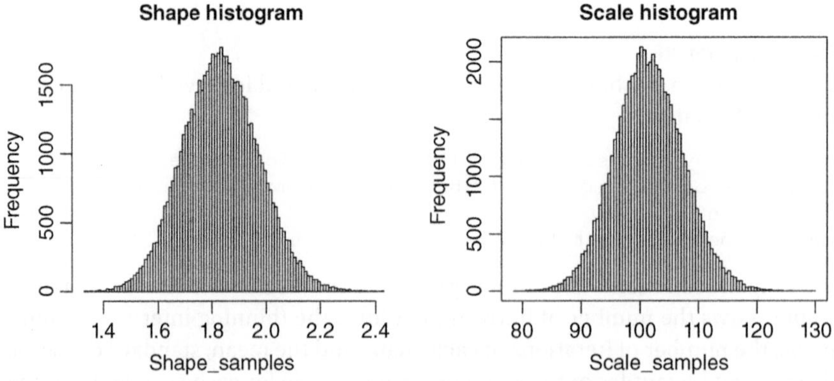

Figure 3.8 Histograms of parameter shape and scale posterior samples.

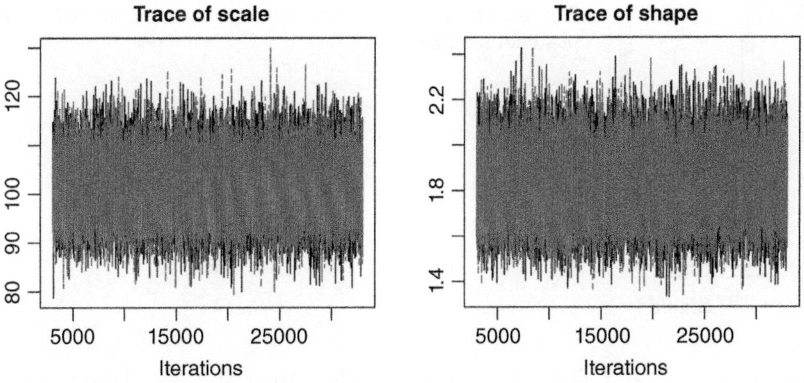

Figure 3.9 Trace plots of parameter scale and shape posterior samples.

```
# Trace plots
par(mfrow=c(1,2))
traceplot(codaList)
```

If the interest is only on the trace plot of one-parameter posterior samples, e.g. the shape parameter, then the following command can be used instead,

```
# traceplot(codaList[,"shape"])
```

3.4.3.3 Autocorrelation Plots

An autocorrelation plot, provided by the command `autocorr.plot`, indicates the autocorrelation within a chain for a single parameter. High autocorrelations indicate slow convergence. In such cases, users may choose to thin out the posterior samples in a chain for post processing and analysis. Figure 3.10 shows the autocorrelation plot for the scale and shape posterior samples.

```
autocorr.plot(codaList)
```

3.4.3.4 Cross-Correlation

Correlation between two saved parameters' posterior samples can be analyzed via the `corsscorr` command. For example, the following command shows the correlation

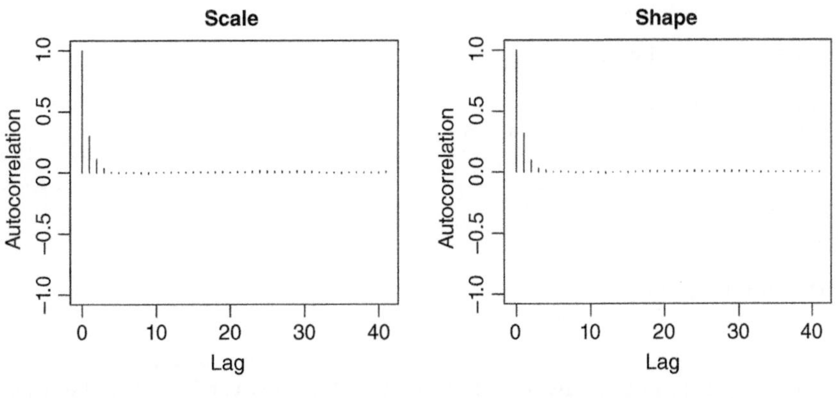

Figure 3.10 Autocorrelation plots of parameter scale and shape posterior samples.

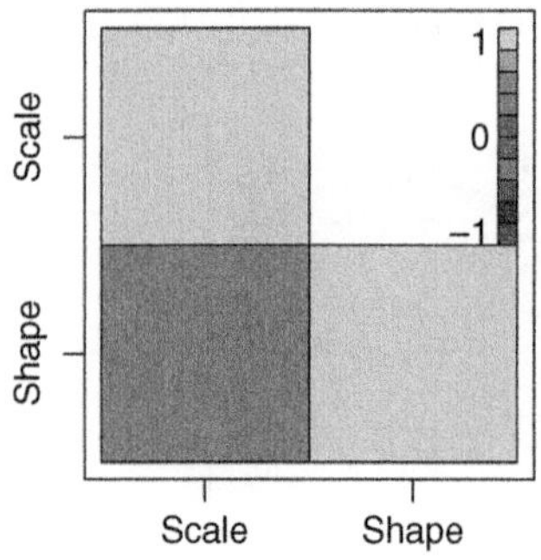

Figure 3.11 Cross-correlation plot of parameter scale and shape posterior samples.

between the saved scale and shape parameter samples. The cross-correlation plot of these two parameters is shown in Figure 3.11.

```
crosscorr(codaList)

> crosscorr(codaList)
            scale     shape
scale 1.0000000 0.3002166
shape 0.3002166 1.0000000

crosscorr.plot(codaList)
```

3.4.3.5 Gelman–Rubin Diagnostic and Plots

Gelman–Rubin diagnostics, given by command `gelman.diag`, can only be calculated when there are multiple chains. When convergence is achieved, the Gelman–Rubin diagnostic approaches 1 (Ntzoufras 2009). This method is based on comparing within- and between-sample variability in the cases when there are two or more chains. The shrinkage factor is defined as the ratio of the pooled variance to the within sample variability. A detailed discussion of the method can be found in Ntzoufras (2009).

```
# Gelman-Rubin diagnostic
# Gelman and Rubin's convergence diagnostic
# Approximate convergence is diagnosed when the upper limit is close
to 1
gelman.diag(codaList)
```

The results are shown below.

```
> gelman.diag(codaList)
Potential scale reduction factors:

      Point est. Upper C.I.
scale          1          1
shape          1          1

Multivariate psrf

1
```

The Gelman–Rubin plots are given by the command `gelman.plot`, and the results are shown in Figure 3.12.

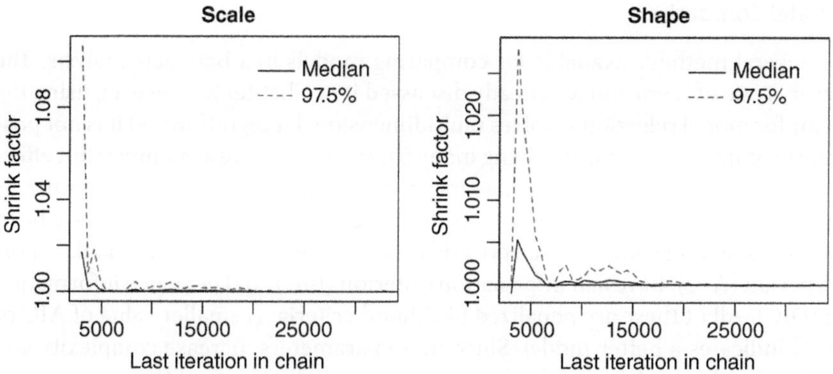

Figure 3.12 Gelman plots of parameter scale and shape posterior samples.

```
    # This plot shows the evolution of Gelman and Rubin's shrink
factor as the number of iterations increases
    # This shows if the shrink factor has really converged, or whether it
is still fluctuating
    gelman.plot(codaList)
```

3.4.4 Sensitivity to the Prior Distributions

Note that when changing the two prior distributions from uniform distributions to gamma distributions, the summary statistics of the posterior distributions shown below are obtained, which are close to the posterior distributions when using the uniform priors.

```
# Vague Gamma distribution for shape and scale prior
shape ~ dgamma(1,1) # mean = a/b; variance = a/(b^2)
scale ~ dgamma(1,0.1)

> summary(codaList, quantiles = c(0.025, 0.05, 0.5, 0.95, 0.975))

Iterations = 3001:33000
Thinning interval = 1
Number of chains = 2
Sample size per chain = 30000

1. Empirical mean and standard deviation for each variable,
   plus standard error of the mean:

        Mean     SD Naive SE Time-series SE
scale 97.896 5.7357 0.0234161      0.0339989
shape  1.785 0.1379 0.0005629      0.0008394

2. Quantiles for each variable:

       2.5%     5%    50%     95%   97.5%
scale 86.96 88.638 97.787 107.492 109.443
shape  1.52  1.562  1.782   2.015   2.062
```

3.4.5 Model Comparison

There are several methods available for comparing models in a Bayesian analysis. The Bayes factor is one of them and we already discussed it in Chapter 2. However, using the Bayes factor for model selection involves multidimensional integration and it is not possible in most complex problems involving many parameters. There are simpler but effective information-based methods available for model comparison. These are commonly known as penalized likelihood criteria (information criteria) and we discuss these here.

Commonly used information-based criteria for model selection include Akaike information criterion (AIC), Bayesian information criterion (BIC), and deviance information criterion (DIC). All of these are penalized likelihood criteria. A smaller value of AIC or BIC or DIC indicates a better model. Since more parameters increase complexity and make it easier for the model to fit the data, there is a penalty for overfitting.

AIC is defined as

$$AIC = -2\log f(y \mid \hat{\theta}_{\text{mle}}) + 2k,$$

where $\hat{\theta}_{\text{mle}}$ is the maximum likelihood estimate of the parameter, $\log f(y \mid \hat{\theta}_{\text{mle}})$ is the maximum of the log-likelihood function, and k is the number of parameters estimated in the model.

BIC, developed by Schwarz (1978), is defined as

$$BIC = -2\log f(y \mid \hat{\theta}_{\text{mle}}) + k\log(n),$$

where $\log f(y \mid \hat{\theta}_{\text{mle}})$ is the maximum of the log-likelihood function, k is the number of estimated parameters in the model, and n is the sample size.

DIC is a generalization of the AIC and is especially suitable for Bayesian hierarchical models.

$$\begin{aligned} DIC &= -2\log f(y \mid \bar{\theta}) + 2p_D \\ &= D(\bar{\theta}) + 2p_D \\ &= \bar{D} + p_D \end{aligned},$$

where $\bar{\theta}$ is the average of θ posterior distributions, i.e. $\bar{\theta} = E_{\theta|y}[\theta]$, D is the deviance, defined by $D(\theta) = -2\log f(y|\theta)$, $\bar{D}$ is the posterior average of the deviance (or mean deviance), defined by $\bar{D} = E_{\theta|y}[D] = -2E_{\theta|y}(\log f(y \mid \theta))$, $D(\bar{\theta})$ is the deviance evaluated at $\bar{\theta}$, i.e. $D(\bar{\theta}) = D(E_{\theta|y}[\theta]) = -2\log f(y \mid \bar{\theta})$, and p_D is the effective number of parameters, indicating the complexity of a model and given by

$$\begin{aligned} p_D &= \bar{D} - D(\bar{\theta}) \\ &= E_{\theta|y}[D] - D(E_{\theta|y}[\theta]) \\ &= -2E_{\theta|y}(\log f(y \mid \theta)) + 2\log f(y \mid \bar{\theta}) \end{aligned}.$$

The deviance is penalized by an increased effective number of parameters. Models with smaller DIC values are preferred. Compared to AIC, DIC calculates the effective number of parameters p_D, instead of the number of estimated parameters k, which is a better way to evaluate complexity in Bayesian hierarchical models. For hierarchical models, the effective number of parameters strongly depends on the variance of the hyperparameters.

In addition, DIC does not require calculation of the maximum likelihood, and computes $\overline{D}$ and $D(\overline{\theta})$ instead, which can be easily calculated from posterior samples generated by MCMC simulations. BUGS-related software such as WinBUGS and the "coda" R package provide functions to directly calculate DIC. Now we use examples to demonstrate how to calculate DIC using R and JAGS code.

DIC is especially suitable for comparison of hierarchical models. Compared to BIC, DIC is easier to use with complex models involving a large number of parameters. The BIC method requires that we know the exact number of parameters estimated in the model. This is not trivial in a hierarchical model with censored times. When comparing different models using DIC, the smaller DIC value indicates a better-fitting model. More discussions can be found in Carlin and Louis (2009) and Gelman et al. (2014).

For the Weibull likelihood in Example 3.1, DIC is estimated from the following code:

```
dic.samples(model=ModelObject,n.iter=10000,thin=1,type='pD')
```

The results indicate that DIC is 1057 with the Weibull likelihood.

```
> dic.samples(model=ModelObject,n.iter=10000,thin=1,type='pD')

Mean deviance:  1055
penalty 1.994
Penalized deviance: 1057
```

Now we change the Weibull likelihood to a normal likelihood and re-run the Bayesian estimation. Our goal is to compare the two likelihood functions to see which one is more suitable in this case. The JAGS model and R script are shown below in **3.4_Normal.JAGS** and **3.6_Normal_Parameters_Bayesian.R.**

```
(3.4_Normal.JAGS)
    model {
    # Likelihood:

    for( i in 1:n) {
    y[i] ~ dnorm(mu, tau)    # likelihood is Normal distribution
    }

    mu ~ dnorm(0,0.000001)   # mean=1, std=1000

    # vague gamma (traditional): mean=0.1/0.1=1 & variance=0.1/(0.1*0.1)
=10
       tau ~ dgamma(0.1,0.1)

    sigma <- 1/sqrt(tau)
    }
```

```
(3.4_Normal_Parameters_Bayesian.R)
    ### Load package rjags (to connect to JAGS from R for Bayesian
analysis)
    library(rjags)

    # Data
    # read data from file
    ttf <- read.table(file="Example3.1Data.txt",header=F,sep="")[,1]
    BayesianData <- list(y = ttf,
```

```
                    n = length(ttf)
)

# Create a JAGS model object
ModelObject <- jags.model(file = "3.4_Normal.JAGS",
                          data=BayesianData,
                          n.chains = 2, n.adapt = 1000
)

# Burn-in stage
update(ModelObject, n.iter=2000)

# DIC calculation in rjags
dic.samples(model=ModelObject ,n.iter=10000,thin=1,type='pD')
```

After running the code, the DIC results are shown below. Since the DIC from the Weibull likelihood is smaller compared to the normal likelihood, we can conclude that the Weibull distribution is a better model in this case.

```
> dic.samples(model=ModelObject,n.iter=10000,thin=1,type='pD')
  |**************************************************| 100%
Mean deviance:  1073
penalty 2.014
Penalized deviance: 1075
```

3.5 Summary

This chapter introduces commonly used Bayesian computation methods, focusing on MCMC algorithms, including the Metropolis-Hastings algorithm and Gibbs sampling. As an example, a Weibull distribution is used to analyze reliability life testing data. We went through different Bayesian computation methods with this example, including the maximum likelihood method, grid approximation for Bayesian inference, and MCMC simulations.

A widely used Bayesian computation software, JAGS, is used to provide MCMC sampling of Bayesian posteriors. R and JAGS codes are introduced step by step to explain the function of creating and running the model, summarizing posterior samples, MCMC diagnostics, and model comparison.

References

Carlin, B. and Louis, T. (2000). *Bayes and Empirical Bayes Methods for Data Analysis*. Text in Statistical Science. New York: Chapman & Hall/CRC.

Carlin, B.P. and Louis, T.A. (2009). *Bayesian Methods for Data Analysis*. Boca Raton, FL: Chapman & Hall/CRC Press.

Duane, S., Kennedy, A.D., Pendleton, B.J., and Roweth, D. (1987). Hybrid Monte Carlo. *Physics Letters B* 195: 216–222.

Gelman, A., Carlin, J.B., Stern, H.S. et al. (2014). *Bayesian Data Analysis*. Boca Raton, FL: Taylor & Francis Group, LLC.

Gilks, W. (1992). Derivative free adaptive rejection sampling for Gibbs sampling. In: *Bayesian Statistics 4* (ed. J.M. Bernardo, J.O. Berger, A.P. Dawid and A.F.A. Smith), 641–665. Oxford: Oxford University Press.

Gilks, W. and Wild, P. (1992). Adaptive rejection sampling for Gibbs sampling. *Applied Statistics* 41 (2): 337–348.

Green, P. (1995). Reversible jump Markov chain Monte Carlo computation and Bayesain model determination. *Biometrika* 82: 711–732.

Green, P.J. and Mira, A. (2001). Delayed rejection in reversible jump Metropolis-Hastings. *Biometrika* 88: 1035–1053.

Hastings, W. (1970). Monte Carlo sampling methods using Markov chains and their applications. *Biometrika* 57: 97–109.

JAGS (2016), http://mcmc-jags.sourceforge.net/ (accessed 04 May 2017).

Lunn, D., Spiegelhalter, D., Thomas, A., and Best, N. (2009). The BUGS project: evolution, critique and future directions. *Statistics in Medicine* 28: 3049–3067.

Lunn, D., Jackson, C., Best, N. et al. (2013). *The BUGS Book: A Practical Introduction to Bayesian Analysis*. Boca Raton, FL: CRC Press, Taylor & Francis Group.

Metropolis, N., Rosenbluth, A.W., Rosenbluth, M.N. et al. (1953). Equation of state calculations by fast computing machines. *Journal of Chemical Physics* 21 (6): 1087–1092.

Neal, R.M. (2003). Slice sampling. *Annals of Statistics* 31 (3): 705–741.

Neal, R.M. (2010) MCMC using Hamiltonian dynamics. In *Handbook of Markov Chain Monte Carlo* (ed. S. Brooks, A. Gelman, G. Jones, and X. L. Meng). Chapman & Hall/CRC Press, Boca Raton, FL, 113–162.

Ntzoufras, I. (2009). *Bayesian Modeling Using WinBugs*. Hoboken, NJ: Wiley.

Plummer, M. (2015) *JAGS Version 4.0.0 user manual.*

Schwarz, G.E. (1978). Estimating the dimension of a model. *Annals of Statistics* 6 (2): 461–464.

Spiegelhalter, D., Thomas, A., Best, N. and Lunn, D. (2003) *WinBUGS User Manual*, Version 1.4, MRC Biostatistics Unit, Institute of Public Health and Department of Epidemiology and Public Health, Imperial College School of Medicine, UK, https://www.mrc-bsu.cam.ac.uk/wp-content/uploads/manual14.pdf (accessed 06 September 2018).

Stan Development Team (2016) *Stan Modeling Language Users Guide and Reference Manual*. Version 2.14.0, http://mc-stan.org (accessed 04 May 2017).

University of Cambridge (n.d.) *The BUGS Project*. https://www.mrc-bsu.cam.ac.uk/software/bugs/ (accessed 04 May 2017).

4

Reliability Distributions (Bayesian Perspective)

This chapter introduces common discrete and continuous probability distributions used in reliability applications from a Bayesian perspective. Examples and R codes are provided to show the estimation of posterior distributions of the parameters of these distributions.

4.1 Introduction

There are two major types of data: attribute data and variable data. Attribute data, or categorical data, are qualitative data (e.g. color, pass/fail, yes/no, etc.). Variable data are quantitative data (e.g. dimension, force, weight, etc.). Attribute data can be converted to count (or discrete) data for analysis, for example the number of units in a sample of 50 passing a certain quality requirement. Compared to discrete data, continuous data usually contain more information and should be a preferred choice during data collection and analysis if possible.

Pass/fail data is one type of commonly used attribute data, especially in acceptance testing or assurance testing, such as design verification testing and qualification testing. In these tests, reliability is defined as the percentage of products conforming to the requirement of interest. Results are recorded as either pass or fail. A predetermined quantity of products (e.g. 22, 59, or 299 parts), based on a desired level of confidence and reliability, is tested against a requirement to ensure that all are conforming so that the desired level of confidence and reliability can be achieved. A binomial distribution is often used to estimate the proportion of successes or failures in a sample of products tested.

Continuous time-to-failure data is usually recorded in component reliability analysis, design characterization, or design verification testing. Results are recorded as numerical data. Here reliability is captured as the probability of surviving beyond a specific time. Probability distributions such as the exponential distribution, the Weibull distribution, and the lognormal distribution are commonly used to model time-to-failure/lifetime data.

In this chapter, we will present examples and R codes for various inferences involving these commonly used distributions in reliability engineering practice. In these examples, we use non-informative prior distributions for parameter estimation and reliability analysis. By using non-informative priors for these examples, the codes present more generic cases, especially where the prior knowledge on the parameters

Practical Applications of Bayesian Reliability, First Edition. Yan Liu and Athula I. Abeyratne.

Companion website: www.wiley.com/go/bayesian20

of interest is absent. If in engineering practice there is previous information on certain parameters, the reader can modify the prior distributions in the R codes to reflect such information.

The methods introduced in this chapter are generally used for component level or subsystem level reliability assessment. This provides a foundation for system reliability assessment based on models including reliability block diagrams and fault trees, which will be discussed in the following chapters. As mentioned earlier, system level reliability is usually estimated by aggregating reliability from component level data with the overall uncertainty quantified. In a Bayesian framework, since posterior distributions are true probability statements about unknown parameters, they may be easily propagated through the system reliability models.

4.2 Discrete Probability Models

4.2.1 Binomial Distribution

A binomial distribution is often used to describe the probability of getting x successes out of n trials. In engineering practice this is widely used in acceptance testing for quality assurance, such as incoming or in-process inspection of material or products, design verification, final inspection of manufactured products, etc. The probability mass function, $f(x)$, of a binomial distribution is given by

$$f(x) = \binom{n}{x} p^x (1-p)^{n-x}, \tag{4.1}$$

where

p is the probability of getting a success in each trial, $0 \leq p \leq 1$
$n > 0$ is the number of trials
$x \geq 0$ is the number of successes
$\binom{n}{x} = \frac{n!}{x!\ (n-x)!}$ is the binomial coefficient.

The cumulative distribution function (CDF), $F(k)$, of observing up to k successes in n trials is given by

$$F(k) = \mathrm{P}(X \leq k) = \sum_{x=0}^{k} \binom{n}{x} p^x (1-p)^{n-x}.$$

The expected number of successes, $E(X)$, in n trials is given by

$$E(X) = np.$$

The variance of number of successes in n trials is given by

$$V(X) = np(1-p).$$

Figure 4.1 shows the probability mass functions of a few binomial distributions (number of trials = 10, probability of success = 0.1, 0.5, 0.9, respectively). The R script to generate Figure 4.1 is in **`4.2.1_Binomial_Distributions.R`**.

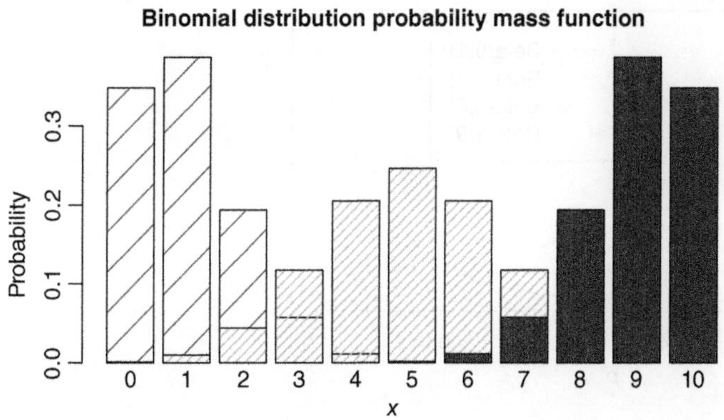

Figure 4.1 Probability mass functions of binomial distributions. Size = 10; probability = 0.1 (left bars), 0.5 (middle bars), and 0.9 (right bars).

```
(4.2.1_Binomial_Distributions.R)
    ## probability mass functions of Binomial distributions

    x <- seq(0,10,by=1)
    binom_1 <- dbinom(x, size=10, prob=0.1)
    binom_2 <- dbinom(x, size=10, prob=0.5)
    binom_3 <- dbinom(x, size=10, prob=0.9)
    barplot(binom_1, col = "red", density=5, names=x, xlab = "x",
ylab="Probability", main = "Binomial distribution probability
mass function")
    barplot(binom_2, col = "green", density=20, add = TRUE)
    barplot(binom_3, col = "blue", density=95, add = TRUE)
```

A commonly used prior distribution of the success/conforming probability p in the binomial distribution is the conjugate prior: beta distribution. A beta probability density is a two-parameter continuous function defined on the unit interval [0, 1], which has the probability density function (PDF) of

$$f(p) = \frac{p^{\alpha-1}(1-p)^{\beta-1}}{B(\alpha, \beta)}, \tag{4.2}$$

where $B(\alpha, \beta) = \frac{\Gamma(\alpha)\ \Gamma(\beta)}{\Gamma(\alpha+\beta)}$

α and β are positive real numbers
$\Gamma(x)$ is the gamma function
$\Gamma(x) = \int_0^\infty s^{x-1}e^{-s}ds, x > 0$ (when $x = 1, 2, \ldots, \Gamma(x) = (x-1)!$).

The mean of the beta probability density function shown in Eq. (4.2) is $\frac{\alpha}{\alpha+\beta}$ and the variance is $\frac{\alpha\beta}{(\alpha+\beta)^2(\alpha+\beta+1)}$.

Figure 4.2 shows the probability density functions of few beta distributions (*Beta* (9, 1), *Beta* (1, 1), *Beta* (5, 5), *Beta* (1, 9)). The R script to generate Figure 4.2 is shown below (included in **`4.2.1_Beta_Distributions.R`**).

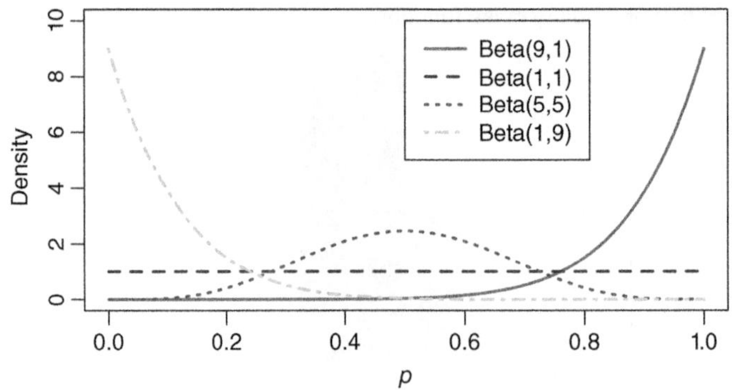

Figure 4.2 Probability density functions of four beta distributions: *Beta* (9, 1), *Beta* (1, 1), *Beta* (5, 5), and *Beta* (1, 9).

```
(4.2.1_Beta_Distributions.R)
## PDF of Beta distributions

p <- seq(0,1,length=500)
beta_1 <- dbeta(p,9,1)  # calculate the PDF of a Beta distribution
beta_2 <- dbeta(p,1,1)
beta_3 <- dbeta(p,5,5)
beta_4 <- dbeta(p,1,9)
plot(p,beta_1,type="l",ylab="Density",xlab="p",lty=1,lwd=2,col="red",
ylim=c(0,10)) # plot PDF of Beta(9,1)
lines(p,beta_2,lty=2,lwd=2,col="blue") # add PDF of Beta(1,1)
lines(p,beta_3,lty=3,lwd=2,col="purple") # add PDF of Beta(2,2)
lines(p,beta_4,lty=4,lwd=2,col="green") # add PDF of Beta(1,9)
legend(.5,10,c("Beta(9,1)","Beta(1,1)","Beta(5,5)","Beta(1,9)"),
lty=c(1,2,3,4),lwd=c(2,2,2,2),col=c("red","blue","purple","green"))
# add legend
```

When the likelihood function is a binomial distribution $X \sim Binomial\,(n, p)$, the Jeffreys prior of p is $Beta\ (0.5,\ 0.5)$ (refer to Appendix D for details), which is a non-informative prior.

As discussed in Section 2.4, when x successes out of n trials are observed, and a beta distribution $Beta(\alpha,\ \beta)$ is used to represent the prior belief on the device reliability p, then the posterior distribution of p is also a beta distribution $Beta\ (x + \alpha,\ n - x + \beta)$.

4.2.2 Poisson Distribution

A Poisson distribution is widely used to express the probability of a certain number of events in a fixed interval of time/space. The Poisson distribution is typically used to model the number of defects in a product.

The Poisson probability mass function is

$$f(x \mid \lambda) = \frac{\lambda^x e^{-\lambda}}{x!}, \tag{4.3}$$

where $x = 0, 1, 2, \ldots$

λ is equal to the mean and the variance of the distribution.

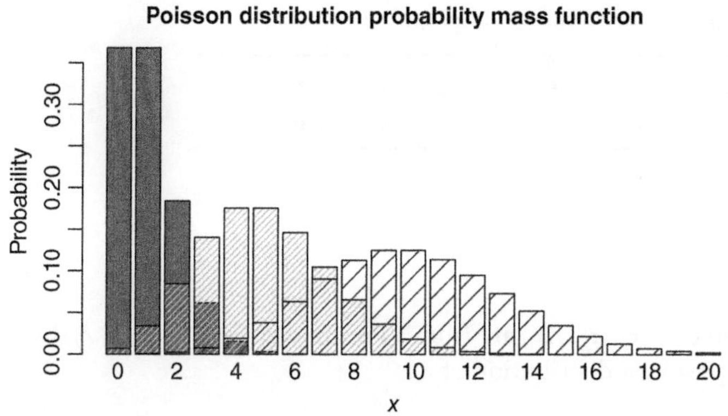

Figure 4.3 Probability mass functions of three Poisson distributions: $\lambda = 1$ (high density), 5 (medium density), and 10 (low density).

The CDF, $F(x)$, of a Poisson distribution with mean λ is

$$F(x) = P(X \le x) = \sum_{k=0}^{x} f(x \mid \lambda) = \sum_{k=0}^{x} \frac{\lambda^k e^{-\lambda}}{k!}.$$

The Poisson distribution can be used as an approximation to the binomial distribution under certain conditions. If the probability of an event, p, is low (≤ 0.05), and the number of trials, n, is large (≥ 20), and $np < 10$, the probability of observing x events in n trials can be approximated with the Poisson probability mass function

$$f(x \mid \lambda) = \frac{\lambda^x e^{-\lambda}}{x!},$$

where $\lambda = np$.

Figure 4.3 shows the probability mass functions of a few Poisson distributions ($\lambda = 1$, 5, 10). The R script to generate Figure 4.3 is as follows (included in **4.2.2_Poisson_Distributions.R**).

```
(4.2.2_Poisson_Distributions.R)
## probability mass functions of Poisson distributions

x <- seq(0,20,by=1)
pois_1 <- dpois(x, lambda=1, log = FALSE) # Poisson PMF
pois_2 <- dpois(x, lambda=5, log = FALSE)
pois_3 <- dpois(x, lambda=10, log = FALSE)

barplot(pois_1, col = "red", density=95, names=x, xlab = "x", ylab=
"Probability", main = "Poisson distribution probability mass function")
barplot(pois_2, col = "green", density=30, add = TRUE)
barplot(pois_3, col = "blue", density=10, add = TRUE)
```

The conjugate prior of a Poisson distribution parameter λ is the gamma distribution. Suppose that the prior distribution of λ is gamma with the shape parameter a and the inverse scale parameter (or rate parameter) b, that is, $f(\lambda) \sim Gamma\ (a, b)$.

Table 4.1 Count of particles greater than 5 μm ml^{-1} in 20 specimens.

5	5	2	12	4	6	3	1	8	10
9	4	3	6	6	4	6	5	6	7

Thus the PDF of λ is given by

$$f(\lambda \mid a, b) = \frac{b^a \lambda^{a-1} e^{-b\lambda}}{\Gamma(a)}. \tag{4.4}$$

Assuming the observed data $x_1, x_2, \ldots x_n$ are from a Poison distribution $f(x_i \mid \lambda)$, based on Bayes' Theorem, the posterior distribution of λ is

$$f(\lambda \mid x_1, x_2, \ldots x_n) = \frac{f(x_1,\ x_2,\ \ldots\ x_n \mid \lambda) f(\lambda)}{f(x_1,\ x_2,\ \ldots\ x_n)} = \frac{f(x_1,\ x_2,\ \ldots\ x_n \mid \lambda) f(\lambda)}{\int f(x_1,\ x_2,\ \ldots\ x_n \mid \lambda) f(\lambda) \mathrm{d}\lambda}$$

$$= \frac{\left(\prod_{i=1}^{n} \frac{\lambda^{x_i} e^{-\lambda}}{x_i!}\right) \frac{b^a \lambda^{a-1} e^{-b\lambda}}{\Gamma(a)}}{\int \left(\prod_{i=1}^{n} \frac{\lambda^{x_i} e^{-\lambda}}{x_i!}\right) \frac{b^a \lambda^{a-1} e^{-b\lambda}}{\Gamma(a)} \mathrm{d}\lambda} = \frac{\left(\prod_{i=1}^{n} \lambda^{x_i} e^{-\lambda}\right) \lambda^{a-1} e^{-b\lambda}}{\int \left(\prod_{i=1}^{n} \lambda^{x_i} e^{-\lambda}\right) \lambda^{a-1} e^{-b\lambda} \mathrm{d}\lambda} = \frac{\lambda^{a+\sum_{i=1}^{n} x_i - 1} e^{-(b+n)\lambda}}{\int \lambda^{a+\sum_{i=1}^{n} x_i - 1} e^{-(b+n)\lambda} \mathrm{d}\lambda}$$

$$= \frac{\frac{(b+n)^{a+\sum_{i=1}^{n} x_i}}{\Gamma\left(a+\sum_{i=1}^{n} x_i\right)} \lambda^{a+\sum_{i=1}^{n} x_i - 1} e^{-(b+n)\lambda}}{\int \frac{(b+n)^{a+\sum_{i=1}^{n} x_i}}{\Gamma\left(a+\sum_{i=1}^{n} x_i\right)} \lambda^{a+\sum_{i=1}^{n} x_i - 1} e^{-(b+n)\lambda} \mathrm{d}\lambda} = \frac{(b+n)^{a+\sum_{i=1}^{n} x_i}}{\Gamma\left(a+\sum_{i=1}^{n} x_i\right)} \lambda^{a+\sum_{i=1}^{n} x_i - 1} e^{-(b+n)\lambda}.$$

Using the conjugate prior, the posterior distribution of λ is

$$f(\lambda \mid x_1,\ x_2,\ \ldots\ x_n) \sim Gamma\left(a + \sum_{i=1}^{n} x_i,\ b + n\right). \tag{4.5}$$

Example 4.1 One product requirement specifies that the average count of particles greater than 5 μm from the specimen shall not exceed 20 ml^{-1}. Sampling test results with 20 specimens are shown in Table 4.1 (data are randomly sampled from a Poisson distribution with mean = 5). What is the probability of meeting this requirement with 95% confidence?

The R script to calculate and plot the posterior distribution of λ, and to calculate the reliability (probability of meeting this requirement) with 95% credible interval is shown in **4.2.2_Poisson_GammaPrior.R.** Here we use *Gamma* (1, 0.1) as a non-informative prior distribution of λ.

```
(4.2.2_Poisson_GammaPrior.R)
########################################################################
## Find the posterior distribution of lambda in a Poisson             ##
## distribution                                                       ##
## Find conformance rate                                              ##
## Conjugate prior: Gamma distribution                                ##
########################################################################
```

```
set.seed(12345)

# data
y <- c(5,5,2,12,4,6,3,1,8,10,9,4,3,6,6,4,6,5,6,7)

# calculate the summation of yi
y_sum <- sum(y)

# calculate the number of elements in y
y_length <- length(y)

# shape and rate parameters in the Gamma prior
a=1   # shape
b=0.1   # rate

# Sample size sampled from the posterior distribution
sample_number <- 10000

lambda <- seq(0,15,length=500)
lambda_prior <- dgamma(lambda, a, rate=b) # mean=a/b, variance=a/b^2
likelihood <- dgamma(lambda, y_sum+1, rate=y_length) # rescaled like-
lihood
lambda_posterior <- dgamma(lambda, a+y_sum, rate=b+y_length)

# Create density plots
#jpeg("Poisson_Gamma.jpeg", width = 6, height = 4, units = 'in',
res = 600)  # save the plot as jpeg format
plot(lambda,likelihood,type="l",ylab="Density",xlab="lambda",
lty=1,lwd=2,col="red",ylim=c(0,1)) # plot likelihood
lines(lambda,lambda_posterior,lty=2,lwd=2,col="blue") # add posterior
lines(lambda,lambda_prior,lty=3,lwd=2,col="purple") # add prior
legend(10,0.9,c("Prior","Likelihood","Posterior"),lty=c(3,1,2),
lwd=c(2,2,2),col=c("purple","red","blue")) # add legend
#dev.off()

## Prior distribution of lambda: Gamma(a,b)
# lambda_prior_samples <- rgamma(sample_number, a, rate=b)

# posterior distribution of lambda:
lambda_posterior_samples <- rgamma(sample_number, a+y_sum,
rate=b+y_length)
print(paste("lambda posterior mean is:", signif(mean(lambda_posterior_
samples), 6)))

# Analytically computed value of the posterior mean of lambda
# using its distribution - Gamma(a+y_sum, b+y_length)
print(paste("Analytically computed value of posterior mean is:",
signif((a+y_sum)/(b+y_length), 6)))
```

The above R script estimates the mean of λ. The first estimate is based on a simulated sample from its posterior distribution. Since the posterior distribution is gamma with shape parameter $a + y\_sum$ and rate parameter $b + y\_length$, the mean can be computed analytically as well. The second estimate provides the analytical estimate. The simulated value is very close to the analytical value. The following results were obtained by running the code.

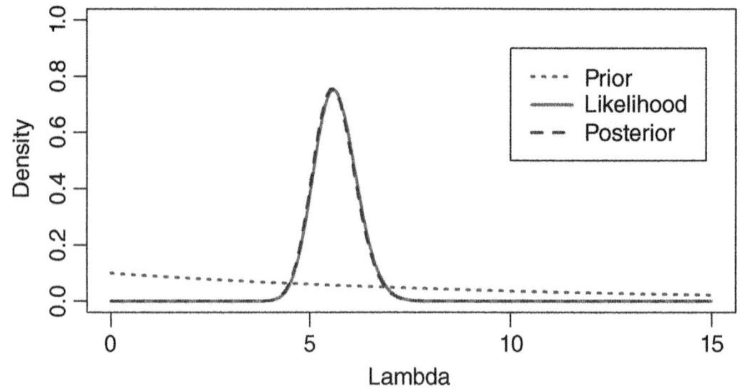

Figure 4.4 λ of a Poisson distribution with a vague gamma prior. Dotted line, prior; solid line, likelihood; dashed line, posterior.

```
[1] "lambda posterior mean is: 5.62909"
[1] "Analytically computed value of posterior mean is: 5.62189"


##############################################
## estimate reliability (conformance rate) ##
##############################################

reliability <- rep(NA, sample_number)

# If there are lambda particles per milliliter,
# find the probability of having 20 or less particles per milliliter.
for(i in 1:sample_number) reliability[i] = ppois(20,
lambda=lambda_posterior_samples[i])    # lower tail

# Calculate the mean and 95% credible interval of reliability (con-
formance rate)
print(paste("reliability mean is:", signif(mean(reliability), 6)))
print(paste("reliability 95% CI lower bound is:",
signif(quantile(reliability, 0.05), 6)))
```

The R script above estimates reliability (probability of meeting the specification). After running the R script, the results obtained are shown below.

```
[1] "reliability mean is: 0.999999"
[1] "reliability 95% CI lower bound is: 0.999995"
```

Based on the results above, there is a 95% probability that reliability is greater than 0.9999. The λ prior and posterior distributions are shown in Figure 4.4.

4.3 Continuous Models

4.3.1 Exponential Distribution

An exponential distribution is often used to model time to failure data if the hazard rate (instantaneous failure rate) is a constant. This means the failure of interest is expected to be random during a product's "useful life." For example, in a system, some electronic components have a mean life much longer than the timeframe the entire system is intended to be used in the field. The lifetime of these electronic components can be modeled using an exponential distribution.

The PDF of an exponential distribution with rate parameter λ (number of failures per unit of time) is

$$f(x \mid \lambda) = \lambda e^{-\lambda x}. \tag{4.6}$$

The CDF, $F(x)$, is

$$F(x) = P(X \leq x) = \int_0^x f(t)\mathrm{d}t = \int_0^x \lambda e^{-\lambda x}\mathrm{d}x = 1 - e^{-\lambda x}. \tag{4.7}$$

The mean, $E(X)$, of the exponential distribution is

$$E(X) = \int_0^\infty x f(x)\mathrm{d}x = \int_0^\infty \lambda x e^{-\lambda x}\mathrm{d}x = \frac{1}{\lambda}.$$

The variance of the exponential distribution is

$$V(X) = \frac{1}{\lambda^2}.$$

The hazard rate or instantaneous failure rate of the exponential distribution is

$$h(x) = \frac{f(x)}{1 - F(x)} = \frac{\lambda e^{-\lambda x}}{e^{-\lambda x}} = \lambda. \tag{4.8}$$

Therefore, a component with an exponential failure time distribution has a constant hazard.

The cumulative hazard of the exponential distribution is

$$H(x) = \int_0^x h(t)\mathrm{d}t = \int_0^x \lambda \mathrm{d}t = \lambda x. \tag{4.9}$$

For any continuous probability distribution, the relationship between the reliability function, $R(x)$, and the cumulative hazard function, $H(x)$, is given by

$$R(x) = e^{-H(x)}. \tag{4.10}$$

Combining (4.9) and (4.10) we can get the reliability function of the exponential distribution as

$$R(x) = e^{-\lambda x}.$$

This can be directly obtained from (4.7) as well.

The exponential distribution has an important property called memorylessness. This property states that the survival time of a new component beyond time t is the same as an old component that is being used for duration of time s will survive an additional time t. This property can be demonstrated as shown below:

$$\begin{aligned} P(X > t + s \mid X > s) &= \frac{P(X > t + s, X > s)}{P(X > s)} \\ &= \frac{P(X > t + s)}{P(X > s)} = \frac{e^{-\lambda(t+s)}}{e^{-\lambda s}} \\ &= e^{-\lambda t} = P(X > t). \end{aligned}$$

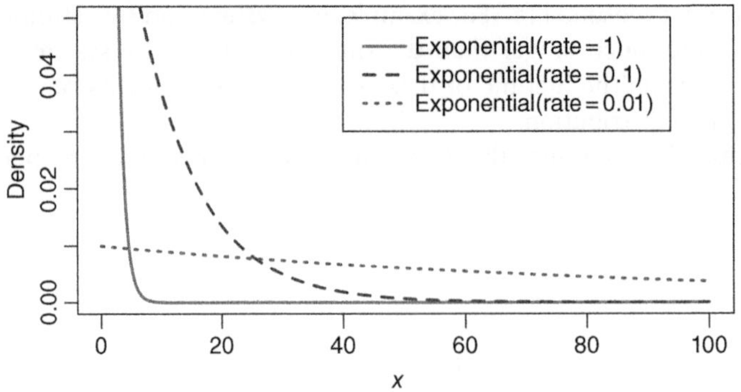

Figure 4.5 Probability density functions of three exponential distributions: $\lambda = 1, 0.1$, and 0.01.

Incandescent light bulbs tend to have this memoryless property and therefore their lifetime can be modeled with an exponential distribution.

There is an interesting relationship between the exponential distribution and the Poisson distribution. Suppose that the time to an event can be described by an exponential probability density with a rate λ. Suppose that events occur independently from each other with the same exponential rate λ. As an example, we can consider customers arriving at a bank ATM machine. Over a time period t the number of customers $N(t)$ arriving at the ATM can be shown to have a Poisson distribution with an average rate of λt. This type of counting process where inter-arrival times are exponentially distributed with the same exponential rate is called a homogeneous Poisson process.

Figure 4.5 shows the probability density function of a few exponential distributions ($\lambda = 1, 0.1, 0.01$). The R script to generate Figure 4.5 is shown in **4.3.1_Exponential_Distributions.R**.

```
(4.3.1_Exponential_Distributions.R)
## PDF of Exponential distributions
x <- seq(0,100,length=500)
exp_1 <- dexp(x, rate = 1, log = FALSE)
exp_2 <- dexp(x, rate = 0.1, log = FALSE)
exp_3 <- dexp(x, rate = 0.01, log = FALSE)

plot(x,exp_1,type="l",ylab="Density",xlab="x",lty=1,lwd=2,col="red",
ylim=c(0,0.05)) # plot 1st PDF
lines(x,exp_2,lty=2,lwd=2,col="blue") # add 2nd PDF
lines(x,exp_3,lty=3,lwd=2,col="purple") # add 3rd PDF
legend(40,0.05,c("Exponential(rate=1)","Exponential(rate=0.1)",
"Exponential(rate=0.01)"),lty=c(1,2,3),lwd=c(2,2,2),col=c("red","blue",
"purple")) # add legend
```

The conjugate prior distribution for λ is the gamma distribution. Refer to Eq. (4.4) for the gamma distribution PDF. Assume that the prior distribution of λ is

$$f(\lambda) \sim Gamma\ (a, b),$$

i.e. the prior PDF of λ is

$$f(\lambda \mid a, b) = \frac{b^a \lambda^{a-1} e^{-b\lambda}}{\Gamma(a)}. \tag{4.11}$$

Table 4.2 Months-to-failure data.

24.9	1.9	29.6	27.9	24.9	27.9	4.0	12.9	37.9	13.6
16.1	67.8	34.8	22.1	6.2	19.3	9.1	17.4	10.1	49.7

Assuming the observed data $x_1, x_2, \ldots, x_n$ are from an exponential distribution $f(x_i \mid \lambda)$, based on Bayes' theorem, the posterior distribution of λ is

$$
\begin{aligned}
f(\lambda \mid x_1, x_2, \ldots x_n)
&= \frac{f(x_1, x_2, \ldots x_n \mid \lambda) f(\lambda)}{f(x_1, x_2, \ldots x_n)} = \frac{f(x_1, x_2, \ldots x_n \mid \lambda) f(\lambda)}{\int f(x_1, x_2, \ldots x_n \mid \lambda) f(\lambda) d\lambda} \\
&= \frac{\left(\prod_{i=1}^{n} \lambda e^{-\lambda x_i}\right) \frac{b^a \lambda^{a-1} e^{-b\lambda}}{\Gamma(a)}}{\int \left(\prod_{i=1}^{n} \lambda e^{-\lambda x_i}\right) \frac{b^a \lambda^{a-1} e^{-b\lambda}}{\Gamma(a)} d\lambda} = \frac{\left(\prod_{i=1}^{n} \lambda e^{-\lambda x_i}\right) \lambda^{a-1} e^{-b\lambda}}{\int \left(\prod_{i=1}^{n} \lambda e^{-\lambda x_i}\right) \lambda^{a-1} e^{-b\lambda} d\lambda} \\
&= \frac{\lambda^{a+n-1} e^{-\left(b+\sum_{i=1}^{n} x_i\right)\lambda}}{\int \lambda^{a+n-1} e^{-\left(b+\sum_{i=1}^{n} x_i\right)\lambda} d\lambda} = \frac{\frac{\left(b+\sum_{i=1}^{n} x_i\right)^{a+n}}{\Gamma(a+n)} \lambda^{a+n-1} e^{-\left(b+\sum_{i=1}^{n} x_i\right)\lambda}}{\int \frac{\left(b+\sum_{i=1}^{n} x_i\right)^{a+n}}{\Gamma(a+n)} \lambda^{a+n-1} e^{-\left(b+\sum_{i=1}^{n} x_i\right)\lambda} d\lambda} \\
&= \frac{\left(b + \sum_{i=1}^{n} x_i\right)^{a+n}}{\Gamma(a+n)} \lambda^{a+n-1} e^{-\left(b+\sum_{i=1}^{n} x_i\right)\lambda}.
\end{aligned}
$$

Based on the equation above, the posterior distribution of λ is

$$f(\lambda \mid x_1, x_2, \ldots x_n) \sim Gamma\left(a + n, b + \sum_{i=1}^{n} x_i\right). \tag{4.12}$$

Example 4.2 Months-to-failure data of 20 units of a product are shown in Table 4.2 (data are randomly sampled from an exponential distribution with a rate of 1/20). What is the product reliability at 5 months (point estimate and 95% credible interval)?

The R script to calculate and plot the posterior distributions of λ, and to calculate reliability with 95% credible interval lower bound is shown in **4.3.1_Exponential_GammaPrior.** Here we use *Gamma* (1, 0.1) as a non-informative prior distribution of λ.

```
(4.3.1_Exponential_GammaPrior)
##################################################################
## Find the posterior distribution of rate parameter lambda ##
## in an Exponential distribution                          ##
## Conjugate prior: Gamma distribution                     ##
##################################################################

set.seed(12345)

# data
y <- c(24.9, 1.9, 29.6, 27.9, 24.9, 27.9, 4.0, 12.9, 37.9, 13.6, 16.1,
67.8, 34.8, 22.1, 6.2, 19.3, 9.1, 17.4, 10.1, 49.7)
```

```
     # calculate the summation of yi
     y_sum <- sum(y)

     # calculate the number of elements in y
     y_length <- length(y)

     # shape and rate parameters in the Gamma prior
     a=1   # shape
     b=0.1   # rate

     # Sample size sampled from the posterior distribution
     sample_number <- 10000

     lambda <- seq(0,0.1,length=500)
     lambda_prior <- dgamma(lambda, a, rate=b) # mean=a/b, variance=a/b^2
     likelihood <- dgamma(lambda, y_length+1, rate=y_sum) # rescaled
likelihood
     lambda_posterior <- dgamma(lambda, a+y_length, rate=b+y_sum)

     # Create density plots
     #jpeg("Exponential_Gamma.jpeg", width = 6, height = 4, units = 'in',
res = 600)  # save the plot as jpeg format
     plot(lambda,likelihood,type="l",ylab="Density",xlab="lambda",lty=1,
lwd=2,col="red",ylim=c(0,50)) # plot likelihood
     lines(lambda,lambda_posterior,lty=2,lwd=2,col="blue") # add posterior
     lines(lambda,lambda_prior,lty=3,lwd=2,col="purple") # add prior
     legend(0.06,40,c("prior","Likelihood","Posterior"),lty=c(3,1,2),
lwd=c(2,2,2),col=c("purple","red","blue")) # add legend
     #dev.off()

     ## Prior distribution of lambda: Gamma(a,b)
     # lambda_prior_samples <- rgamma(sample_number, a, rate=b)

     # posterior distribution of lambda:
     lambda_posterior_samples <- rgamma(sample_number, a+y_length,
rate=b+y_sum)
print(paste("lambda posterior mean is:", signif(mean(lambda_posterior_
samples), 6)))
```

The R script above estimates the mean of the posterior distribution of λ. After running the code, the following results are obtained:

```
[1] "lambda posterior mean is: 0.0459512"
```

```
     ########################################
     ## estimate reliability at 5 months ##
     ########################################

     reliability <- rep(NA, sample_number)

     # Find the probability of failure time being greater than 5.
     for(i in 1:sample_number) reliability[i] = 1 - pexp(5, rate =
lambda_posterior_samples[i])   # lower tail
```

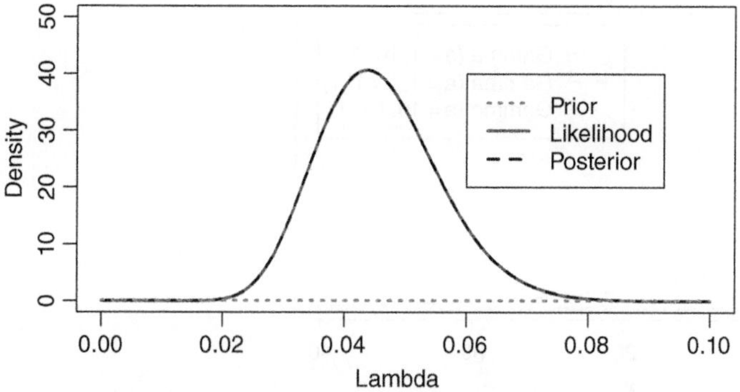

Figure 4.6 Rate parameter λ of an exponential distribution with vague gamma prior. Dotted line, prior; solid line, likelihood; dashed line, posterior.

```
    # Calculate the mean and 95% credible interval of reliability
at 5 months
    print(paste("reliability mean is:", signif(mean(reliability), 6)))
    print(paste("reliability 95% CI is:", signif(quantile(reliability,
0.025), 6), signif(quantile(reliability, 0.975), 6)))
```

The R script above estimates reliability at 5 months. After running the R script, the following results are obtained:

```
[1] "reliability mean is: 0.795716"
[1] "reliability 95% CI is: 0.712434 0.867628"
```

Based on the results above, at 5 months the reliability has a mean of 0.80. The reliability two-sided 95% credible interval is (0.71, 0.87). λ prior and posterior distributions are shown in Figure 4.6.

4.3.2 Gamma Distribution

In Bayesian statistics, a gamma distribution is often used as a conjugate prior distribution for rate parameters, e.g. rate parameters in a Poisson distribution or an exponential distribution (see Sections 4.2.2 and 4.3.1). The PDF of a gamma distribution with shape parameter a and rate parameter b is shown in Eq. (4.4). The mean of a gamma distribution is a/b and the variance is a/b^2.

There is another commonly used parameterization of the gamma distribution where the shape parameter a and the inverse of the rate parameter b are used. In this parameterization, $\theta = 1/b$ is called the scale parameter. The PDF of the gamma distribution with shape-scale parameterization is given by

$$f(x \mid a, \theta) = \frac{1}{\Gamma(a)} \frac{x^{a-1}}{\theta^a} e^{-\frac{x}{\theta}}, x > 0,$$

where $a > 0$ is the shape parameter and $\theta > 0$ is the scale parameter.

With this parameterization, the mean and the standard deviation are $a\theta$ and $a\theta^2$, respectively.

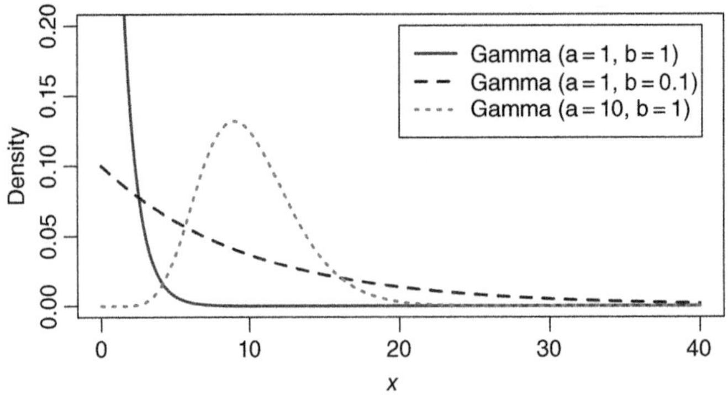

Figure 4.7 Probability density functions of three gamma distributions: $a = 1, b = 1; a = 1, b = 0.1$; $a = 10, b = 1$.

Figure 4.7 shows the probability density functions of a few gamma distributions. The R script to generate Figure 4.7 is as follows (included in **`4.3.2_Gamma_Distributions.R`**).

```
(4.3.2_Gamma_Distributions.R)
x <- seq(0,40,length=500)
gamma_1 <- dgamma(x, 1, rate=1) # mean=a/b, variance=a/b^2
gamma_2 <- dgamma(x, 1, rate=0.1) # mean=a/b, variance=a/b^2
gamma_3 <- dgamma(x, 10, rate=1) # mean=a/b, variance=a/b^2
plot(x,gamma_1,type="l",ylab="Density",xlab="x",lty=1,lwd=2,col="red",
ylim=c(0,0.2))
lines(x,gamma_2,lty=2,lwd=2,col="blue")
lines(x,gamma_3,lty=3,lwd=2,col="purple")
legend(20,0.2,c("Gamma (a=1, b=1)","Gamma (a=1, b=0.1","Gamma (a=10,
b=1)"),lty=c(1,2,3),lwd=c(2,2,2),col=c("red","blue","purple")) # add legend
```

When the shape parameter a is known and the rate parameter b is unknown, the conjugate prior distribution for the rate parameter b is a gamma distribution, *Gamma* (a_0, b_0),

$$f(b \mid a_0, b_0) = \frac{{b_0}^{a_0} b^{a_0-1} e^{-b_0 b}}{\Gamma(a_0)} \propto b^{a_0-1} e^{-b_0 b}.$$

The likelihood function is

$$f(x \mid a, b) = \frac{b^a x^{a-1} e^{-bx}}{\Gamma(a)} \propto b^a e^{-bx}.$$

The posterior distribution is

$$\begin{aligned} p(b \mid a_0, b_0, x) &\propto p(b \mid a_0, b_0) f(x \mid a, b) \propto b^{a_0-1} e^{-b_0 b} b^a e^{-bx} \\ &= b^{a_0+a-1} e^{-b(b_0+x)} \propto Gamma\ (a_0 + a, b_0 + x). \end{aligned} \tag{4.13}$$

4.3.3 Weibull Distribution

The Weibull distribution was introduced in Chapter 3. The PDF of a two-parameter Weibull distribution is

$$f(t \mid \beta, \eta) = \frac{\beta}{\eta}\left(\frac{t}{\eta}\right)^{\beta-1} e^{-(t/\eta)^{\beta}}, \tag{4.14}$$

where

t is the time to failure ($t \geq 0$)
β is the shape parameter ($\beta > 0$)
η is the scale parameter ($\eta > 0$).

The hazard function of the Weibull distribution is

$$h(t \mid \beta, \eta) = \frac{\beta}{\eta}\left(\frac{t}{\eta}\right)^{\beta-1}$$

As mentioned earlier, the Weibull distribution is flexible enough to model a variety of different failure modes: $\beta > 1$ indicates increasing hazard rate or wear-out failures, $\beta < 1$ indicates decreasing hazard rate or early failures, and $\beta = 1$ indicates constant hazard rate or random failures. A Weibull distribution with $\beta = 1$ is also an exponential distribution.

The reliability at time t is

$$R(t \mid \beta, \eta) = e^{-(t/\eta)^{\beta}}.$$

One of the interesting properties of the Weibull distribution is that when $t = \eta$, the cumulative probability of failure is $e^{-1} = 0.632$. That is, 63.2% of the units are expected to fail by the characteristic life regardless of the value of the shape parameter. Therefore, η is also known as the characteristic life of the Weibull distribution.

With the above parameterization, the mean is $\eta\Gamma(1+1/\beta)$ and the variance is

$$\eta^2[\Gamma(1+2/\beta) - (\Gamma(1+1/\beta))^2],$$

where $\Gamma(x)$ is the gamma function given by

$$\Gamma(k) = \int_0^{\infty} x^{k-1} e^{-x} dx.$$

Figure 4.8 shows the probability density function of a few Weibull distributions. The R script to generate Figure 4.8 is as follows (included in **4.3.3_Weibull_Distributions.R**).

```
(4.3.3_Weibull_Distributions.R)
## PDF of Weibull distributions
# with different shape parameters
x <- seq(0,40,length=500)
weib_1 <- dweibull(x, shape = 0.5, scale = 10, log = FALSE)
weib_2 <- dweibull(x, shape = 1, scale = 10, log = FALSE)
weib_3 <- dweibull(x, shape = 2, scale = 10, log = FALSE)
plot(x,weib_1,type="l",ylab="Density",xlab="x",lty=1,lwd=2,col="red",
ylim=c(0,0.2))
lines(x,weib_2,lty=2,lwd=2,col="blue")
lines(x,weib_3,lty=3,lwd=2,col="purple")
```

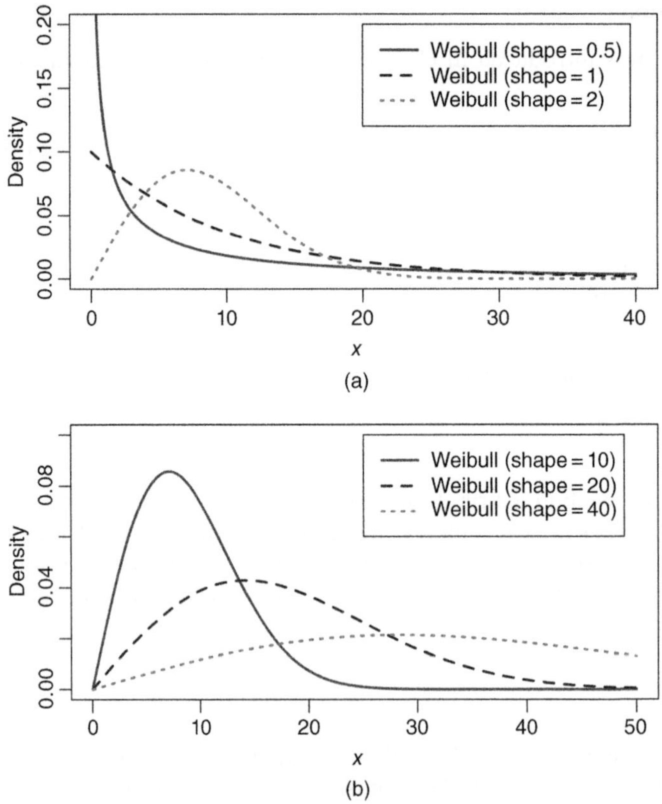

Figure 4.8 Probability density functions of various Weibull distributions. (a) Shape = 0.5, 1, 2; scale = 50. (b) Shape = 2; scale = 10, 20, 40.

```
    legend(20,0.2,c("Weibull (shape=0.5)","Weibull (shape=1)",
"Weibull (shape=2)"),lty=c(1,2,3),lwd=c(2,2,2),col=c("red","blue",
"purple")) # add legend

    # with different scale parameters
    x <- seq(0,50,length=500)
    weib_4 <- dweibull(x, shape = 2, scale = 10, log = FALSE)
    weib_5 <- dweibull(x, shape = 2, scale = 20, log = FALSE)
    weib_6 <- dweibull(x, shape = 2, scale = 40, log = FALSE)
    plot(x,weib_4,type="l",ylab="Density",xlab="x",lty=1,lwd=2,col="red",
ylim=c(0,0.1))
    lines(x,weib_5,lty=2,lwd=2,col="blue")
    lines(x,weib_6,lty=3,lwd=2,col="purple")
    legend(30,0.1,c("Weibull (scale=10)","Weibull (scale=20)",
"Weibull (scale=30)"),lty=c(1,2,3),lwd=c(2,2,2),col=c("red","blue",
"purple")) # add legend
```

4.3.3.1 Fit Data to a Weibull Distribution

In Chapter 1 we introduced censored data. Censored data are commonly seen in reliability practices. In this section, we provide an example of fitting data to a Weibull distribution and provide Bayesian estimation of the distribution parameters, where data is a mixture of right-censored data and uncensored data. Since there are no analytical conjugate prior distributions for the Weibull shape and scale parameters, there are no closed-form solutions to calculate the posterior distributions, thus we use Just Another Gibbs Sampler (JAGS) to estimate posterior distributions from Markov chain Monte Carlo (MCMC) sampling, which was discussed in Chapter 3. The JAGS model and R script are shown in **4.3.3_Weibull.R** and **4.3.3_Weibull.R**, respectively.

Example 4.3 Months-to-failure data of a product are shown in Table 4.3 (data are randomly sampled from a Weibull distribution with a scale of 100 and a shape of 3, and then all data values greater than 100 are marked as right-censored with a censoring limit of 100). Fit a Weibull distribution to the data and compute the Bayesian estimations of the scale and shape parameters.

```
(4.3.3_Weibull.JAGS)
model {
# Likelihood:
lambda <- 1/pow(scale, shape)  # power function pow(x,z) = x^z

for( i in 1:n) {
Censor[i] ~ dinterval(ttf[i], CenLimit[i])
ttf[i] ~ dweib(shape, lambda)
}

# Vague Gamma distributions for shape and scale prior
shape ~ dgamma(1,1) # mean = a/b; variance = a/(b^2)
scale ~ dgamma(1,0.1)
}
```

```
(4.3.3_Weibull.R)
########################################################
##  Fit data to a Weibull distribution                ##
##  Data: right censored & uncensored data            ##
########################################################

require(rjags)

## Data
# time to failure data (NA indicates right censored data)
ttf <- c(NA, 84.8, 87.5, 61.5, 99.3, NA, NA, 60.3, 80.3, NA, 51.7,
         68.5, 99.6, NA, 53.2, 46.6, 26.4, 72.3, 62.9, 70.5, 22.2, NA,
         NA, 75.2,87.2, 47.4, NA, 47.1, 98.2, 67.3)

# Censoring limit
CenLimit <- rep(100.0, length(ttf))

# Censor: 0 means uncensored; 1 means right censored
Censor <- c(1, 0, 0, 0, 0, 1, 1, 0, 0, 1, 0, 0, 0, 1, 0, 0, 0, 0, 0,
0, 0, 1, 1, 0, 0, 0, 1, 0, 0, 0)

# Data used in Bayesian analysis
BayesianData <- list(ttf = ttf,
                  CenLimit = CenLimit, Censor = Censor, n = length(ttf)
)
```

Table 4.3 Time to failure data (with right-censored data) for 30 specimens.

Time to failure (months)	Censor (0, uncensored; 1, right-censored)	Censoring limit
NA	1	100.0
84.8	0	
87.5	0	
61.5	0	
99.3	0	
NA	1	100.0
NA	1	100.0
60.3	0	
80.3	0	
NA	1	100.0
51.7	0	
68.5	0	
99.6	0	
NA	1	100.0
53.2	0	
46.6	0	
26.4	0	
72.3	0	
62.9	0	
70.5	0	
22.2	0	
NA	1	100.0
NA	1	100.0
75.2	0	
87.2	0	
47.4	0	
NA	1	100.0
47.1	0	
98.2	0	
67.3	0	

NA, not available.

```
    # Create a JAGS model object
    ModelObject <- jags.model(file = "4.3.3_Weibull.JAGS",
                              data=BayesianData,
                              n.chains = 2, n.adapt = 1000
    )

    # Burn-in stage
    update(ModelObject, n.iter=2000)

    # Run MCMC and collect posterior samples in coda format for selected
variables
```

```
    codaList <- coda.samples(model=ModelObject, variable.names =
c("shape", "scale"), n.iter = 30000, thin = 1)

    # Show summary statistics for collected posterior samples.
    # Quantiles of the sample distribution can be modified in the
quantiles argument.
    summary(codaList, quantiles = c(0.025, 0.05, 0.5, 0.95, 0.975))
```

If there is no prior knowledge and no conjugate distributions available, readers are free to use any vague distributions and run sensitivity analysis to ensure that the final results are robust against the choice of the prior distributions, as discussed in Chapter 3. In the R script above, we used different gamma distributions as priors of the shape and scale parameters in the Weibull distribution, as they both are positive and gamma distributions flexibly support a range of values between 0 and infinity.

In addition to sensitivity analysis, to ensure that the chosen prior distributions are adequate, one can run Monte Carlo simulations to check the prior predictive distributions to see if they match existing engineering knowledge. In Example 4.4, we check the prior predictive distribution of reliability via Monte Carlo simulations (R script shown in **4.3.3_Weibull_Prior.R**).

`Censor[i]` here is a binary indicator which equals 1 if `ttf[i]` is right-censored and 0 if `ttf[i]` is uncensored. Time to failure `ttf[i]` is unobserved when `ttf[i]` is greater than the censoring limit `CenLimit[i]`. For the right-censored data, `ttf[i]` takes the value `NA`. `dinterval` is a distribution representing interval-censored data. `Censor[i] ~ dinterval(ttf[i], CenLimit[i])` means the following

- `Censor[i] = 0 if ttf[i] ≤ CenLimit[i]` (`ttf[i]` is uncensored)
- `Censor[i] = 1 if ttf[i] > CenLimit[i]` (`ttf[i]` is right-censored).

The way to code interval-censored data can be found in JAGS user manual (Plummer 2015).

Summary statistics of the scale and shape parameters obtained from the MCMC sampling of the posterior distributions are shown below (you are likely to see slightly different results as random number generators are involved). A 95% credible interval of the scale parameter is (74.7, 102.4) and a 95% credible interval of the shape parameter is (1.7, 3.7). Note that 95% credible intervals of the scale and shape parameters contain the true values (scale = 100; shape = 3). For comparison, the frequentist maximum likelihood estimates (MLE) of the scale and shape parameters are 90.9 and 2.9, respectively. The corresponding 95% confidence intervals are (78.5, 105.3) and (2.0, 4.1), respectively. Both the frequentist confidence interval and the Bayesian credible interval contain the true values of the parameters. As mentioned before, the interpretations are different. A Bayesian credible interval is a true probabilistic statement whereas the frequentist confidence interval is a true or false statement. For this example, doing Bayesian analysis does not have added benefit, however in a complex problem situation the results could be completely different as will be demonstrated in later chapters.

```
    > summary(codaList, quantiles = c(0.025, 0.05, 0.5, 0.95, 0.975))

    Iterations = 33001:63000
    Thinning interval = 1
```

```
Number of chains = 2
Sample size per chain = 30000

1. Empirical mean and standard deviation for each variable,
   plus standard error of the mean:

          Mean     SD Naive SE Time-series SE
scale 87.641 7.0010 0.028581       0.045963
shape  2.589 0.4837 0.001975       0.003951

2. Quantiles for each variable:

       2.5%     5%    50%   95%   97.5%
scale 74.70 76.794 87.353 99.57 102.419
shape  1.72  1.845  2.561  3.43   3.618
```

Note that when you run the above code you may not get the same results due to differences in the random seed and the R software versions.

4.3.3.2 Demonstrating Reliability using Right-censored Data Only

For highly reliable products, sometimes only right-censored data are available in reliability testing. In these cases, traditional statistical methods cannot be used to estimate quantities such as "mean time to failure." If treating the right-censored data as pass/fail data, useful information will be lost. Furthermore, it is often challenging to demonstrate high reliability from attribute data due to sample size limitations. For example, to demonstrate 99% reliability with 95% confidence, a sample size of 299 is needed in an acceptance test with zero failures. In Example 4.4, we demonstrate reliability estimation based only on right-censored data.

Example 4.4 An accelerated bench test is used to demonstrate the reliability of a certain device component. This test is designed in such a way that 1 day survival on the bench test translates to 1 year survival under the device use conditions. To demonstrate a new component reliability, 135 parts were tested for different length of times. No failure is observed in the test: 78 parts survived 9 days, 22 parts survived 12 days, 35 parts survived 21 days. The objective is to demonstrate that reliability at 15 days is at least 95%, assuming time-to-failure data fits a Weibull distribution.

Since there are only right-censored data in this case, MLE cannot be used to estimate the two parameters of the Weibull distribution. Therefore, in this situation, the frequentist approach is to assume that the shape parameter β is known and the first failure is imminent (see Abernethy 1998, section 6-2). One can obtain the shape parameter from (i) historical failure data for a similar failure mode, (ii) prior experience, or (iii) engineering knowledge of the physics of the failure mode. With these assumptions a conservative $(1-\alpha)$ 100% lower confidence bound of the scale parameter η is given by

$$\eta = \left[\frac{\sum_{i=1}^{N} t_i^{\beta}}{-\log(\alpha)}\right]^{1/\beta}, \tag{4.15}$$

where t_i is time on the ith unit and N is the total number of units tested.

Actually, the above estimate is a conservative MLE estimate of the scale parameter assuming that the shape parameter is known.

Suppose we know from historical data for a similar component that the shape parameter $\beta = 1$. Then from Eq. 4.15 we get a conservative estimate of the 95% lower confidence bound of η as

$$\eta = \left[\frac{\sum_{i=1}^{135} t_i}{-\log(0.05)} \right] = \frac{1701}{2.995732} = 567.808.$$

The estimate of η given in Eq. (4.15) is a conservative MLE estimate and since MLE estimates are invariant under the transformation, a conservative 95% lower bound of the reliability at 15 days (translates to 15 years under the use conditions) is obtained as $e^{-15/567.8077} = 0.974$. This type of estimate requires the knowledge of the shape parameter. In the absence of such knowledge, we cannot estimate the lower bound of the reliability at 15 years.

Another limitation of the frequentist approach is that it does not take into account the uncertainty in the assumed shape parameter, which could lead to an optimistic or pessimistic reliability estimation. In addition, often in real-life applications there is not much prior knowledge about the shape parameter. To address these issues, we use a Bayesian Weibull model with vague prior distributions to assess the reliability for this case.

Since there is no prior information on the Weibull shape and scale parameters, vague priors of these two parameters are defined using gamma distributions. Several different vague prior distributions can be chosen to assess if results are sensitive to the selection of priors. In this case, the prior distribution of the shape parameter is assumed to have the same mean and standard deviation equal to 1. Since a shape parameter less than 1 indicates early (infant mortality) failures, equal to 1 indicates constant hazard (useful life), and greater than 1 indicates the wear-out phase, this prior distribution allows different types of hazard functions to be considered. We choose the prior distribution of the scale parameter to have both mean and standard deviation equal to 10. In order to be able to reproduce the results, we use the additional parameters ".RNG.name" and ".RNG.seed" in the list of initial values for each chain in the "jags.model." The parameter ".RNG.name" is used to specify the random number generator and the parameter ".RNG.seed" is used to specify the starting value of the "seed."

The JAGS model and R script are shown in **4.3.3_Weibull_RightCensored.JAGS** and **4.3.3_Weibull_RightCensored.R**, respectively.

```
(4.3.3_Weibull_RightCensored.JAGS)
model {
# Likelihood:
lambda <- 1/pow(scale, shape)
reliability <- exp(-pow(15/scale,shape))   # reliability of ith
product at 15 years

for( i in 1:135) {
Censor[i] ~ dinterval(ttf[i], CenLimit[i])
ttf[i] ~ dweib(shape, lambda)
}
```

```
# Vague Gamma distribution for shape and scale prior
shape ~ dgamma(1,1) # mean = a/b; variance = a/(b^2)
scale ~ dgamma(1,0.1)
}
```

(4.3.3_Weibull_RightCensored.R)

```
###################################################################
## Estimate 15 years' reliability based on right censored data ##
## Likelihood: Weibull distribution                              ##
## Vague priors: Gamma distribution                              ##
###################################################################

### Load package rjags (to connect to JAGS from R for
Bayesian analysis)
library(rjags)

# There are 135 right censored data in total.
# 78 data points are censored at 9 years;
# 22 data points are censored at 12 years;
# 35 data points are censored at 21 years

# Censoring time
CenLimit <- c( rep(9,78), rep(12,22), rep(21,35) )

# Censor: 0 means uncensored; 1 means right censored
Censor <- rep(1,135)

ttf <- rep(NA, 135)

# Data used in Bayesian analysis
BayesianData <- list(ttf = ttf,
                   CenLimit = CenLimit, Censor = Censor
)

# Initial values of censored data: different initial values for
different chains
ttfInit1 <- rep(NA,135)
ttfInit2 <- rep(NA,135)

ttfInit1[as.logical(Censor)] = CenLimit[as.logical(Censor)]+1
ttfInit2[as.logical(Censor)] = CenLimit[as.logical(Censor)]+2

# Create a JAGS model object
ModelObject <- jags.model(file = "4.3.3_Weibull_RightCensored.JAGS",
                        data=BayesianData, inits=list(list
(.RNG.name="base::Mersenne-Twister", .RNG.seed=1349, shape=1.5, scale=40,
ttf = ttfInit1),
                        list(.RNG.name="base::Mersenne-Twister",
.RNG.seed=6247, shape=3.5, scale=65, ttf = ttfInit2)),
                        n.chains = 2, n.adapt = 1000
)

# Burn-in stage
update(ModelObject, n.iter=10000)
```

```
    # Run MCMC and collect posterior samples in coda format for selected
variables
    codaList <- coda.samples(model= ModelObject, variable.names =
c("shape", "scale",
                            "reliability"), n.iter = 50000, thin = 1)

    # Summary plots (trace plots and density plots)
    # jpeg("Fig 4.9_summary_plots.jpeg", width = 8, height = 6,
units = 'in', res = 600)  # save the plot as jpeg format
    plot(codaList)
    # dev.off()
```

After running the R script above, the trace and density plots of estimated reliability at 15 years, and the Weibull scale and shape parameters are shown in Figure 4.9. Visual inspection of these plots suggest inadequate mixing in the scale parameter and perhaps in the shape and reliability parameters as well. We need to determine if we have run the MCMC chains long enough to accurately estimate the parameters of interest. The most critical parameter is the reliability at 15 years.

When using a Bayesian model, it is important to verify the convergence of the Markov chains. There are several methods available for checking convergence. The Gelman and Rubin (1992) diagnostic is based on comparing within and between sample variability when multiple chains are implemented in the JAGS model. An ANOVA-type approach is used to construct a test statistic, R, that approaches 1 when the convergence is achieved.

Geweke (1992) proposed a method where the mean value of each parameter was checked separately for its convergence using the sampled values within each chain. His method is based on constructing a Z test using non-overlapping subsamples from each MCMC chain of the parameter. If there are multiple chains, separate Z-tests were performed for each parameter for each chain. The rejection of the Z test is an indication of non-convergence. If the absolute value of the Z score is greater than 1.96, the convergence of the mean is rejected.

A third method of convergence diagnostic test is due to Raftery and Lewis (1992). The method is separately applied to the MCMC output from each chain for each parameter. The focus of this method is to achieve a prespecified level of accuracy for estimating a specified quantile with specified probability. The default quantile, accuracy, and probability in CODA are 2.5%, 0.005, and 0.95, respectively. The user can change these default values to their preference. When this method is applied to MCMC output, several quantities are reported: M, N, N_{min}, and I. These quantities are defined as follows:

M is the number of burn-in iterations

N is the total number of iterations that the chain is required to run

N_{min} is the minimum number of iterations required to estimate the quantile of interest with the prespecified accuracy and the probability, assuming that the data are independent (zero correlation)

I is the dependence factor computed as $I = N/N_{min}$.

I is interpreted as the relative increase in the number of iterations attributable to autocorrelations. If I is close to 1, generated values are closer to being independent. High dependence factors (>5) are problematic and may be due to different factors such as

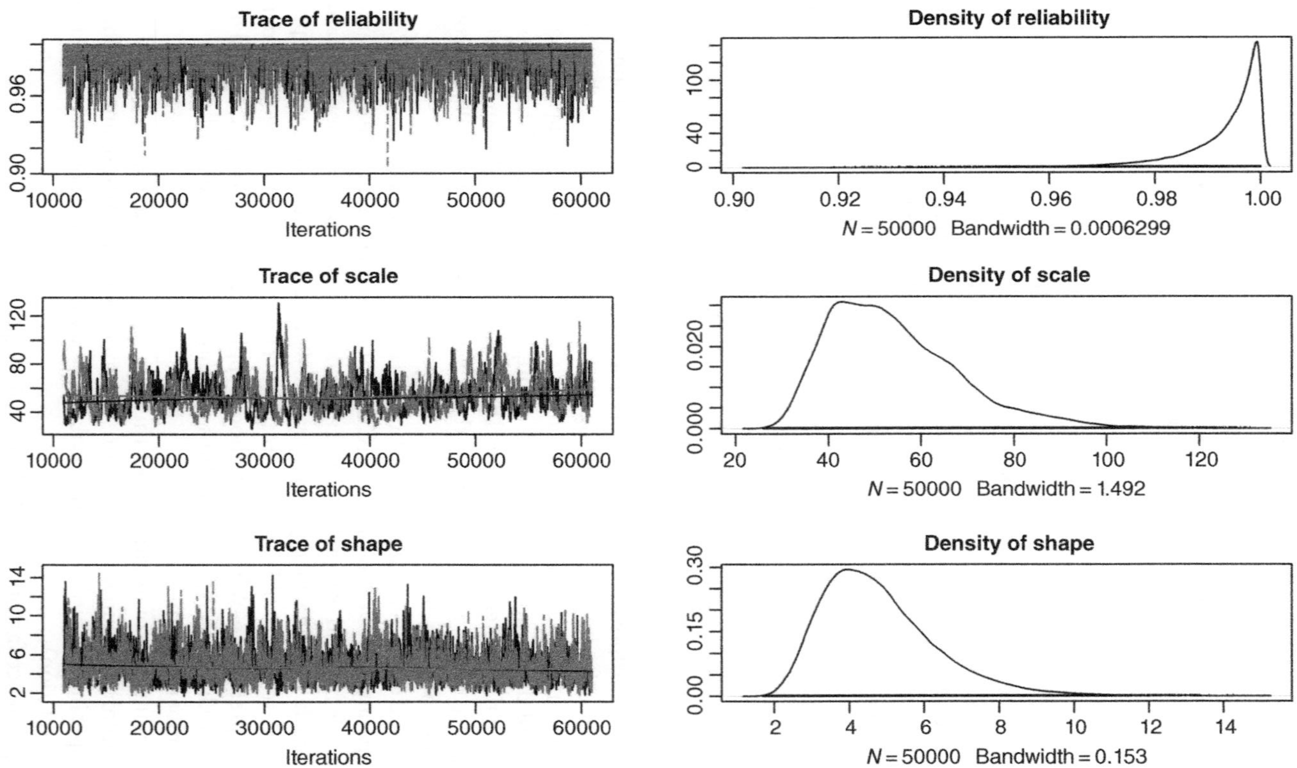

Figure 4.9 Trace plots and density plots of reliability at 15 years, and Weibull scale and shape parameters obtained from 10 000 burn-in followed by 50 000 iterations.

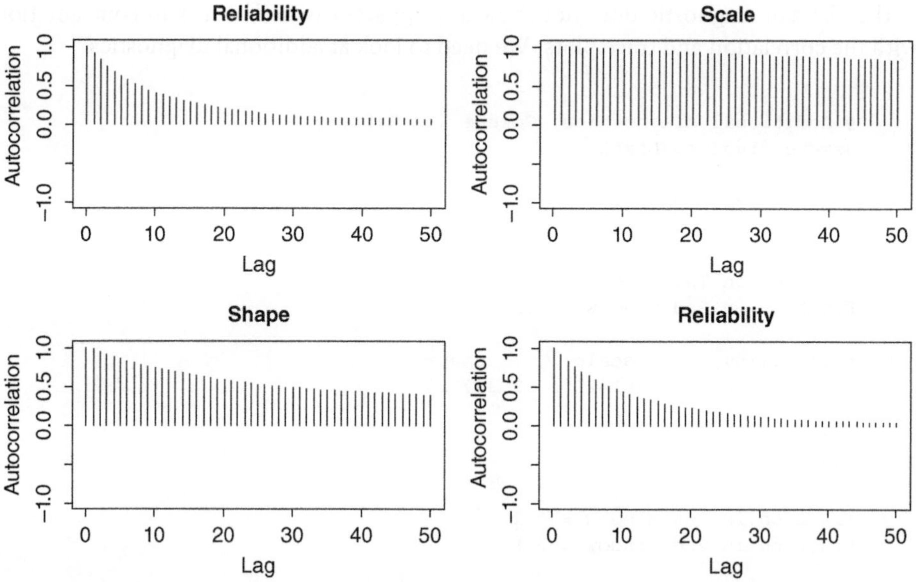

Figure 4.10 Autocorrelation plots with 10 000 burn-in followed by 50 000 iterations.

poor mixing, influential starting values, or high correlations between model coefficients. The I factor also indicates that thinning needs to be used to achieve an independent sequence of MCMC values for that parameter.

A good discussion of these and other diagnostic methods of convergence can be found in Ntzoufras (2009).

Before applying the convergence diagnostic tests to determine whether we need to run the MCMC chains longer, check the autocorrelations. Figure 4.10 shows a plot of autocorrelations for each parameter for each chain. The scale parameter shows the highest autocorrelation followed by the shape parameter. The reliability has the least autocorrelation.

```
    # Autocorrelation plots
    autocorr.plot(codaList,lag.max=50,ask=FALSE)

    # Gelman-Rubin diagnostic
    # Gelman and Rubin's convergence diagnostic
    # Approximate convergence is diagnosed when the upper limit is close
to 1
    gelman.diag(codaList)

    Potential scale reduction factors:

                 Point est. Upper C.I.
    reliability           1       1.00
    scale                 1       1.01
    shape                 1       1.00

    Multivariate psrf

    1.01
```

The Gelman diagnostic does not show any apparent issue but it is in contradiction with the correlation and trace plots. We need to look at additional diagnostics.

```
# Additional diagnostics; Geweke and Raftery
geweke.diag(codaList)

> geweke.diag(codaList)
[[1]]

Fraction in 1st window = 0.1
Fraction in 2nd window = 0.5

reliability       scale       shape
  -0.002332   -1.881429    3.387023


[[2]]

Fraction in 1st window = 0.1
Fraction in 2nd window = 0.5

reliability       scale       shape
    0.37088     0.09619     0.50599
```

The Geweke diagnostic indicates that the mean of the shape parameter in the first chain failed to converge (Z score = 3.39 > 1.96). Geweke plots (not shown) provide information regarding the changes in the Z score as the number of iterations increases. We need to check the Raftery diagnostic to assess the adequacy of the number of iterations of the chains.

```
raftery.diag(codaList)

> raftery.diag(codaList)
[[1]]

Quantile (q) = 0.025
Accuracy (r) = +/- 0.005
Probability (s) = 0.95

             Burn-in  Total  Lower bound  Dependence
             (M)      (N)    (Nmin)       factor (I)
 reliability 24       24496  3746          6.54
 scale       224      254208 3746         67.90
 shape       54       67707  3746         18.10


[[2]]

Quantile (q) = 0.025
Accuracy (r) = +/- 0.005
Probability (s) = 0.95
```

```
             Burn-in  Total   Lower bound  Dependence
             (M)      (N)     (Nmin)       factor (I)
reliability 30        33805   3746          9.02
scale       200       216296  3746         57.70
shape       55        66154   3746         17.70
```

The Raftery diagnostic indicates that the number of iterations required for accurately (within 0.005) estimating the 2.5th percentile (i.e. 97.5% lower bound) of the reliability is 24 496 for the first chain and 33 805 for the second chain. These numbers are much larger for the scale parameter: 254 208 and 216 296 for the first and second chains, respectively. For the shape parameter, the numbers are 67 707 and 66 154, respectively. Next, we will run the code with 300 000 iterations to achieve approximately independent MCMC samples.

```
    ModelObject <- jags.model(file = "4.3.3_Weibull_RightCensored.JAGS",
data=BayesianData, inits=list(list(.RNG.name="base::Mersenne-Twister",
.RNG.seed=1349, shape=1.5, scale=40, ttf = ttfInit1),  list(.RNG.name="base
::Mersenne-Twister", .RNG.seed=6247, shape=3.5, scale=65, ttf = ttfInit2)),
n.chains = 2, n.adapt = 1000 )

    # Burn-in stage
    update(ModelObject, n.iter=10000)
    # Now run MCMC and collect posterior samples in coda format
    codaList <- coda.samples(model=ModelObject, variable.names = c("shape",
"scale", "reliability"), n.iter = 300000, thin = 1)

    # Summary plots (trace plots and density plots)
    plot(codaList)
```

When the above code is run, it generates the trace plots and density plots shown in Figure 4.11. Visual inspection of these plots suggests adequate mixing of the two chains for all three parameters. We need to do the convergence diagnostic tests to confirm that the number of iterations (300 000) is adequate for accurately estimating the parameters of interest.

```
gelman.diag(codaList)
> gelman.diag(codaList)
Potential scale reduction factors:

            Point est. Upper C.I.
reliability          1       1.00
scale                1       1.01
shape                1       1.01

Multivariate psrf

1
```

The results of the Gelman diagnostic are satisfactory, indicating the convergence of both chains for all three parameters. Next we run the Geweke diagnostic test.

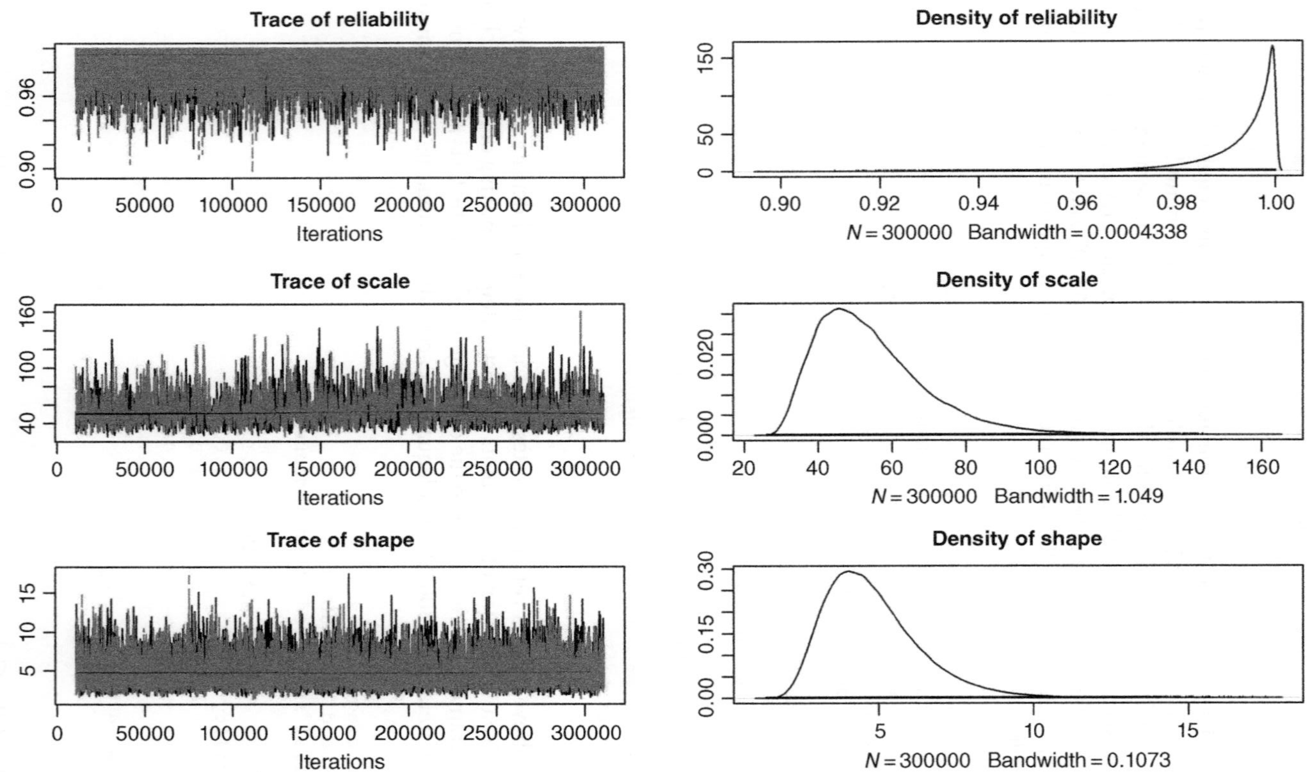

Figure 4.11 Trace plots and density plots of reliability at 15 years, and Weibull scale and shape parameters obtained from 10 000 burn-in followed by 300 000 iterations.

```
geweke.diag(codaList)
> geweke.diag(codaList)
[[1]]

Fraction in 1st window = 0.1
Fraction in 2nd window = 0.5

reliability       scale       shape
     0.4031     -0.6003      0.6309

[[2]]

Fraction in 1st window = 0.1
Fraction in 2nd window = 0.5

reliability       scale       shape
    -1.1923     -0.8485      1.4077
```

The Geweke diagnostic indicates that the means of the three parameters tend to converge for each chain. Finally, we run the Raftery diagnostic.

```
raftery.diag(codaList)
> raftery.diag(codaList)
[[1]]

Quantile (q) = 0.025
Accuracy (r) = +/- 0.005
Probability (s) = 0.95

             Burn-in  Total  Lower bound  Dependence
             (M)      (N)    (Nmin)       factor (I)
 reliability 36       42147  3746          11.3
 scale       368      404478 3746         108.0
 shape       160      207280 3746          55.3


[[2]]

Quantile (q) = 0.025
Accuracy (r) = +/- 0.005
Probability (s) = 0.95

             Burn-in  Total  Lower bound  Dependence
             (M)      (N)    (Nmin)       factor (I)
 reliability 32       39952  3746          10.7
 scale       368      399786 3746         107.0
 shape       128      152000 3746          40.6
```

The Raftery diagnostic suggests that the number of iterations in the chain (300 000) is sufficient for estimating the 2.5th percentile of both reliability and the shape parameters with the prespecified level of accuracy (within ± 0.005) and probability (0.95). However, the number of iterations may not be enough for estimating the scale parameter. Since our main focus is reliability, we can stop with the current number of iterations. The factor I for each parameter in each chain suggests that we need thinning to get independent samples. The function "ess" in the R package "mcmcse" can be used to determine the effective sample size from the MCMC samples generated by each chain. Using these sample sizes, gather independent samples from each chain for each parameter. Subsequently, the samples are combined to compute summary statistics for each parameter.

```
    # Use "ess" function in the "mcmcse" package to determine
effective sample size from
    # the posterior samples for shape, scale and reliability parameters.

    install.packages("mcmcse")
    library(mcmcse)

    A <- as.matrix(codaList [[1]])
    EffSS1 <- ess(A, g=NULL) # Compute effective sample sizes for all
three parameters from the 1st mcmc chain
    print("Effective sample sizes estimated from chain 1:", q=FALSE)
    print(EffSS1)
    print("Effective sample sizes estimated from chain 2:", q=FALSE)
    B <- as.matrix(codaList[[2]])
    EffSS2 <- ess(B, g=NULL) # Compute effective sample sizes for all
three parameters from the 2nd mcmc chain
    print(EffSS2)
```

When the above code is run the following results are obtained:

```
    [1] Effective sample sizes estimated from chain 1:
    > print(EffSS1)
    reliability       scale       shape
     11652.1507    937.3518   2130.3285
    > print("Effective sample sizes estimated from chain 2:", q=FALSE)
    [1] Effective sample sizes estimated from chain 2:
    > B <- as.matrix(codaList[[2]])
    > EffSS2 <- ess(B, g=NULL) # Compute effective sample sizes for all
three parameters from the 2nd mcmc chain
    > print(EffSS2)
    reliability       scale       shape
     11513.9159    883.4218   2172.2628
```

The effective sample sizes from each chain for the reliability parameter are much larger than the minimum number (3746) required by the Raftery diagnostic results shown earlier. For the shape parameter, the effective sample size for each chain is not larger than the minimum number required but when the two chains are combined, there is a sufficiently large sample. However, for the scale parameter, even the combined sample does not exceed the minimum number (3746) required. Since the focus is on reliability, there is no need to further increase the number of iterations in the two chains. The following codes are used to extract approximate random samples from each chain and then to get summary statistics for each parameter.

```
    n11 <- ceiling(EffSS1[1]) #get the effective sample size for the
"reliability" parameter from the 1st mcmc chain
    n12 <- ceiling(EffSS1[2]) #get the effective sample size for the
"scale" parameter from the 1st mcmc chain
    n13 <- ceiling(EffSS1[3]) #get the effective sample size for the
"shape" parameter from the 1st mcmc chain
```

```
    n21 <- ceiling(EffSS2[1]) #get the effective sample size for the
"reliability" parameter from the 2nd mcmc chain
    n22 <- ceiling(EffSS2[2]) #get the effective sample size for the
"scale" parameter from the 2nd mcmc chain
    n23 <- ceiling(EffSS2[3]) #get the effective sample size for the
"shape" parameter from the 2nd mcmc chain

    # Do necessary thinning to reduce autocorrelation of
"reliability" within the 1st chain as follows
    ChainLnth <- dim(A)[1]
    intvl <- ceiling(ChainLnth/n11)
    Indx <- seq(1,ChainLnth,intvl)
    Reliability_RS1 <- A[,1][Indx]

    # Do necessary thinning to reduce autocorrelation of
"reliability" within the 2nd chain as follows
    intvl <- ceiling(ChainLnth/n21)
    Indx <- seq(1,ChainLnth,intvl)
    Reliability_RS2 <- B[,1][Indx]

    # Combine the two samples for reliability
    Reliability_RS <- c(Reliability_RS1,Reliability_RS2)

    #plot the autocorrelations upto Lag50 for the "reliability" using the
data selected from the 1st chain
    Acf(Reliability_RS1, lag.max = 50,type = "correlation", plot = TRUE)
    #plot the autocorrelations upto Lag50 for the "reliability" using the
data selected from the 2nd chain
    Acf(Reliability_RS2, lag.max = 50,type = "correlation", plot = TRUE)

    # Do necessary thinning to reduce autocorrelation of "scale" within
the 1st chain
    intvl <- ceiling(ChainLnth/n12)
    Indx <- seq(1,ChainLnth,intvl)
    Scale_RS1 <- A[,2][Indx]

    # Do necessary thinning to reduce autocorrelation of "scale" within
the 2nd chain
    intvl <- ceiling(ChainLnth/n22)
    Indx <- seq(1,ChainLnth,intvl)
    Scale_RS2 <- B[,2][Indx]

    # Combine the two samples for "scale"
    Scale_RS <- c(Scale_RS1,Scale_RS2)

    #plot the autocorrelations upto Lag50 for the "scale" using the data
selected from the 1st chain
    Acf(Scale_RS1, lag.max = 50,type = "correlation", plot = TRUE)
    #plot the autocorrelations upto Lag50 for the "scale" using the data
selected from the 2nd chain
    Acf(Scale_RS2, lag.max = 50,type = "correlation", plot = TRUE)

    # Do necessary thinning to reduce autocorrelation of "shape" within
the 1st chain
    intvl <- ceiling(ChainLnth/n13)
    Indx <- seq(1,ChainLnth,intvl)
```

```
    Shape_RS1 <- A[,3][Indx]

    # Do necessary thinning to reduce autocorrelation of "shape" within
the 2nd chain
    intvl <- ceiling(ChainLnth/n23)
    Indx <- seq(1,ChainLnth,intvl)
    Shape_RS2 <- B[,3][Indx]

    # Combine the two samples for reliability
    Shape_RS <- c(Shape_RS1,Shape_RS2)

    #plot the autocorrelations upto Lag50 for the "scale" using the data
selected from the 1st chain
    Acf(Shape_RS1, lag.max = 50,type = "correlation", plot = TRUE)
    #plot the autocorrelations upto Lag50 for the "scale" using the data
selected from the 2nd chain
    Acf(Shape_RS2, lag.max = 50,type = "correlation", plot = TRUE)

    # Compute basic statistics and quantiles for the three
parameters; reliability, scale, and shape

    summary(Reliability_RS)
    quantile(Reliability_RS, probs = c(0.025, 0.05, 0.5, 0.95, 0.975))

    summary(Scale_RS)
    quantile(Scale_RS, probs = c(0.025, 0.05, 0.5, 0.95, 0.975))

    summary(Shape_RS)
    quantile(Shape_RS, probs = c(0.025, 0.05, 0.5, 0.95, 0.975))
```

When the above code is run, the resulting plots (not shown here) show negligible autocorrelation for each parameter and for each chain. This is an indication that the selected samples are roughly independent and therefore can be used to estimate the parameters. A summary of the parameter estimates is given below.

```
> summary(Reliability_RS)
   Min. 1st Qu.  Median    Mean 3rd Qu.    Max.
 0.9198  0.9910  0.9960  0.9933  0.9986  1.0000
> quantile(Reliability_RS, probs = c(0.025, 0.05, 0.5, 0.95, 0.975))
     2.5%        5%       50%       95%     97.5%
0.9705896 0.9772029 0.9960209 0.9998531 0.9999445

> summary(Scale_RS)
   Min. 1st Qu.  Median    Mean 3rd Qu.    Max.
  29.91   43.59   51.41   54.44   62.20  146.60
> quantile(Scale_RS, probs = c(0.025, 0.05, 0.5, 0.95, 0.975))
    2.5%       5%      50%      95%    97.5%
33.48045 35.75122 51.41029 82.36527 90.80747

> summary(Shape_RS)
   Min. 1st Qu.  Median    Mean 3rd Qu.    Max.
  1.786   3.700   4.571   4.793   5.627  11.850
> quantile(Shape_RS, probs = c(0.025, 0.05, 0.5, 0.95, 0.975))
    2.5%       5%      50%      95%    97.5%
   2.509171 2.736350 4.571279 7.730044 8.330848
```

These results show that the 15 years' reliability has a mean value of 0.993 and 97.5% credible interval lower bound of 0.971. Finally, the results obtained for the parameters using approximately independent samples can be compared with those obtained using all MCMC samples, as shown below.

```
    # Summary statistics for Markov Chain Monte Carlo chains
    # Quantiles of the sample distribution can be modified in the
quantiles argument
    summary(codaList, quantiles = c(0.025, 0.05, 0.5, 0.95, 0.975))

    > summary(codaList, quantiles = c(0.025, 0.05, 0.5, 0.95, 0.975))

    Iterations = 11001:311000
    Thinning interval = 1
    Number of chains = 2
    Sample size per chain = 3e+05

    1. Empirical mean and standard deviation for each variable,
       plus standard error of the mean:

                    Mean        SD  Naive SE Time-series SE
    reliability  0.9933  0.007983 1.031e-05      5.155e-05
    scale       54.4543 15.250701 1.969e-02      4.764e-01
    shape        4.8055  1.531431 1.977e-03      2.054e-02

    2. Quantiles for each variable:

                   2.5%      5%     50%     95%   97.5%
    reliability  0.971  0.9771  0.9961  0.9999  0.9999
    scale       33.523 35.3468 51.5932 83.3078 92.0308
    shape        2.550  2.7848  4.5606  7.6711  8.4624
```

The summary results obtained using all MCMC samples compare well with those obtained from approximately independent samples. At least in this example, there is no real advantage of going through convergence diagnostics and then carefully selecting roughly independent samples to generate the estimates of the selected parameters. However, results could have been different in a more complex problem with many parameters. So, it is always a good idea to do larger number of iterations after burn-in than a smaller number of iterations.

We can also check the prior distribution of reliability at 15 years to see if the prior distributions chosen for the shape and scale parameters result in a prior distribution of reliability that is also vague and matches the designers' belief. This can be done using direct Monte-Carlo simulation using the assumed vague priors on the shape and scale parameters. The R script to estimate the prior distribution of 15 years' reliability is shown in **`4.3.3_Weibull_Prior.R`**.

Results show the reliability prior has a mean and median around 0.24, with the 97.5th percentile being 0.71. This prior distribution on reliability is viewed as vague and conservative. The mean and the 95% credible interval of reliability posterior distribution are much larger than the corresponding values in the reliability prior distribution. This R script can be easily modified to check the reliability prior with a few different prior distributions of scale and shape parameters to assess sensitivity.

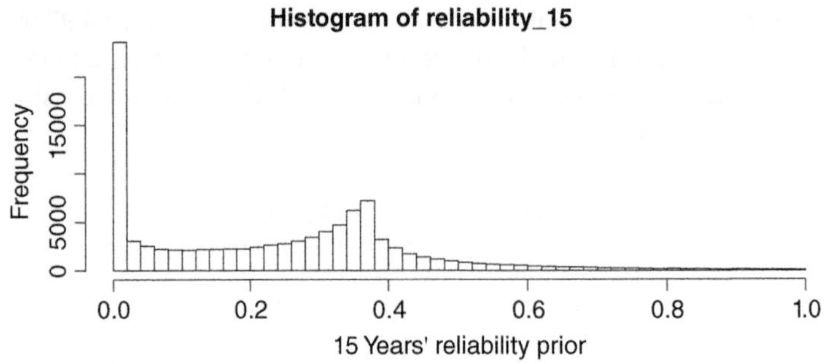

Figure 4.12 Histogram of 15 years' reliability prior distribution.

```
(4.3.3_Weibull_Prior.R)
    # Do a Monte-Carlo simulation to see if the selected vague priors
    # on the scale and shape parameters of the Weibull model
    # in Example 4.3.3 cause a conservative prior distribution of
reliability

    # Generate 100,000 samples for shape and scale priors
    # each from the Vague Gamma distributions
    shape <- rgamma(n=100000, shape=1, rate=1) # mean = a/b;
variance = a/(b^2)
    scale <- rgamma(n=100000,shape=1,rate=0.1)
    #compute relibility at 15 years
    reliability_15 <- exp(-(15/scale)**(shape))

    mean(reliability_15)
    quantile(reliability_15, c(0.025, 0.50, 0.975))

    mean(shape)
    quantile(shape, c(0.025, 0.50, 0.975))

    hist(reliability_15, xlab="15 years' reliability prior", breaks=50)
```

Summary statistics of simulated 15 years' reliability prior and shape prior distributions are shown below. The histogram of 15 years' reliability prior distribution is shown in Figure 4.12. These summary results show that the assumptions about the prior distributions of shape and the reliability are fairly conservative compared to what was achieved in the above Bayesian analysis.

```
> mean(reliability_15)
[1] 0.2349964
> quantile(reliability_15, c(0.025, 0.50, 0.975))
         2.5%           50%          97.5%
8.121193e-273  2.429172e-01  7.079728e-01

> mean(shape)
[1] 0.998279
> quantile(shape, c(0.025, 0.50, 0.975))
      2.5%        50%      97.5%
0.02488366 0.69317034 3.68038939
```

4.3.4 Normal Distribution

The PDF of a normal distribution with mean μ and variance σ^2 is

$$f(x \mid \mu,\ \sigma^2) = \frac{1}{\sqrt{2\pi\sigma^2}} e^{-\frac{(x-\mu)^2}{2\sigma^2}}. \tag{4.16}$$

The normal density is a bell-shaped curve that is symmetric around its mean. A random variable X with normal PDF having mean μ and standard deviation σ can be converted to the standard normal density with mean zero and standard deviation 1 by the transformation $Y = (X - \mu)/\sigma$.

The normal distribution is not commonly used to model time to failure data. However, it is often used in other engineering applications, where quality characteristics such as dimensions, weights, volumes etc. are used.

The CDF of the normal distribution is given by

$$\varphi(x \mid \mu, \sigma^2) = \int_{-\infty}^{x} f(t \mid \mu, \sigma^2)\mathrm{d}t = \int_{-\infty}^{x} \frac{1}{2\pi\sigma^2} e^{-\frac{(t-\mu)^2}{2\sigma^2}} \mathrm{d}t. \tag{4.17}$$

There is no closed-form solution to compute the CDF of a normal distribution, however numerical integration methods are used to obtain good approximations.

When the variance is known, the conjugate prior of the unknown mean parameter μ is a normal distribution. Assuming the prior distribution of μ is normal with meanμ_0 and variance ${\sigma_0}^2$, the PDF of μ is given by

$$f(\mu \mid \mu_0,\ {\sigma_0}^2) = \frac{1}{\sqrt{2\pi{\sigma_0}^2}} e^{-\frac{(\mu-\mu_0)^2}{2{\sigma_0}^2}}.$$

Assuming the observed data $x_1,\ x_2,\ \ldots\ x_n$ are from a normal distribution$f(x_i \mid \mu,\ \sigma^2)$, based on Bayes' theorem, the posterior distribution of μ is

$$\begin{aligned} &f(\mu \mid x_1,\ x_2,\ \ldots\ x_n) \\ &= \frac{g(x_1,\ x_2,\ \ldots\ x_n \mid \mu, \sigma^2)h(\mu)}{k(x_1,\ x_2,\ \ldots\ x_n)} = \frac{g(x_1,\ x_2,\ \ldots\ x_n \mid \mu, \sigma^2)h(\mu)}{\int g(x_1,\ x_2,\ \ldots\ x_n \mid \mu, \sigma^2)h(\mu)\mathrm{d}\mu} \\ &= \frac{\left(\prod_{i=1}^{n} \frac{1}{\sqrt{2\pi\sigma^2}} e^{-\frac{(x_i-\mu)^2}{2\sigma^2}}\right) \frac{1}{\sqrt{2\pi{\sigma_0}^2}} e^{-\frac{(\mu-\mu_0)^2}{2{\sigma_0}^2}}}{\int \left(\prod_{i=1}^{n} \frac{1}{\sqrt{2\pi\sigma^2}} e^{-\frac{(x_i-\mu)^2}{2\sigma^2}}\right) \frac{1}{\sqrt{2\pi{\sigma_0}^2}} e^{-\frac{(\mu-\mu_0)^2}{2{\sigma_0}^2}} \mathrm{d}\mu} = \frac{\left(\prod_{i=1}^{n} e^{-\frac{(x_i-\mu)^2}{2\sigma^2}}\right) e^{-\frac{(\mu-\mu_0)^2}{2{\sigma_0}^2}}}{\int \left(\prod_{i=1}^{n} e^{-\frac{(x_i-\mu)^2}{2\sigma^2}}\right) e^{-\frac{(\mu-\mu_0)^2}{2{\sigma_0}^2}} \mathrm{d}\mu}. \end{aligned} \tag{4.18}$$

The numerator of Eq. (4.18) becomes

$$\begin{aligned} &\exp\left(-\sum_{i=1}^{n} \frac{(x_i-\mu)^2}{2\sigma^2} - \frac{(\mu-\mu_0)^2}{2{\sigma_0}^2}\right) \propto \exp\left(-\sum_{i=1}^{n} \frac{(-2x_i\mu+\mu^2)}{2\sigma^2} - \frac{(\mu^2-2\mu\mu_0)}{2{\sigma_0}^2}\right) \\ &= \exp\left[-\frac{1}{2\sigma^2{\sigma_0}^2}\left(\sum_{i=1}^{n}(-2x_i\mu+\mu^2){\sigma_0}^2 + (\mu^2-2\mu\mu_0)\sigma^2\right)\right] \\ &= \exp\left\{-\frac{1}{2\sigma^2{\sigma_0}^2}\left[(n{\sigma_0}^2+\sigma^2)\mu^2 - 2\left({\sigma_0}^2\sum_{i=1}^{n} x_i + \sigma^2\mu_0\right)\mu\right]\right\} \end{aligned}$$

$$\propto \exp\left[-\frac{(n\sigma_0{}^2+\sigma^2)}{2\sigma^2\sigma_0{}^2}\left(\mu-\frac{\left(\sigma_0{}^2\sum_{i=1}^{n}x_i+\sigma^2\mu_0\right)}{(n\sigma_0{}^2+\sigma^2)}\right)^2\right]$$

$$\propto \frac{1}{\sqrt{2\pi\frac{\sigma^2\sigma_0{}^2}{(n\sigma_0{}^2+\sigma^2)}}}\exp\left[-\frac{1}{2\frac{\sigma^2\sigma_0{}^2}{(n\sigma_0{}^2+\sigma^2)}}\left(\mu-\frac{\left(\sigma_0{}^2\sum_{i=1}^{n}x_i+\sigma^2\mu_0\right)}{(n\sigma_0{}^2+\sigma^2)}\right)^2\right]$$

$$\sim Normal\left(\frac{\left(\sigma_0{}^2\sum_{i=1}^{n}x_i+\sigma^2\mu_0\right)}{(n\sigma_0{}^2+\sigma^2)},\ \frac{\sigma^2\sigma_0{}^2}{(n\sigma_0{}^2+\sigma^2)}\right). \tag{4.19}$$

Equation (4.18) becomes

$$f(\mu \mid x_1,\ x_2,\ \ldots\ x_n)=\frac{f\left(\mu \mid \frac{\left(\sigma_0{}^2\sum_{i=1}^{n}x_i+\sigma^2\mu_0\right)}{(n\sigma_0{}^2+\sigma^2)},\ \frac{\sigma^2\sigma_0{}^2}{(n\sigma_0{}^2+\sigma^2)}\right)}{\int f\left(\mu \mid \frac{\left(\sigma_0{}^2\sum_{i=1}^{n}x_i+\sigma^2\mu_0\right)}{(n\sigma_0{}^2+\sigma^2)},\ \frac{\sigma^2\sigma_0{}^2}{(n\sigma_0{}^2+\sigma^2)}\right)d\mu}$$

$$=f\left(\mu \mid \frac{\left(\sigma_0{}^2\sum_{i=1}^{n}x_i+\sigma^2\mu_0\right)}{(n\sigma_0{}^2+\sigma^2)},\ \frac{\sigma^2\sigma_0{}^2}{(n\sigma_0{}^2+\sigma^2)}\right).$$

In summary, when the variance is known, the posterior distribution of the mean is also a normal distribution using the conjugate prior

$$f(\mu \mid x_1,\ x_2,\ \ldots\ x_n) \sim Normal\left(\frac{\sigma_0{}^2\sum_{i=1}^{n}x_i+\sigma^2\mu_0}{n\sigma_0{}^2+\sigma^2},\ \frac{\sigma^2\sigma_0{}^2}{n\sigma_0{}^2+\sigma^2}\right)$$

$$= Normal\left(\frac{\sigma_0{}^2}{\left(\sigma_0{}^2+\frac{\sigma^2}{n}\right)}\bar{x}+\frac{\frac{\sigma^2}{n}}{\left(\sigma_0{}^2+\frac{\sigma^2}{n}\right)}\mu_0,\ \frac{\sigma^2\sigma_0{}^2}{n\sigma_0{}^2+\sigma^2}\right). \tag{4.20}$$

Let $\omega=\frac{\sigma_0^2}{\left(\sigma_0^2+\frac{\sigma^2}{n}\right)}$ then $(1-\omega)=1-\frac{\sigma_0^2}{\left(\sigma_0^2+\frac{\sigma^2}{n}\right)}=\frac{\frac{\sigma^2}{n}}{\left(\sigma_0^2+\frac{\sigma^2}{n}\right)}$.

Therefore, the posterior distribution of the mean has the form $\omega\bar{x}+(1-\omega)\mu_0$, which is a weighted average of the prior distribution mean μ_0 and the data average $\bar{x}$.

In the next example we will examine a case where both the mean μ and variance σ^2 are unknown. The R script for this example is in **`4.3.4_Normal.R`**, which is given

Table 4.4 Pull strength (lbf) test data for 75 lead/connectors.

15.06	15.91	16.07	11.92
15.34	11.11	13.63	13.03
13.97	17.25	15.38	14.68
12.52	16.83	15.68	13.27
18.9	16.29	13.74	16.83
14.96	15.68	17.43	14.11
15.09	18.98	19.89	13.33
15.87	12.65	12.77	19.44
18.01	14.71	18.05	11.45
17.79	16.82	12.81	13.26
18.67	18.77	15.57	15.34
9.55	14.81	16.27	12.19
17.37	17.34	15.72	10.64
10.12	11.09	12.66	8.92
14.54	16.38	15.21	11.23
13.55	17.13	16.55	
14.81	15.72	15.46	
15.4	13.34	14.41	
15.75	17.04	11.92	
15.66	13.99	18.82	

below. The JAGS model **3.4_Normal.JAGS** introduced in Chapter 3 is reused in this case. We will use vague priors on both the mean and the precision (1/variance) of the normal distribution while implementing MCMC method for estimating distributions of mean μ and variance σ^2.

Example 4.5 Leads are connected to a device through a component called a connector. The leads are inserted into the ports in the connector and then the set screws are tightened. A bench test of 75 lead/connectors were conducted to assess the pull strength and recorded in pounds of force (lbf) required to pull the lead out of the connector. The results of this test are shown in Table 4.4 and are also included in the text file Ex4.5_Normal.txt. Compute the lower bound of the one-sided 95% credible interval of the B5 life (i.e. the fifth percentile) of the pull strength. Historical data suggest that pull strength test data follow a normal distribution.

```
(4.3.4_Normal.R)
##############################
# Fit a Normal distribution using non-informative mean and variance
##############################

### Load package rjags (to connect to JAGS from R for Bayesian analy-
sis)
library(rjags)

# read data from file
data <- read.table("Ex4.5_Normal.txt", header=F, sep="")
```

```
y <- data[,1]

# Data for Bayesian analysis
BayesianData <- list(y = y, n=75)

# Initial values
listial1 <- list( mu = 5, tau = 1 )
listial2 <- list( mu = 7, tau = 0.5 )
InitialValues <- list(listial1, listial2)

# Create a JAGS model object
ModelObject <- jags.model(file = "3.4_Normal.JAGS",
                         data=BayesianData, inits=InitialValues,
                         n.chains = 2, n.adapt = 1000
)

# Burn-in stage
update(ModelObject, n.iter=10000)

# Run MCMC and collect posterior samples in coda format for selected
variables
codaList <- coda.samples(model=ModelObject, variable.names = c("mu",
"sigma",
                      "tau"), n.iter = 100000, thin = 1)

# Summary statistics for Markov Chain Monte Carlo chains
# Quantiles of the sample distribution can be modified in the quan-
tiles argument
summary(codaList, quantiles = c(0.025, 0.05, 0.5, 0.95, 0.975))
```

Summary statistics of the posterior distributions of mean μ and σ are shown below.

```
> summary(codaList, quantiles = c(0.025, 0.05, 0.5, 0.95, 0.975))

Iterations = 10001:110000
Thinning interval = 1
Number of chains = 2
Sample size per chain = 1e+05

1. Empirical mean and standard deviation for each variable,
   plus standard error of the mean:

         Mean      SD  Naive SE Time-series SE
mu    14.9659 0.28598 6.395e-04      6.395e-04
sigma  2.4706 0.20624 4.612e-04      4.618e-04
tau    0.1672 0.02748 6.145e-05      6.193e-05

2. Quantiles for each variable:

         2.5%      5%     50%     95%   97.5%
mu    14.4050 14.4977 14.9653 15.4377 15.5293
sigma  2.1078  2.1584  2.4562  2.8316  2.9168
tau    0.1175  0.1247  0.1658  0.2147  0.2251
```

The above results show that the one-sided lower 95% credible interval for the fifth percentile of the distribution of pull strength test is 10.13 lbf.

4.3.5 Lognormal Distribution

If a random variable Y has a normal distribution, then $X = \exp(Y)$ has a lognormal distribution. A lognormal distribution has the PDF

$$f(x) = \frac{1}{x\sigma\sqrt{2\pi}} e^{-\frac{(\ln x - \mu)^2}{2\sigma^2}}. \tag{4.21}$$

If $Y = \log(X - k)$, where $X > k$ is normally distributed with mean μ and standard deviation σ, then the distribution of X said to have a three-parameter lognormal distribution with threshold k. A three-parameter lognormal distribution has the PDF

$$f(x) = \frac{1}{(x-k)\sigma\sqrt{2\pi}} e^{-\frac{(\ln(x-k) - \mu)^2}{2\sigma^2}}. \tag{4.22}$$

The lognormal distribution is frequently used to model semiconductor degradation failure mechanisms, material fatigue failures, propagation of cracks, personal income, etc. As in the case of the normal distribution, the lognormal distribution is completely characterized by its mean θ and the standard deviation γ. Let the random variable X have a lognormal distribution with mean θ and standard deviation γ. Then, the distribution of $Y = \log(X)$ is normal with mean μ and standard deviation σ. The lognormal parameters θ and γ can be expressed in terms of μ and σ, and vice versa. These relationships are given below:

$$\mu = \log\left(\frac{\theta}{\sqrt{1+\frac{\gamma^2}{\theta^2}}}\right) \text{ and } \sigma^2 = \log\left(1 + \frac{\gamma^2}{\theta^2}\right), \tag{4.23}$$

$$\theta = e^{\left(\mu + \frac{\sigma^2}{2}\right)} \text{ and } \gamma^2 = (e^{\sigma^2} - 1)e^{(2\mu + \sigma^2)}. \tag{4.24}$$

Figure 4.13 shows the probability density functions of a few lognormal distributions. The R script to generate Figure 4.13 is as shown below (included in **`4.3.5_Lognormal_Distributions.R`**).

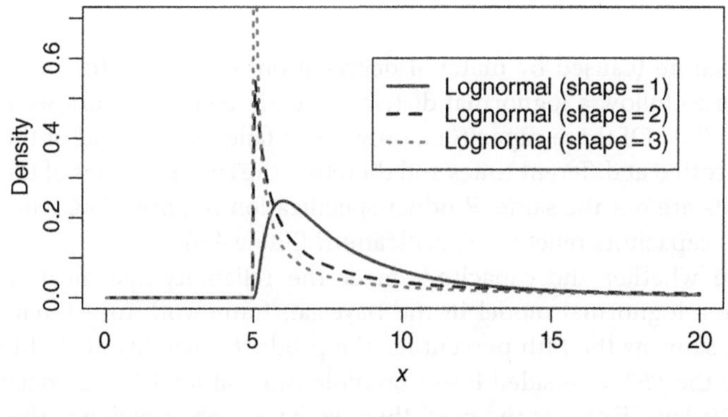

Figure 4.13 Probability density functions of three lognormal distributions with shape = 1, 2, 3 and scale = 1.

Table 4.5 Time to failure data for 16 capacitors.

Time (days)	Status	Time (days)	Status
20.5	Susp	50.5	Susp
30.0	Susp	80.0	Fail
28.0	Fail	55.0	Susp
35.0	Fail	100.0	Fail
32.1	Susp	60.0	Susp
54.0	Fail	67.0	Susp
70.0	Fail	90.0	Susp
49.0	Susp	140.0	Fail

Note: "Susp" denotes the unfailed units.

```
(4.3.5_Lognormal_Distributions.R)
## PDF of lognormal distributions

# 2-parameter lognormal distribution
# dlnorm(x, meanlog = 0, sdlog = 1, log = FALSE)
# 3-parameter lognormal distribution
# dlnorm3(x,shape=1,scale=1,thres=0,log=FALSE)
install.packages("FAdist")
library(FAdist)
x <- seq(0,20,length=500)
lnorm3_1 <- dlnorm3(x, shape=1,scale=1,thres=5,log=FALSE)
lnorm3_2 <- dlnorm3(x, shape=2,scale=1,thres=5,log=FALSE)
lnorm3_3 <- dlnorm3(x, shape=3,scale=1,thres=5,log=FALSE)
plot(x,lnorm3_1,type="l",ylab="Density",xlab="x",lty=1,lwd=2,col="red",
ylim=c(0,0.7))
lines(x,lnorm3_2,lty=2,lwd=2,col="blue")
lines(x,lnorm3_3,lty=3,lwd=2,col="purple")
legend(12,0.6,c("Lognormal (shape=1)","Lognormal (shape=2)","Lognormal
(shape=3)"),lty=c(1,2,3),lwd=c(2,2,2),col=c("red","blue","purple")) # add
legend
```

Example 4.6 Time to failure (caused by material degradation) of a capacitor in an accelerated test is known to follow a lognormal distribution. Sixteen capacitors were tested to determine reliability. Of these capacitors, only seven failed by the end of the test. The capacitors were tested at different times and therefore the running times of the unfailed (suspended) units are not the same. Product specification requires 95% reliability at 10 days. Do these capacitors meet the specification? (Table 4.5).

In order to determine whether the capacitors meet the reliability specification requirement, we will use a lognormal model in the Bayesian framework to estimate B5 life. The B5 life is the same as the fifth percentile. The product reliability at B5 life is 95%. We will compute the 95% one-sided lower credible interval for B5 and check if it is still greater than 10 days. If this is the case, then we have high confidence that the product is meeting the specification requirement. For Bayesian simulations we will assume vague priors on the mean μ and standard deviation σ of the lognormal distribution. The JAGS model and R scripts are shown in **4.3.5_Lognormal.JAGS** and **4.3.5_ Lognormal.R**, respectively.

```
(4.3.5_Lognormal.JAGS)
model {
for( i in 1:n) {
#Handling the censored data
Censor[i] ~ dinterval(ttf[i], CenLimit[i])
# lognormal likelihood using mean mu and precision tau (=1/sigma^2) in
the log-scale,
ttf[i] ~ dlnorm(mu,tau)
}

mu ~ dnorm(0,0.000001)  # mean=1, std=1000
tau ~ dgamma(0.1,0.1) # vague gamma (traditional): mean=1 & vari-
ance=10
sigma <- 1/sqrt(tau)
}
```

```
(4.3.5_Lognormal.R)
# Bayesian lognormal model for analyzing right censored data
### Load package rjags (to connect to JAGS from R for Bayesian analy-
sis)
library(rjags)

# time to failure data (NA indicates right censored data)
ttf <- c(NA, NA,28.0,35.0,NA,54.0,70.0,NA,NA,80.0,NA,100.0,NA,NA,
NA,140.0)
# Censoring limit
CenLimit <- c(20.5,30.0,28.0,35.0,32.1,54.0,70.0,49.0,50.5,80.0,55.0,
100.0,60.0,67.0,90.0,140.0)
# Censor: 0 means uncensored; 1 means right censored
Censor <- c(1,1,0,0,1,0,0,1,1,0,1,0,1,1,1,0)
Censor_TF <- as.logical(Censor)

# Initial values of censored data:
ttfInit1 <- rep(NA,length(ttf))
ttfInit2 <- rep(NA,length(ttf))
ttfInit1[Censor_TF] = CenLimit[Censor_TF]+1
ttfInit2[Censor_TF] = CenLimit[Censor_TF]+2.5

# Initial values. Note: To be able to reproduce the results we fix
the values of .RNG.name and .RNG.seed
Initial1 <- list(.RNG.name="base::Mersenne-Twister", .RNG.seed=1237,
mu = -1, tau = 1, ttf=ttfInit1)
Initial2 <- list(.RNG.name="base::Mersenne-Twister", .RNG.seed=2645,
mu = -2, tau = 1.5, ttf=ttfInit2)
InitialValues <- list(Initial1, Initial2)

# Data used in Bayesian analysis
BayesianData <- list(ttf = ttf, CenLimit=CenLimit, Censor=Censor,
n=length(ttf))

# Create a JAGS model object
ModelObject <- jags.model(file = "4.3.5_Lognormal.JAGS",
                          data= BayesianData, inits= InitialValues,
                          n.chains = 2, n.adapt = 1000
)
```

```
    # Burn-in stage
    update(ModelObject, n.iter=10000)

    # Run MCMC and collect posterior samples in coda format for selected
variables
    codaList <- coda.samples(model= ModelObject, variable.names = c("mu",
"sigma"),
                             n.iter = 100000, thin = 1)

    # Summary statistics for Markov Chain Monte Carlo chains
    # Quantiles of the sample distribution can be modified in the
quantiles argument
    summary(codaList, quantiles = c(0.025, 0.05, 0.5, 0.95, 0.975))
```

When the above code is run we get the following results:

```
    > summary(codaList, quantiles = c(0.025, 0.05, 0.5, 0.95, 0.975))

    Iterations = 11001:111000
    Thinning interval = 1
    Number of chains = 2
    Sample size per chain = 1e+05

    1. Empirical mean and standard deviation for each variable,
       plus standard error of the mean:

            Mean     SD  Naive SE Time-series SE
    mu    4.4706 0.2498 0.0005586       0.001539
    sigma 0.6573 0.2213 0.0004949       0.001370

    2. Quantiles for each variable:

            2.5%     5%    50%   95% 97.5%
    mu    4.0778 4.1351 4.4399 4.907 5.046
    sigma 0.3822 0.4075 0.6111 1.057 1.202
```

The summary results show that on the log-scale the mean "mu" and the standard deviation "sigma" have mean values of 4.47 and 0.6573, respectively. The estimated lower bounds of the 95% credible interval for these parameters are 4.14 and 0.4075, respectively. The B5-life (fifth percentile) of the time to distribution is computed as follows:

```
    sims <- as.matrix(codaList)
    mu <- sims[,"mu"]
    sigma <- sims[,"sigma"]
    ## Estimating B5 (5th percentile) and one-sided lower 95%
Credible Interval
    # Mean value on the data-scale
    B5_Life <- qlnorm(p=rep(0.05,length(mu)), meanlog = mu, sdlog=sigma)
    #mean of B5-life
    print(paste("The point estimate of B5-life (95% reliability) is",
round(mean(B5_Life),2)),quote=FALSE)
    #Compute one-sided lower bound of 95% Credible Interval
    B595Low <- quantile(B5_Life, prob=0.05)
    print(paste("One-sided 95% lower bound of B5-Life (95%
reliability) is", round(B595Low,2)),quote=FALSE)
```

When the above code is run, we get the following results:

```
[1] The point estimate of B5-life (95% reliability) is 31
[1] One-sided 95% lower bound of B5-Life (95% reliability) is 16.69
```

On average, the capacitors have 95% reliability at 31 days. Also, the results show that there is a 95% chance that 95% or more reliability can be achieved at 16.69 days. We can therefore conclude that the specification requirement of 95% reliability at 10 days is achieved.

4.4 Model and Convergence Diagnostics

If there are no historical data suggesting which probability model to apply for the failure mechanism being considered, then consider several competing models and use the deviance information criterion (DIC) (described in Chapter 3) to choose the most appropriate model. The smaller the DIC value the better the model fits the data. The other model selection criteria that can be used for model selection include Bayesian information criterion and Akaike information criterion, as discussed in the previous chapter. We discussed several parameter convergence diagnostic methods that included graphical and analytical methods. The graphical methods include trace plots, density plots, and autocorrelation plots. The analytical methods include the Gelman, Geweke, and Raftery methods.

References

Abernethy, R.B. (1998). *The New Weibull Handbook*. North Palm Beach, 3e. R.B. Abernethy FL 33408–4328.

Gelman, A. and Rubin, D. (1992). Inference from iterative simulation using multiple sequences. *Statistical Science* 7: 457–511.

Geweke, J. (1992). Evaluating the accuracy of sampling-based approaches to calculating posterior moment. In: *Bayesian Statistics*, vol. 4 (ed. J. Bernado, J. Berger, A. Davis and A. Smith), 169–194. Oxford: Claredon Press.

Ntzoufras, I. (2009). *Bayesian Modeling Using WinBugs*. Hoboken, NJ: Wiley.

Plummer, M. (2015) *JAGS Version 4.0.0 user manual.*

Raftery, A. and Lewis, S. (1992). How many iterations in the Gibbs sampler? In: *Bayesian Statistics, vol. 4* (ed. J. Bernado, J. Berger, A. Dawid and A. Smith), 763–774. Oxford: Claredon Press.

5

Reliability Demonstration Testing

There are two types of reliability demonstration tests: substantiation tests and reliability tests. The purpose of a substantiation test is to demonstrate that a redesigned part or component has either eliminated or significantly improved a known failure mode. In this case, the distribution of the existing failure mode is assumed to be known (e.g. the Weibull scale parameter η and the shape parameter β are known). This is a classical statistical concept where we assume that the parameters of a distribution are fixed. A frequentist test can be developed with a specified number of test units or specified test duration to show that the new design is significantly better than the old design.

The purpose of a reliability test is to demonstrate that a certain reliability objective has been achieved. A reliability objective may be stated as a desired reliability goal (e.g. 95%, 99%, etc.) at a specified test duration (e.g. 5 days or 150 000 cycles, etc.) or mean time to failure (MTTF). Again, in this case a frequentist test can be developed with a fixed number of test units or fixed test duration to demonstrate that the specified reliability objective is met. In reliability or substantiation testing involving a fixed number of test units, how long each unit needs to be tested must be determined. A test plan specifies how many units to test and for how long. It also specifies a stopping rule which can be stated in terms of number of failures or the amount of time to accumulate on each test unit. The test plan also specifies a success criterion. The test is passed if the success criterion is met. As an example, a zero-failure test plan allows no failures in all units tested for the duration specified in the plan. If one or more units fail, the test is failed. When developing a statistically designed test plan, a high level of confidence (typically 90% or 95%) needs to be specified. This means that if the failure mode in question is not fixed or significantly improved or the specified reliability goal has not been met the test will fail with the specified level of confidence.

These classical concepts can be extended to the Bayesian thinking process by treating distributional parameters as random variables and thereby assuming prior distributions. Prior distributions of the parameters can be obtained using available historical data, literature search, or expert opinion. One can also consider noninformative or vague priors on the parameters. Sample sizes for substantiation or reliability (with specific goal) test plans can be developed using the posterior distribution of the reliability for a given level of confidence (probability). We will present several examples in this chapter to discuss both classical and Bayesian solutions for estimating sample sizes for substantiation and reliability testing.

Practical Applications of Bayesian Reliability, First Edition. Yan Liu and Athula I. Abeyratne.

Companion website: www.wiley.com/go/bayesian20

5.1 Classical Zero-failure Test Plans for Substantiation Testing

In new product design and development, it may be very important to predetermine how many units should be tested and for how long to demonstrate that the current design is substantially more reliable than the predicate design. If the number of available units is fixed, then how long each unit needs to be tested must be determined. On the other hand, if the test duration needs to be constrained, then how many units should be tested must be determined. The failure mechanism of the current design may be similar to a previous design, but the reliability has been significantly improved. If there is literature or knowledge about the underlying distribution of the failure mechanism of the previous design, the number of units that needs to be tested could be reduced. As an example, suppose the failure mechanism is such that the time to failure follows a Weibull distribution with a known shape parameter. In certain industries, zero failure test plans are considered common practice.

Example 5.1 Suppose that the reliability of a certain mechanical component of an automobile is tested using an accelerated bench test. The current design of the component has a time to failure distribution that can be modeled by a Weibull distribution with scale parameter $\eta = 60$ hours and shape parameter $\beta = 3$. A redesigned component is expected to significantly improve the component's life. If we want to demonstrate that this is the case with high confidence (90%), how many redesigned components should be tested on the bench in a zero-failure test plan? Assume that we want to test each component to 30 hours on the bench.

In this example, it is important to note that the failure mechanism of the current design is well understood so that the time to failure is given by a Weibull distribution with known scale and shape parameters. Further, a redesigned component is expected to have the same shape parameter but improved characteristic life. There is a closed-form solution to this problem which does not require Bayesian analysis under the assumptions that the shape and scale parameters of the current design are constants. These are very strict assumptions. The Bayesian approach does not make these assumptions and would provide a more realistic solution.

In the frequentist solution, under the null hypothesis we assume that the new and current designs are the same. If the null hypothesis is true, then the test plan we develop should fail the test with high confidence (90%). Let n be the number of components needed to achieve the specified level of confidence. The test plan requires that each component be tested to 30 hours or failure, whichever comes first. If we see 1 or more failures, the overall test will be failed. Under the null hypothesis, the probability of test success is given by $[R(t_0)]^n$, where $t_0 = 30$ hours and $R(t_0)$ is the reliability of the current design at t_0. Since the current design has a time to failure distribution represented by a Weibull probability model with $\eta = 60$ hours and $\beta = 3$, we have

$$R(t_0) = e^{-(t_0/\eta)^\beta}.$$

The probability that there are zero failures in a sample of n units tested to time t_0 each is given by

$$[R(t_0)]^n = [e^{-(t_0/\eta)^\beta}]^n.$$

Therefore, the probability that the test fails (1 or more units fail) under the null hypothesis (new design is the same as old design, no improvement to the life of the component) is given by

$$1 - [R(t_0)]^n = 1 - [e^{-(t_0/\eta)\beta}]^n.$$

Since we want the test to fail under the null hypothesis (new design is the same or worse than current design), set this failure probability equal to the level of confidence. Thus, we have

$$\begin{aligned}
1 - [e^{-(t_0/\eta)\beta}]^n &= 0.90 \\
[e^{-(t_0/\eta)\beta}]^n &= 1 - 0.90 \\
-n(t_0/\eta)^\beta &= \log(0.1) \\
n &= -(\eta/t_0)^\beta \log(0.1) \\
n &= (60/30)^3 \log(0.1) = 18.42.
\end{aligned}$$

So, the required sample size for the test is 19 (round up) components.

The above calculations can be generalized as follows. Suppose the underlying failure mechanism is represented by a Weibull distribution with scale and shape parameters η and β, respectively. Then the sample size n for a zero-failure substantiation test with fixed test duration t_0 is given by

$$n = -(\eta/t_o)^\beta \log(\alpha), \tag{5.1}$$

where $\alpha = (1 - \textit{confidence level})$.

On the other hand, if there are only n units available for the test, then the test duration t_0 for each unit is given by

$$t_0 = \eta\left[-\frac{1}{n}\log(\alpha)\right]^{1/\beta}. \tag{5.2}$$

Zero-failure test plans can be developed for other situations where the underlying failure mechanism tends to produce different time to failure distributions such as normal, lognormal, gamma, etc.

5.2 Classical Zero-failure Test Plans for Reliability Testing

The goal of a reliability test is to demonstrate that a certain reliability objective has been achieved. The reliability objective may be stated as a desired reliability goal γ_0 at a specified time duration t_{γ_0}. Typically, γ_0 is stated as a probability such as 0.90, 0.95, etc. The reliability goal, t_{γ_0}, can be interpreted as the $(100 * (1 - \gamma_0))$th percentile of the failure time distribution. We want to develop a zero-failure test plan such that if the test is passed, then the reliability goal is achieved with high confidence $(1 - \alpha)$. Mathematically, this can be stated as

$$P(T_{\gamma_0} > t_{\gamma_0} \mid \textit{test is passed}) \geq (1 - \alpha),$$

where T_{γ_0} is the time at which the survival probability is γ_0.

A classical zero-failure test plan for reliability testing can be developed in the same way as a substantiation test. For simplicity, suppose the underlying failure mechanism is represented by a Weibull distribution with scale and shape parameters η and β, respectively. The goal of the test is to show that the product reliability at t_{γ_0} is at least γ_0 with confidence $(1-\alpha)$. Suppose each unit in the sample is tested to time t_*, then the sample size n of the zero-failure test plan can be obtained as follows.

First, convert the reliability goal to that of the scale parameter, assuming that the shape parameter stays the same:

$$\gamma_0 = e^{-(t_{\gamma_0}/\eta)^\beta}$$

$$\left(\frac{t_{\gamma_0}}{\eta}\right)^\beta = -\log(\gamma_0)$$

$$\eta = \frac{t_{\gamma_0}}{(-\log(\gamma_0))^{1/\beta}}$$

The required sample size n can now be obtained by substituting the value of η in Eq. (5.1):

$$\begin{aligned} n &= -(\eta/t_*)^\beta \log(\alpha) \\ &= -(t_{\gamma_0}/t_*(-\log(\gamma_0))^{1/\beta})^\beta \times \log(\alpha) \\ &= -\log(\alpha) \times t_{\gamma_0}^\beta/(t_*^\beta(-\log(\gamma_0)), \end{aligned} \tag{5.3}$$

where $\alpha = (1 - \textit{confidence level})$.

On the other hand, if there are only n units available for the test, then the test duration t_* for each unit is given by

$$t_* = \left(\frac{\log(\alpha)}{\log(\gamma_0)}\right)^{1/\beta} \times \frac{t_{\gamma_0}}{n^{1/\beta}}. \tag{5.4}$$

Zero-failure reliability test plans can be developed for other situations where the underlying failure mechanism tends to produce different time to failure distributions such as normal, lognormal, gamma, etc.

Example 5.2 Suppose that in Example 5.1 we want to demonstrate 99% reliability at 40 hours and that we intend to test each unit for 75 hours. How many units should we test to be 90% confident that the desired reliability goal has been achieved?

Recall that the shape parameter β of the Weibull model is 3. The following R code is written to implement Eq. (5.3) to carry out the necessary calculations.

```
(5.2_ZeroFail_RealTestPlan_Classical.R)
## Calculating Sample Size for Zero-fail reliability Test plan using
Weibull Model #

# Desired reliability
Gama0 <- 0.99
# Time at which the reliability is desired
TGam0 <- 40
# Desired test duration
T0 <- 70
# Get the Desired confidence level
Conf = 0.90
```

```
  # Get the shape parameter
  Beta <- 3
  # Compute required sample size, N, for zero-failure substantiation test
as follows
  N <- ceiling(-log(1-Conf)*(TGam0^Beta)/(T0^Beta*(-log(Gama0))))
print(paste("Required sample size for Zero-failure reliability test
plan=",N),quote=FALSE)
```

The following result is obtained by running the above code:

```
[1] Required sample size for Zero-failure reliability test plan= 35
```

5.3 Bayesian Zero-failure Test Plan for Substantiation Testing

In the classical test plan considered in Example 5.1 we assumed that Weibull scale parameter η and shape parameter β of the current design are known. This is a very strong assumption, and rarely do we know the exact values of those parameters. In the Bayesian approach we do not assume that they are constants rather random variables with some probability distributions. This is a much more reasonable assumption than assuming parameter values are known. Usually, we have reliability bench test data for the current design and we could use them in a Bayesian framework to estimate the posterior distributions of Weibull scale and shape parameters. Subsequently, these posterior distributions can be used to estimate the necessary sample size for a zero-failure substantiation test plan for testing the new components. This approach can also be used in situations where time to failure data has other distributions such as normal, log-normal, etc.

Suppose the lifetime data of a system or component follows a Weibull distribution with scale parameter η and shape parameter β. Let t be the failure time, then the probability density function is given by

$$f(t \mid \beta, \eta) = \frac{\beta}{\eta}\left(\frac{t}{\eta}\right)^{\beta-1} e^{-(t/\eta)^{\beta}} \tag{5.5}$$

and the reliability function is given by

$$R(t \mid \beta, \eta) = e^{-(t/\eta)^{\beta}}. \tag{5.6}$$

Suppose $g(\eta, \beta \mid \boldsymbol{t})$ is the posterior distribution of η and β given the data $\boldsymbol{t}$. In a classical substantiation testing situation, it is assumed that the shape parameter β is known and then a test plan is developed to demonstrate that the scale parameter (or characteristic life) η of the new system or component is significantly improved compared to the previous design. These classical test plans are discussed in Meeker and Escobar (1998) and Abernethy (1998). Hamada et al. (2008) in Chapter 10 provide a good discussion of various Bayesain reliability test plans.

We will consider two different Bayesian solutions to this problem: (i) assume β is known and model the variability in η through its posterior distribution obtained from historical data and (ii) assume both β and η are unknown and model their variability through their posterior distributions obtained from historical data. In the next example we will consider both solutions to a given set of data. In the case of the Weibull distribution, we can state this objective in terms of the scale parameter. Suppose η_0 is the mean value of η of the current design. We want to develop a test plan so that if the test is passed then the new design has a scale parameter that is greater than η_0 with high probability, $(1-\alpha)$. Typically, α is selected in the range of $0<\alpha<0.2$. The sample size can be estimated as follows.

Let N be the number of posterior samples obtained from the posterior distribution of the scale parameter η.

$$P(\eta > \eta_0 \mid test\ is\ passed) \geq (1-\alpha). \tag{5.7}$$

The left side of Eq. (5.5) can be rewritten as

$$P(\eta > \eta_0 \mid test\ is\ passed) = \frac{P(test\ is\ passed\ and\ \eta > \eta_0)}{P(test\ is\ passed)}$$

$$= \frac{\int_{\eta_0}^{\infty} e^{-n(t_0/\eta)^\beta} f(\eta)d\eta}{\int_0^{\infty} e^{-n(t_0/\eta)^\beta} f(\eta)d\eta} \approx \frac{\sum_{i=1}^{N} I_{(\eta_i>\eta_0)} e^{-n(t_0/\eta_i)^\beta}}{\sum_{i=1}^{N} e^{-n(t_0/\eta_i)^\beta}} \geq (1-\alpha). \tag{5.8}$$

where $f(\eta)$ is the posterior density of the Weibull parameter η and $I_{(\eta_i>\eta_0)}$ is the indicator function, which is equal to 1 if $\eta_i > \eta_0$ and zero otherwise.

An iterative procedure can be used to obtain the minimum sample size that satisfies inequality (5.8).

Example 5.3 Suppose in Example 5.1 that we know that the Weibull shape parameter β is 3 but the value of the scale parameter is unknown. Assume that historical reliability test data for the current design are available from a design qualification study. These data are provided in Table 5.1. We want to develop a zero-failure test plan to demonstrate that the characteristic life (the scale parameter) η of the new design is greater than 70 with high probability (say, 0.9). We wish to test each component to 45 hours on the bench. How many components of the new design should be tested?

We use design qualification data to obtain the posterior distribution of the scale parameter η and then compute the required sample size by varying n from 1 to 500 and selecting the smallest n that satisfies the inequality 5.8.

The R code **5.3.1_Bayesian_Subs_Test_SS_Known_Beta.R** is written to implement the inequality (5.8) to estimate the sample size. In addition, this code also generates sample size versus test duration data for three confidence levels: 90%, 95%, and 99%. Figure 5.1 shows the graphs of this relationship for three confidence levels. If each unit is tested to 45 hours, then the required sample sizes for the zero-failure substantiation test for confidence levels 90%, 95%, and 99% are 20, 28, and 44, respectively.

(5.3.1_Bayesian_Subs_Test_SS_Known_Beta.R)

```
  ## Bayesian approach for estimating the sample size for a zero-failure
substantiation
  ## test plan for known shape parameter.

  library("rjags")
  # Read the dataset for Example 5.3.1
  df <- read.table(file="5.3.1_SubsTest_data.txt",header=TRUE)
```

Table 5.1 Historical time (hours) to failure data for 30 components of the current design.

Hours to failure	Censor (1, right-censored; 0, uncensored)	Hours to failure	Censor (1, right-censored; 0, uncensored)
25.5	0	46.2	0
80.0	1	25.4	0
42.6	0	71.5	0
67.4	0	80.0	1
68.4	0	80.0	1
38.6	0	80.0	1
50.5	0	76.4	0
70.1	0	66.7	0
72.7	0	55.2	0
30.8	0	80.0	1
63.4	0	23.7	0
69.6	0	70.0	0
74.5	0	77.5	0
76.7	0	80.0	1
43.3	0	53.9	0

```
  # Jags require that right censored times to be labeled as "NA"
  TimeToFail <- ifelse(df$Censored==1,NA,df$T_Fail)
  #summary(TimeToFail)
  MaxTime <- max(df$T_Fail)
  # In jags syntax, a limiting vector of times is required for both cen-
sored and
  # uncensored times. All uncensored times can be limited by the maxi-
mum of the
  # time vector, while censored times are used as they are
  CensLimVec <- ifelse(df$Censored==1,df$T_Fail,MaxTime) #Get the limit
of censored times
  IsCensored <- df$Censored
  N <- length(TimeToFail)

  # Create JAGS model:
  modeltext <- "
  model{
  for(i in 1:N){
  # The censoring part:
  IsCensored[i] ~ dinterval(TimeToFail[i], CensLimVec[i])
  TimeToFail[i] ~ dweib(beta, lambda) # likelihood is Weibull distribu-
tion with beta=3
  }
  beta <- 3 # assume fixed shape parameter
  lambda ~ dgamma(0.02, 0.1) # Cause a vague prior on eta (Weibull scale
parameter)
  eta <- 1/pow(lambda,1/beta)
```

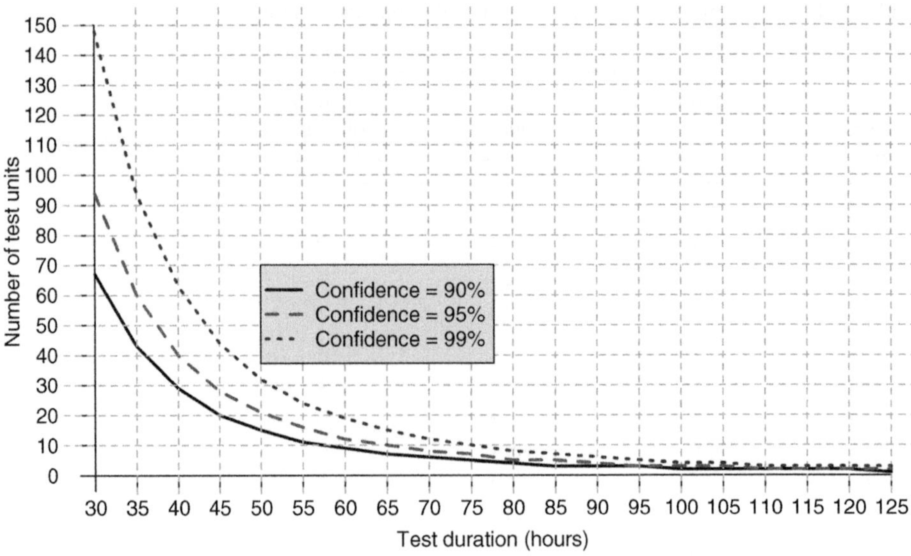

Figure 5.1 Number of test units vs. test duration for the zero-failure Bayesain substantiation test for the Weibull model with known shape parameter.

```
  #data# N, IsCensored, CensLimVec, TimeToFail
  #monitor# eta
  }
  "
  # Write model to a file:
  writeLines(modeltext,con="5.3.1WeibModel.txt")

  # initialize censored values in TimeToFail:
  TimeToFail.init1 <- ifelse(is.na(TimeToFail), 81, NA)
  TimeToFail.init2 <- ifelse(is.na(TimeToFail), 82, NA)

  # In order to preserve the reproducibility of the results we fix the seed
and the random
  # number generators
  list1 <- list(lambda=.1, TimeToFail=TimeToFail.init1,".RNG.name"="base:
:Mersenne-Twister",
              ".RNG.seed" = 231467)
  list2 <- list( lambda=.01, TimeToFail=TimeToFail.init2,".RNG.name"="base:
:Wichmann-Hill",
               ".RNG.seed" = 132984)

  InitList <- list(list1, list2)

  ModelFit <- jags.model(file = "5.3.1WeibModel.txt",
                        data=list(TimeToFail=TimeToFail,
IsCensored=IsCensored,
                        CensLimVec=CensLimVec, N=N), inits=InitList,
                        n.chains = 2, n.adapt = 1000
  )

  # Burn-in stage
  update(ModelFit, n.iter=5000)
```

```
  # Run MCMC and collect posterior samples in coda format for selected
variables
  CodaList <- coda.samples(model=ModelFit, variable.names = c("eta"),
                              n.iter=50000, thin = 1)

  # View trace plot to check for convergence of MCMC chains
  plot(CodaList)

  # View Autocorrelation plots
  autocorr.plot(CodaList,lag.max=50,ask=FALSE)

  # View Gelman and Rubin's convergence diagnostic
  # Approximate convergence is diagnosed when the upper limit is close to 1
  gelman.diag(CodaList)

  # View Summary statistics for Markov Chain Monte Carlo chains
  summary(CodaList, quantiles = c(0.025, 0.05, 0.5, 0.95, 0.975))

  # Create a single data vector by combining the two MCMC chains for the
Weibull scale
  # parameter Eta
  Eta <- c(CodaList[[1]],CodaList[[2]])

  # Get the value of the scale parameter that the new design is expected
to exceed with
  # specified level of confidence
  Eta0 <- 70
  # Get the known shape parameter of the Weibull distribution
  b0 <- 3

  # Create a function to compute the smallest sample size for the
zero-failure
  # substantiation test for a given test duration and confidence
level assuming
  # a fixed shape parameter
  DurVsSS <- function(Conf,tin,Eta,Eta0,b0) {
    Dur <- numeric(20)
    SS <- numeric(20)
    for (k in 1:20) {
      t0 <- (k-1)*5+tin
      for (n in 1:500) {
        ss <- NA
        Num <- sum(exp(-n*(t0/Eta[Eta>Eta0])^b0))
        Den <- sum(exp(-n*(t0/Eta)^b0))
        Value <- Num/Den
        if(Value > Conf) {
          ss <- n
          break
        }
      }
      Dur[k] = t0
      SS[k] = n
    }
    return(list(Dur=Dur,SS=SS))
  }
```

```
  # Get the initial test duration
  tin <- 30
  ###  Case 1:
  # Get the desired confidence level for the zero-failure substantia-
tion test plan
  Conf <- 0.90

  SSConf90 <- DurVsSS(Conf=Conf,tin=tin,Eta=Eta,Eta0=Eta0,b0=b0)

  X1 <- SSConf90$Dur
  Y1 <- SSConf90$SS

  ###  Case 2:
  # Get the desired confidence level for the zero-failure substantia-
tion test plan
  Conf <- 0.95

  SSConf95 <- DurVsSS(Conf=Conf,tin=tin,Eta=Eta,Eta0=Eta0,b0=b0)
  X2 <- SSConf95$Dur
  Y2 <- SSConf95$SS

  ###  Case 3:
  # Get the desired confidence level for the zero-failure substantia-
tion test plan
  Conf <- 0.99

  SSConf99 <- DurVsSS(Conf=Conf,tin=tin,Eta=Eta,Eta0=Eta0,b0=b0)
  X3 <- SSConf99$Dur
  Y3 <- SSConf99$SS

  # Get needed sample sizes for different confidence levels assuming each
  # unit is tested to 45 hours
  Tdur <- 45
  N1 <- Y1[X1==Tdur]
  N2 <- Y2[X2==Tdur]
  N3 <- Y3[X3==Tdur]
  cat("If each component is tested to", t0, "hours, the required Sample ",
      "\nsize for the zero-failure test plan for 90%, 95% and 99% Confi-
dence",
      "\nlevels are = ",N1,", ",N2,", ","and ", N3, " respectively",sep="")

  # Plot separate Sample Size Versus Test duration curves for confidence
  # levels 90%, 95% and 99%

  jpeg("SSVsTestDur_Cruves_BayesSubsTest_KnownBeta.jpeg", width=7, height=5,
       units='in', res = 800)
  plot(x=X1,y=Y1, xlim=c(30, 125), ylim=c(0,150),axes=FALSE,
       xlab="Test Duration (Hours)",ylab="Number of Test Units",
       main="Number of Test Units Vs. Test Duration",
       mgp=c(2.4, 0.8, 0), type="l",lwd=2, lty=1, col="black")
  # Get custom x and y axes
  axis(side=1, at=c(30,35,40,45,50,55,60,65,70,75,80,85,90,95,100,105,110,
115,120,125),
       labels=NULL, pos=0,lty=1, col="black", las=1,cex.axis=0.6)
  axis(side=2, at=c(0,10,20,30,40,50,60,70,80,90,100,110,120,130,140,150),
labels=NULL,
```

```
      pos=30, lty=1, col="black", las=1,cex.axis=0.6)
  # Get custom grid lines
  abline(h=c(10,20,30,40,50,60,70,80,90,100,110,120,130,140,150),lty=2,
col="grey")
  abline(v=c(35,40,45,50,55,60,65,70,75,80,85,90,95,100,105,110,115,120,
125),lty=2,
        col="grey")
  lines(x=X2, y=Y2, type="l", lwd=2, lty=2, col="red")
  lines(x=X3, y=Y3, type="l", lwd=2, lty=3, col="blue")

  legend(x=50, y=70, c("Confidence = 90%", "Confidence = 95%", "Confi-
dence = 99%"),
                        col = c("black","red", "blue"),text.col="black",
        cex=0.8, lty=c(1,2, 3), lwd=c(2,2,2), merge=TRUE, bg="gray90")

  dev.off()
```

When the above code is run the following results and Figure 5.1 are generated:

```
Iterations = 5001:55000
Thinning interval = 1
Number of chains = 2
Sample size per chain = 50000

   1. Empirical mean and standard deviation for each variable,
      plus standard error of the mean:

            Mean           SD      Naive SE Time-series SE
        71.81439      4.98764       0.01577        0.01941

   2. Quantiles for each variable:

    2.5%     5%    50%    95% 97.5%
   63.03 64.25 71.46 80.60 82.55

   If each component is tested to 45 hours, the required Sample
   size for the zero-failure test plan for 90%, 95% and 99% Confidence
   levels are = 20, 28, and 44 respectively
```

If we use the classical method (Eq. (5.1)) to estimate the required sample size for a zero-failure substantiation test with a test duration of 45 hours, $\eta = 70$ and $\beta = 3$, then we get sample sizes of 9, 12, and 18 for the confidence levels 90%, 95%, and 99%, respectively. The classical sample sizes are much smaller than those obtained from Bayesian approach. The main reason for this sample size discrepancy is that the Bayesian method directly incorporates the uncertainty of estimation of η into the calculation of the sample size through its posterior distribution. The classical approach treats η as a constant.

Eq. (5.8) can be easily modified to compute the required sample size for a zero-failure substantiation test plan when both the scale parameter η and the shape parameter β of the Weibull model are unknown. Let N be a posterior sample from the joint posterior distribution of η and β then the modified equation is given by

$$P(\eta > \eta_0 \mid test\ is\ passed, \beta) = \frac{P(test\ is\ passed, \eta > \eta_0, \beta)}{P(test\ is\ passed)}$$

$$= \frac{\int_0^\infty \int_{\eta_0}^\infty e^{-n(t_0/\eta)^\beta} f(\eta,\beta) d\eta\, d\beta}{\int_0^\infty \int_0^\infty e^{-n(t_0/\eta)^\beta} f(\eta,\beta) d\eta\, d\beta}$$

$$\approx \frac{\sum_{i=1}^{N} I_{(\eta_i>\eta_0)} e^{-n(t_0/\eta_i)^{\beta_i}}}{\sum_{i=1}^{N} e^{-n(t_0/\eta_i)^{\beta_i}}} \geq (1-\alpha) \tag{5.9}$$

where $f(\eta, \beta)$ is the posterior joint density of the Weibull parameters η and β, and $I_{(\eta_i>\eta_0)}$ is the indicator function, which is equal to 1 if $\eta_i > \eta_0$ and zero otherwise. Inequality (5.9) can be solved iteratively to obtain the desired sample size n for a specified test duration t_0 and vice versa.

Example 5.4 Suppose in Example 5.1 that we do not know the exact values of the Weibull scale and shape parameters. Assume that historical reliability test data for the current design are available from a design qualification study as given in Table 5.1. We want to develop a zero-failure test plan to demonstrate that the characteristic life (the scale parameter) η of the new design is greater than 70 with high probability (say, 0.9). We wish to test each component to 45 hours on the bench. How many components of the new design should be tested?

The R code **5.3.2_Bayesian_Subs_Test_SS_Unk_Beta.R** is written to implement the inequality (5.9) to estimate the sample size for a zero-failure substantiation test under the assumption that time to failure distribution is Weibull. In addition, this code also generates sample size versus test duration data for three confidence levels: 90%, 95%, and 99%. Figure 5.2 shows the graphs of this relationship for three confidence levels. If each unit is tested to 45 hours, then the required sample size for the

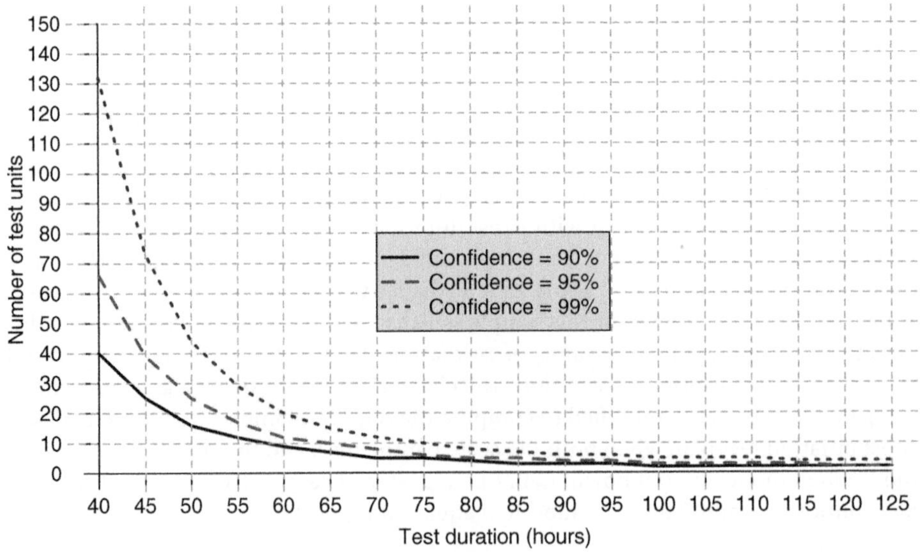

Figure 5.2 Number of test units vs. test duration for the zero-failure Bayesain substantiation test for the Weibull model with unknown shape parameter.

zero-failure substantiation test for confidence levels 90%, 95%, and 99% with unknown β are 25, 39, and 73, respectively.

(5.3.2_Bayesian_Subs_Test_SS_Unk_Beta.R)

```
## Bayesian approach for estimating the sample size for a zero-failure
substantiation
## test plan for unknown shape parameter.

library("rjags")
# Read the dataset from Example 5.3.1
df <- read.table(file="5.3.1_SubsTest_data.txt",header=TRUE)

# Jags require that right censored times to be labeled as "NA"
TimeToFail <- ifelse(df$Censored==1,NA,df$T_Fail)
#summary(TimeToFail)
MaxTime <- max(df$T_Fail)
# In jags syntax, a limiting vector of times is required for both
censored and
# uncensored times. All uncensored times can be limited by the
maximum of the
# time vector, while censored times are used as they are
CensLimVec <- ifelse(df$Censored==1,df$T_Fail,MaxTime) #Get the limit
of censored times
IsCensored <- df$Censored
N <- length(TimeToFail)

# Create JAGS model:
modeltext <- "
model{
for(i in 1:N){
# The censoring part:
IsCensored[i] ~ dinterval(TimeToFail[i], CensLimVec[i])
TimeToFail[i] ~ dweib(beta, lambda) # likelihood is Weibull
distribution with beta=3
}
lambda ~ dgamma(0.02, 0.1) # Cause a vague prior on eta (Weibull scale
parameter)
beta ~ dgamma(2,0.5) # mean = a/b; variance = a/(b^2) #Vague prior on beta
eta <- 1/pow(lambda,1/beta)
#data# N, IsCensored, CensLimVec, TimeToFail
#monitor# beta, eta
}
"
# Write model to a file:
writeLines(modeltext,con="5.3.2WeibModel.txt")

# initialize censored values in TimeToFail:
TimeToFail.init1 <- ifelse(is.na(TimeToFail), 81, NA)
TimeToFail.init2 <- ifelse(is.na(TimeToFail), 82, NA)

# In order to preserve the reproducibility of the results we fix the seed
and the random
# number generators
list1 <- list(lambda=.1, beta=1,TimeToFail=TimeToFail.init1,
              ".RNG.name"="base::Mersenne-Twister",".RNG.seed" = 231467)
list2 <- list(lambda=.01, beta=1.5,TimeToFail=TimeToFail.init2,
```

```
                ".RNG.name"="base::Wichmann-Hill", ".RNG.seed" = 132984)

  InitList <- list(list1, list2)

  ModelFit <- jags.model(file = "5.3.2WeibModel.txt",
                         data=list(TimeToFail=TimeToFail,
IsCensored=IsCensored,
                         CensLimVec=CensLimVec, N=N), inits=InitList,
                         n.chains = 2, n.adapt = 1000
  )

  # Burn-in stage
  update(ModelFit, n.iter=5000)

  # Run MCMC and collect posterior samples in coda format for selected
variables
  # Note: It was found that there is high autocorrelation in mcmc sam-
ple of "beta"
  # and therefore, heavy thinning is required to get an approx. indepen-
dent
  # samples
  CodaList <- coda.samples(model=ModelFit, variable.names = c("eta","beta"),
                           n.iter=1000000, thin = 100)
  #str(CodaList)
  # View trace plot to check for convergence of MCMC chains
  plot(CodaList)

  # View Gelman and Rubin's convergence diagnostic
  # Approximate convergence is diagnosed when the upper limit is close to 1
  gelman.diag(CodaList)

  # Convert mcmc output to matrix format
  MCMC_Out <- as.matrix(CodaList)

  # Create separate vectors containing MCMC samples for eta and beta

  Eta <- MCMC_Out[,"eta"]
  Beta <- MCMC_Out[,"beta"]
  # To get a better view of the autocorrelation plots use the func-
tion "acf"
  acf(Beta, lag.max = 50,type = "correlation", plot = TRUE)
  acf(Eta, lag.max = 50,type = "correlation", plot = TRUE)

  # Use "ess" function in the "mcmcse" package to determine effective sam-
ple size from
  # the posterior samples for shape, scale and reliability parameters.
  library(mcmcse)

  EffSS1 <- ess(Eta, g=NULL)
  print("Effective sample size estimated for Eta:", q=FALSE)
  print(EffSS1)
  print("Effective sample size estimated for Beta:", q=FALSE)
  EffSS2 <- ess(Beta, g=NULL)
  print(EffSS2)

  # View Summary statistics for Markov Chain Monte Carlo chains
```

```
summary(CodaList, quantiles = c(0.025, 0.05, 0.5, 0.95, 0.975))


# Create separate vectors containing MCMC samples for eta and beta

Eta <- MCMC_Out[,"eta"]
Beta <- MCMC_Out[,"beta"]

# Get the value of the scale parameter that the new design is expected to
exceed with
# specified level of confidence
Eta0 <- 70

# Create a function to compute the smallest sample size for the
zero-failure
# substantiation test for a given test duration and confidence level

DurVsSS <- function(Conf,tin,Eta,Beta,Eta0,b0) {
  Dur <- numeric(18)
  SS <- numeric(18)
  for (k in 1:18) {
    t0 <- (k-1)*5+tin
    for (n in 1:500) {
      ss <- NA
      Num <- sum(exp(-n*(t0/Eta[Eta>Eta0])^Beta[Eta>Eta0]))
      Den <- sum(exp(-n*(t0/Eta)^Beta))
      Value <- Num/Den
      if(Value > Conf) {
        ss <- n
        break
      }
    }
    Dur[k] = t0
    SS[k] = n
  }
  return(list(Dur=Dur,SS=SS))
}

# Get the initial test duration
tin <- 40
### Case 1:
# Get the desired confidence level for the zero-failure substantia-
tion test plan
# with 90% confidence
Conf <- 0.90
SSConf90 <- DurVsSS(Conf=Conf,tin=tin,Eta=Eta,Beta=Beta,Eta0=Eta0,b0=b0)

X1 <- SSConf90$Dur
Y1 <- SSConf90$SS

### Case 2:
# Get the desired confidence level for the zero-failure substantia-
tion test plan
# with 95% confidence
Conf <- 0.95
```

```
  SSConf95 <- DurVsSS(Conf=Conf,tin=tin,Eta=Eta,Beta=Beta,Eta0=Eta0,b0=b0)
  X2 <- SSConf95$Dur
  Y2 <- SSConf95$SS

  ###  Case 3:
  # Get the desired confidence level for the zero-failure substantia-
tion test plan
  # with 99% confidence
  Conf <- 0.99

  SSConf99 <- DurVsSS(Conf=Conf,tin=tin,Eta=Eta,Beta=Beta,Eta0=Eta0,b0=b0)
  X3 <- SSConf99$Dur
  Y3 <- SSConf99$SS

  # Get needed sample sizes for different confidence levels assuming each
  # unit is tested to 45 hours
  Tdur <- 45
  N1 <- Y1[X1==Tdur]
  N2 <- Y2[X2==Tdur]
  N3 <- Y3[X3==Tdur]
  cat("If each component is tested to ", Tdur, " hours, the required Sam-
ple size for",
      "\nthe zero-failure test plan with unknown Beta for 90%, 95% and 99%
Confidence",
      "\nlevels are = ",N1,", ",N2,", ","and ", N3, " respectively",sep="")

  # Plot separate Sample Size Versus Test duration curves for confidence
  # levels 90%, 95% and 99%

  jpeg("SSVsTestDur_Cruves_BayesSubsTest_UnKBeta.jpeg", width=7, height=5,
       units='in', res = 800)
  plot(x=X1,y=Y1, xlim=c(40, 125), ylim=c(0,150),axes=FALSE,
       xlab="Test Duration (Hours)",ylab="Number of Test Units",
       main="Number of Test Units Vs. Test Duration",
       mgp=c(2.4, 0.8, 0), type="l",lwd=2, lty=1, col="black")
  # Get custom x and y axes
  axis(side=1, at=c(40,45,50,55,60,65,70,75,80,85,90,95,100,105,110,115,
120,125),
       labels=NULL, pos=0,lty=1, col="black", las=1,cex.axis=0.6)
  axis(side=2, at=c(0,10,20,30,40,50,60,70,80,90,100,110,120,130,140,150),
labels=NULL,
       pos=40, lty=1, col="black", las=1,cex.axis=0.6)
  # Get custom grid lines
  abline(h=c(10,20,30,40,50,60,70,80,90,100,110,120,130,140,150),lty=2,
col="grey")
  abline(v=c(45,50,55,60,65,70,75,80,85,90,95,100,105,110,115,120,125),
lty=2,
         col="grey")
  lines(x=X2, y=Y2, type="l", lwd=2, lty=2, col="red")
  lines(x=X3, y=Y3, type="l", lwd=2, lty=3, col="blue")

  legend(x=70, y=80, c("Confidence = 90%", "Confidence = 95%", "Confi-
dence = 99%"),
                        col = c("black","red", "blue"),text.col="black",
         cex=0.8, lty=c(1,2, 3), lwd=c(2,2,2), merge=TRUE, bg="gray90")

  dev.off()
```

When the above code is run, the following results and Figure 5.2 are generated. As expected, the required sample sizes are much larger. This is because the uncertainties of both β and η are now directly incorporated into the calculation of sample sizes through their posterior distributions.

```
[1] Effective sample size estimated for Eta:
[1] 20092.1
[1] Effective sample size estimated for Beta:
[1] 3471.26

Iterations = 6100:1006000
Thinning interval = 100
Number of chains = 2
Sample size per chain = 10000

1. Empirical mean and standard deviation for each variable,
   plus standard error of the mean:

        Mean     SD Naive SE Time-series SE
beta   3.311 0.5846 0.004134       0.009423
eta   71.840 4.7363 0.033490       0.033491

2. Quantiles for each variable:

        2.5%     5%    50%    95%  97.5%
beta   2.265  2.407   3.28  4.314  4.538
eta   63.190 64.612  71.55 79.974 81.984
If each component is tested to 45 hours, the required Sample size for
the zero-failure test plan with unknown Beta for 90%, 95% and 99% Confi-
dence
levels are = 25, 39, and 73 respectively
```

5.4 Bayesian Zero-failure Test Plan for Reliability Testing

As mentioned in Section 5.2, the goal of a reliability test is to demonstrate that a certain reliability objective has been achieved. Reliability objective may be stated as a desired reliability goal, γ_0, at a specified test duration t_{γ_0}. Typically, γ_0 is stated as a probability such as 0.90, 0.95, etc. The reliability goal t_{γ_0} can be interpreted as the $(100 * (1-\gamma_0))$th percentile of failure time distribution. We want to develop a zero-failure Bayesain test plan such that if the test is passed, then the reliability goal is achieved with high probability, $(1-\alpha)$, which is sometimes called the confidence level. Mathematically, these requirements can be stated as

$$P(T_{\gamma_0} > t_{\gamma_0} \mid \textit{test is passed}) \geq (1-\alpha), \tag{5.10}$$

where T_{γ_0} is the time at which the survival probability is γ_0.

Suppose the underlying failure mechanism is represented by a Weibull distribution with scale and shape parameters of η and β, respectively. Let t_{γ_0}, γ_0, and $(1-\alpha)$ be as defined above. Suppose each unit in the sample is tested to time t_*, then the sample size n of the zero-failure test plan can be obtained as follows.

For a Weibull model we have

$$\gamma_0 = e^{-(T_{\gamma_0}/\eta)^\beta}$$

$$-\log(\gamma_0) = \left(\frac{T_{\gamma_0}}{\eta}\right)^{\beta}$$
$$T_{\gamma_0} = \eta(-\log(\gamma_0))^{1/\beta}.$$

By substituting the value of T_{γ_0} from the above formula into the left side of Eq. (5.10) we get

$$\begin{aligned}
P(T_{\gamma_0} > t_{\gamma_0} \mid \textit{test is passed}) &= P(\eta(-\log(\gamma_0))^{1/\beta} > t_{\gamma_0} \mid \textit{test is passed}) \\
&= P(\eta > t_{\gamma_0}/(-\log(\gamma_0))^{1/\beta} \mid \textit{test is passed}) \\
&= P(\eta > \eta_0^{\beta} \mid \textit{test is passed}) \\
&= \frac{P(\eta > \eta_0^{\beta}, \textit{test is passed})}{P(\textit{test is passed})} \\
&= \frac{\int_0^{\infty}\int_{\eta_0^{\beta}}^{\infty} e^{-n(t_*/\eta)^{\beta}} f(\eta,\beta)d\eta\, d\beta}{\int_0^{\infty}\int_0^{\infty} e^{-n(t_*/\eta)^{\beta}} f(\eta,\beta)d\eta\, d\beta} \\
&\approx \frac{\sum_{i=1}^{N}\sum_{j=1}^{N} e^{-n(t_*/\eta_i)^{\beta_j}} I_{(\eta_i > \eta_0^{\beta_j})}}{\sum_{i=1}^{N}\sum_{j=1}^{N} e^{-n(t_*/\eta_i)^{\beta_j}}} \geq (1-\alpha),
\end{aligned} \tag{5.11}$$

where N is the total number of MCMC samples, $\eta_0^{\beta_i} = t_{\gamma_0}/(-\log(\gamma_0))^{1/\beta_i}$, $f(\eta, \beta)$ is the posterior distribution of the Weibull parameters (η,β), and $I_{(\eta_i > \eta_0^{\beta_i})}$ is an indicator function having value 1 if $\eta_i > \eta_0^{\beta_i}$ and zero otherwise.

Eq. (5.11) can be solved iteratively to obtain the desired sample size n for a zero-failure Weibull reliability test plan for a specified test duration t_* and vice versa.

Note that, unlike a classical approach, in a Bayesian framework sample size cannot be computed for any arbitrary level of reliability. Reliability is based on the posterior distributions of the parameters obtained from the available data. Therefore, reliability expectations must be consistent with the posterior reliability distribution. The necessary sample size to demonstrate a specific reliability goal can be obtained by solving inequality (5.11). The R code that was used for the sample size calculations in Example 5.4 can be modified to do the necessary calculations. For our problem, t_{γ_0}=40, γ_0=0.90, and t_*=75.

5.5 Summary

In this chapter we introduced the concepts of substantiation and reliability testing, and provided methods for developing test plans to meet specified objectives. Zero-failure test plans are the most commonly used in some industries. Classical and Bayesian methods were discussed for developing zero-failure test plans for substantiation and reliability testing. Examples were provided for these sampling plans assuming that the underlying failure time distributions are Weibull.

Sample size estimates obtained from classical methods were smaller compared to those obtained from Bayesain methods, but they do not consider the uncertainty associated with estimation of the model parameters. Bayesian methods naturally account for these uncertainties through the posterior distributions.

References

Abernethy, R.B. (1998). *The New Weibull Handbook*, 3e (ed. R.B. Abernethy), 33408–34328. Abernethy.

Hamada, M.S., Wilson, A., Reese, C.S., and Martz, H. (2008). *Bayesian Reliability*. Springer Science & Media LLC.

Meeker, W.Q. and Escobar, L.A. (1998). *Statistical Methods for Reliability Data*. New York: Wiley.

6

Capability and Design for Reliability

This chapter introduces Bayesian methods for design capability analysis and design for reliability analysis. Design capability analysis and design for reliability analysis have been widely practiced in industry, using methods including Monte Carlo simulations. Typically, traditional methods using Monte Carlo simulations provide a point estimate of reliability. However, these methods can be extended to incorporate the uncertainty associated with the parameter point estimates and to obtain approximate distribution of reliability. In this approach, often we have to rely on the large sample approximations of the distributions of the parameter estimates. In a Bayesian framework, uncertainty of the parameters can also be taken into account and be propagated to the final outcome of reliability estimation. With Bayesian analysis, posterior distributions of the parameter estimates can be easily used to obtain the distribution of any quantity of interest that depends on the parameters. The method we use for estimating the distribution of reliability is called two-level nested Monte Carlo simulations.

6.1 Introduction

Predictive methods are often used to ensure that products meet design requirements during the development phase to avoid or minimize late findings and "firefighting." Design capability analysis and design for reliability analysis are common predictive methods to ensure first-pass success during design verification tests. These methods are usually based on first principle models or empirical models, which are also called transfer functions. Examples of a first principle transfer function include tolerance stack up, stress-strength interference, etc. Examples of an empirical transfer function include equations from linear regression, design of experiments, response surface design, etc. (Maass and McNair 2010). Once transfer functions have been developed, Monte Carlo simulations can be run using the transfer functions to provide reliability estimates.

The objective of design capability analysis is to ensure that there is a high probability of the design meeting critical requirements. To achieve this, the probability of meeting the design requirement(s) is estimated based on transfer functions via Monte Carlo simulations. Transfer functions are usually developed utilizing test data gathered at the development phase. Typically, a Six Sigma level of performance limits the defect rate to no more than 3.4 ppm.

Practical Applications of Bayesian Reliability, First Edition. Yan Liu and Athula I. Abeyratne.

Companion website: www.wiley.com/go/bayesian20

Design for reliability refers to the efforts to minimize the risk of failure due to certain failure modes during development, based on physics of failure models, reliability life testing, and/or accelerated life testing. Monte Carlo simulations can be run with physics of failure models. In these efforts, the main objective is to estimate the probability of success (reliability) under intended use conditions.

In these Monte Carlo analyses, unknown parameters of probability distributions are usually treated as fixed (using the point estimators). Probability distribution models are often used to describe physical variations, e.g. dimension or weight variation. Sample data is used to estimate a population distribution for physical properties (e.g. dimension). Distribution parameters are usually estimated from methods such as maximum likelihood estimation (MLE). With point estimates of distribution parameters, reliability is estimated as a fixed value. We call these analyses single-level Monte Carlo simulations.

In Section 6.3, we use two common scenarios as examples to illustrate single-level Monte Carlo simulations. In Section 6.4, we extend the analyses in these two examples and provide methods and codes to estimate reliability credible intervals based on two-level nested Monte Carlo simulations and Bayesian estimation of the parameters. Different from the analysis in Section 6.3, distribution parameters are re-analyzed using the Bayesian approach. Since in a Bayesian framework the parameters and physical properties are both statistical distributions, two-level nested Monte Carlo simulations vary the physical property at the inner loop and vary the parameters at the outer loop. As a result, estimated population reliability is summarized as a distribution from the sample data instead of a fixed number.

6.2 Monte Caro Simulations with Parameter Point Estimates

Monte Carlo simulation is a method to obtain numerical results by random sampling. Here are the steps to perform Monte Carlo simulations.

1) Build a model (transfer function) between an output y and inputs $x_1, x_2, \ldots, x_n$.
2) Assign a probability distribution to each of the inputs $x_1, x_2, \ldots, x_n$.
3) Run Monte Carlo simulation. At each iteration, random numbers are generated for the inputs based on the assigned probability distributions at step 2. Output y is calculated based on the model defined at step 1 and input values at each iteration.
4) After many iterations, summarize the output distribution of y by computing statistics such as mean, median, percentiles etc.

6.2.1 Stress-strength Interference Example

Stress-strength interference is commonly seen in reliability analysis. If the strength of a product is greater than the stress, the product will survive (a reliable product), otherwise it will fail (an unreliable product). For a group of products, their strength or stress will vary. Typically, the data of one sample is used to represent the stress distribution of a product and the data of another sample is used to represent the strength distribution of the product. Generally, we want to have as many reliable products as possible. In other words, there should be no or minimal interference between the stress distribution and the strength distributions.

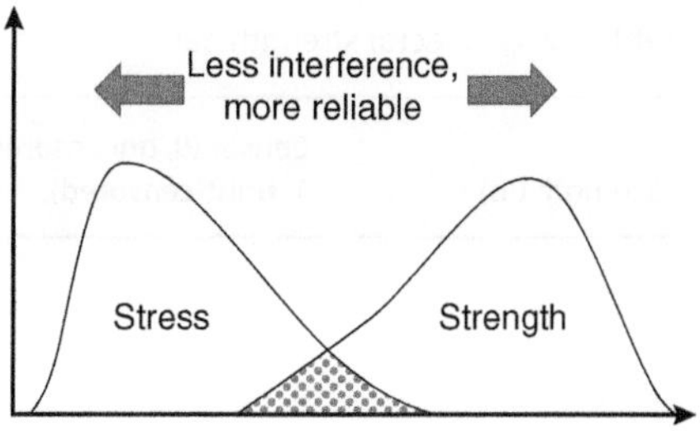

Figure 6.1 Illustration of stress-strength interference.

An illustration of stress-strength interference is shown in Figure 6.1. The dotted area is the overlap between the two distributions. The smaller the overlap/interference, the smaller the probability of failure, or the higher the reliability. The goal of design for reliability is to minimize the amount of overlap between the two distributions. This can be achieved by shifting the stress curve to the left, and/or shifting the strength curve to the right, and/or decreasing the variance of stress and/or strength or both.

To achieve high reliability, first we need to quantify the probability of overlap between the stress and the strength distributions. However, it is often challenging to calculate this overlap probability analytically. To estimate reliability or failure rate, Monte Carlo simulations are usually used instead. In Example 6.1 we use an R script to run Monte Carlo simulations to estimate reliability in the presence of stress-strength interference. Here reliability is defined as the probability of strength being stochastically greater than stress.

Example 6.1 Connector stress and strength sample data are collected from two separate characterization tests. The stress data is a sample of the amount of stress conditions that the connectors experience in the field based on customer usage data (shown in Table 6.1). The strength data is a sample of forces resulting in connector failures in bench tests (shown in Table 6.2). Note that there are right-censored data in the strength measurement (four connectors survived when tested at 15 lbs). The objective is to estimate the reliability (the probability of strength being stochastically larger than stress via Monte Carlo simulations).

First, we analyze the data using MLE. Minitab® version17 to determine the best fitting distribution to the strength and the stress data, separately. From the analysis (Individual Distribution Identification tool in Minitab) version17, it is shown that a lognormal distribution fits the stress data and a Weibull distribution fits the strength data. So, in this case, we use a two-parameter lognormal distribution to fit the stress data and a

Table 6.1 Connector stress data.

2.53	3.06	1.70	4.53	1.94	1.30
2.76	2.16	5.77	4.77	1.81	2.84
1.89	2.20	4.35	1.68	1.53	3.85
3.85	1.90	5.30	1.85	1.60	3.32
3.62	1.96	3.61	2.32	0.47	
3.89	2.09	2.63	2.11	1.06	

Table 6.2 Connector strength data.

Strength (lb)	Censor (0, uncensored; 1, right-censored)	Censoring limit (lb)
7.52	0	
NA	1	15.00
8.44	0	
6.67	0	
11.48	0	
11.09	0	
NA	1	15.00
5.85	0	
13.27	0	
13.09	0	
12.73	0	
11.08	0	
NA	1	15.00
8.41	0	
12.34	0	
8.77	0	
6.47	0	
10.51	0	
7.05	0	
10.90	0	
12.38	0	
7.78	0	
14.61	0	
NA	1	15.00
10.99	0	
11.35	0	
4.72	0	
6.72	0	
11.74	0	
8.45	0	
13.26	0	
13.89	0	
12.83	0	
6.49	0	

NA, not available.

two-parameter Weibull distribution to fit the connector strength data. It is important to note that this approach of finding the best fitting distribution should only be used in situations where there is no prior knowledge of the nature of the distributions of interest. There are many situations where the failure mechanism of a certain phenomenon of interest is well documented in literature and therefore which probability distribution to use is known. In this case one should use a known probability model instead of looking for the best fitting model.

If a random variable $y = \log(x)$ is normally distributed with mean μ and standard deviation σ, then the distribution of x becomes a two-parameter lognormal distribution with location parameter μ and scale parameter σ. Fitting the stress data in Table 6.1 to a two-parameter lognormal distribution using Minitab version17, the MLE method resulted in lognormal location and scale parameter estimates of

- *location*: $\widehat{\mu} = 0.884642$
- *scale*: $\widehat{\sigma} = 0.502301$.

Fitting the strength data in Table 6.2 to a two-parameter Weibull distribution using Minitab version17, the MLE method resulted in Weibull shape and scale parameter estimates of

- *shape*: 3.56588
- *scale*: 12.0002.

With the MLE parameter estimates, we use R script, **6.2.1_StressStrengthMonteCarlo.R** to run Monte Carlo simulations to estimate reliability (probability of strength being greater than stress) in Example 6.1. To help readers understand the R script, we divide the codes to a few sections and explain the function of each section.

(6.2.1_StressStrengthMonteCarlo.R)

```
##############################################################
## Monte Carlo simulation                                   ##
## to estimate probability of strength greater than stress  ##
##############################################################

set.seed(12345)
iter <- 1000000

# Stress: Lognormal (location 0.884642; scale 0.502301)
stress <- rlnorm(iter, meanlog = 0.884642, sdlog = 0.502301)

# Strength: Weibull (shape 28.9703; scale 12.0971)
# strength <- rweibull(iter, shape = 28.9703, scale = 12.0971)
strength <- rweibull(iter, shape = 3.56588, scale = 12.0002)

# overlapping the histograms of stress and strength
# Histogram Grey Color
# Create plot
hist(stress, col=rgb(0.1,0.1,0.1,0.5), main = "Overlapping
Histogram: Stress and Strength", xlab="lb", xlim=range(0,22), breaks=30 )
hist(strength, col=rgb(0.8,0.8,0.8,0.5), add=T)
box()
```

The script above plots the overlapping histogram of stress vs. strength from Monte Carlo simulations. This histogram is shown in Figure 6.2. Note that there is a small overlap of the stress and strength distributions.

```
        # strength - stress
        delta <- strength - stress
        hist(delta, main = "Histogram: Strength-Stress", xlab="Strength-
Stress (lb)", xlim=range(-5,20), breaks=50 )
        box()
```

The script above plots the histogram of strength minus stress. This histogram is shown in Figure 6.3.

```
        # sum(delta>0) counts how many elements in vector delta is >0
        reliability <- sum(delta>0)/iter  # or use: reliability <- mean
(delta>0)
        print(paste("reliability is:", reliability))
```

The script above calculates reliability as the proportion of the difference between strength and stress being positive. The reliability result is shown below.

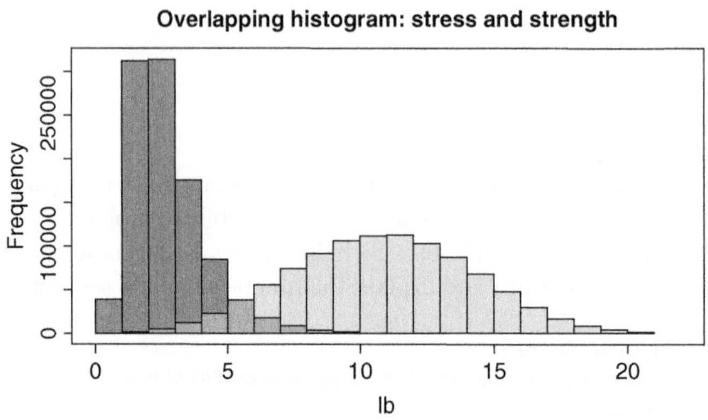

Figure 6.2 Histogram of stress (dark gray) vs. strength (light gray) (lb).

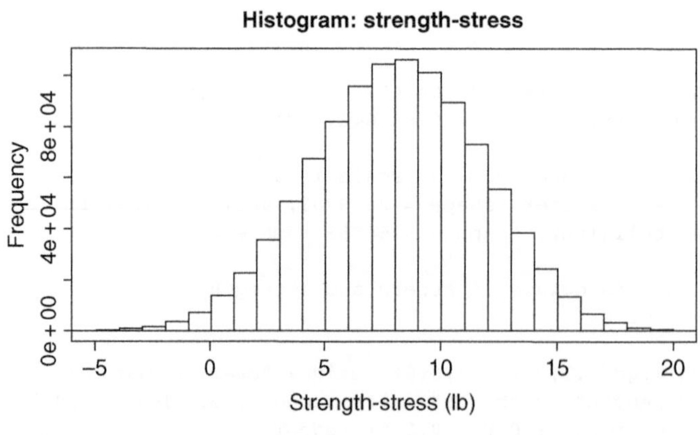

Figure 6.3 Histogram of strength – stress (lb).

```
[1] "reliability is: 0.985189"
```

The result shows that in this case reliability is 0.985 estimated from Monte Carlo simulations (using R version 3.3.0). This value, 0.985, is the reliability point estimate based on single-level Monte Carlo simulations. In Section 6.3.1 we will add a second loop of Monte Carlo simulations to estimate the reliability confidence interval, using Bayesian estimations of the distribution parameters.

6.2.2 Tolerance Stack-up Example

Tolerance stack-up analysis is one common scenario in design capability analysis. For example, in mechanical design, manufacturing processes result in variability of dimensions. When variances of different dimensions add up, under worst-case scenarios the design performance may not meet the design intent. The objective of design capability analysis is to quantify the probability of design failure (not meeting design requirements) due to manufacturing variations.

In Example 6.2, we estimate the probability of meeting a requirement based on a first principle model of tolerance stack-up analysis. This example can be modified to apply to other similar cases. Monte Carlo simulations are used to estimate the probability of failure.

Example 6.2 A connector is designed such that the electrodes on the top parts (solid blocks in Figure 6.4) contact the electrodes on the bottom parts (blocks filled with downward diagonal) to ensure the device functions properly. Figure 6.4a shows the connector design when all the dimensions are nominal. Stack-up tolerance analysis results indicate the potential of no electrical contact at the leftmost electrode in Figure 6.4b. Figure 6.4b shows that since there is variation of the part dimensions, when some dimensions are at extreme cases it is possible that the leftmost electrode on the top has no contact with the corresponding electrode on the bottom, leading to an electrical conductivity problem (no electronic contact).

A major source of dimension is assumed to be due to wear-out of the connector mold. If a mold wears out, under the worst-case condition, a sample of connector dimensions from the mold is shown in Table 6.3. There are three connectors in a device, which are

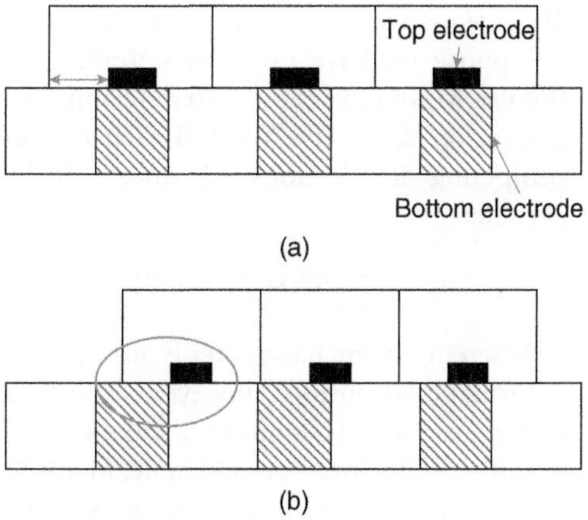

Figure 6.4 Connector design at (a) nominal dimensions and (b) tolerance stack-up scenario.

Table 6.3 Connector dimension measurements (inches) from a wear-out mold.

0.01635	0.01802	0.01826	0.01846	0.01873
0.01708	0.01804	0.01826	0.01846	0.01874
0.01738	0.01805	0.01827	0.01847	0.01877
0.01753	0.01806	0.01827	0.01847	0.01877
0.01766	0.01807	0.01828	0.01848	0.01878
0.01766	0.01807	0.01829	0.01849	0.01879
0.01767	0.01809	0.01830	0.01849	0.01880
0.01771	0.01811	0.01830	0.01851	0.01881
0.01776	0.01812	0.01832	0.01852	0.01882
0.01778	0.01814	0.01832	0.01852	0.01882
0.01781	0.01814	0.01832	0.01854	0.01884
0.01782	0.01816	0.01835	0.01855	0.01885
0.01784	0.01817	0.01835	0.01859	0.01885
0.01785	0.01818	0.01838	0.01861	0.01885
0.01786	0.01818	0.01838	0.01861	0.01887
0.01786	0.01818	0.01841	0.01862	0.01888
0.01790	0.01819	0.01841	0.01863	0.01889
0.01793	0.01820	0.01842	0.01863	0.01891
0.01794	0.01821	0.01842	0.01866	0.01892
0.01796	0.01822	0.01842	0.01868	0.01894
0.01798	0.01823	0.01842	0.01868	0.01894
0.01801	0.01824	0.01844	0.01871	0.01894

independent and come out of the same type of molds. An engineer wants to estimate the probability of proper electrical conductivity at the leftmost connector (defined as reliability in this case), given the presence of mold wear-out. For demonstration purposes, we simplify the problem here and only consider variation of the connector dimension (connector dimension is shown by the double-sided arrow in Figure 6.4a). All other dimensions are assumed to be nominal.

Suppose we have no prior knowledge of the distribution of this dimension data. We use Minitab version17 to determine the best fitting distribution for the connector dimension data. The Weibull distribution is found to be the best fit among many competing distributions. Maximum likelihood estimates of the Weibull distribution parameters are

$$\text{shape} = 54.53, \text{scale} = 0.01851.$$

When the shape parameter is as high as this, the data can very well be represented by a normal distribution. However, since Weibull has a better goodness of fit results, we would use it for subsequent calculations.

Similar to the previous example, first we provide R script for regular one-level Monte Carlo simulations to assess the probability of the leftmost electrode having no proper electrical contacts. The R script is shown in **6.2.2_ToleranceStackupMonteCarlo.R**. Assuming the leftmost connector (we call it connector 3 in the following equation and in the R script) position is calculated as

$$\text{connector 3 position} = \text{connector 1 dimension} + \text{connector 2 dimension} + \text{connector 3 dimension} + 0.0540\ (\text{in.}).$$

If connector 3 position is greater than 0.1105 in., a failure (no electric contact) occurs.

```
(6.2.2_ToleranceStackupMonteCarlo.R)
set.seed(12345)
shape <- 54.53
scale <- 0.01851

iter <- 1000000
connector_3_pos <- rep(0,iter)

for (i in 1:iter) {
  connector_1_dim <- rweibull(1, shape, scale)
  connector_2_dim <- rweibull(1, shape, scale)
  connector_3_dim <- rweibull(1, shape, scale)
  connector_3_pos[i] <- connector_1_dim + connector_2_dim +
connector_3_dim + 0.0540
}

# Histogram of connector 3 position
hist(connector_3_pos, main = " Histogram of connector 3
position", xlab="Inch", xlim=range(0.1,0.115), breaks=30)
box()

# calculate the probability that connector 3 has no
electrical contact
# length() and which() count the number of elements that meet
certain criteria
# in a data set
# Max requirement is 0.1105
connector_3_OOS_prob <- length(which(connector_3_pos>0.1105))/iter

print(paste("probability of failure is:", connector_3_OOS_prob))
```

Monte Carlo simulation results show that the probability of the leftmost electrode having no proper electrical contact is 0.5% (using R version 3.3.0).

```
[1] "probability of failure is: 0.004663"
```

We assume other dimensions are nominal to simplify the calculations for demonstration purposes. If variations of other dimensions are not negligible, the constants in the R script can be replaced by random variables. A histogram of the leftmost connector (connector 3) position is shown in Figure 6.5.

Based on the above results from Monte Carlo simulations, the probability of the leftmost electrode having proper contacts (reliability) is 100 − 0.5% = 99.5%. This value, 0.995, is the reliability point estimate based on single-level Monte Carlo simulations. In Section 6.3.2, we will add a second loop of Monte Carlo simulations to estimate the reliability credible interval, using Bayesian estimations of the distribution parameters.

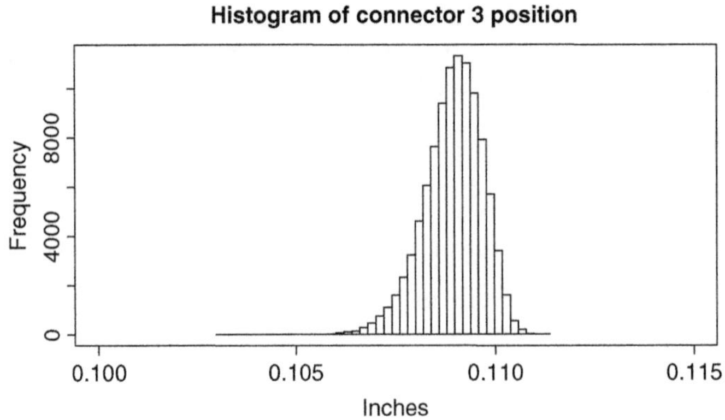

Figure 6.5 Histogram of the leftmost connector (connector 3) position (inches).

6.3 Nested Monte Carlo Simulations with Bayesian Parameter Estimation

As introduced earlier, the Monte Carlo simulations in Section 6.2 only provide reliability point estimates, based on point estimates of distribution parameters (i.e. parameters of the lognormal or Weibull distribution). Since distribution parameters are estimated based on sample data, there is non-negligible uncertainty due to limited sample size. However, the reliability results in Section 6.2 do not include uncertainties due to the estimation of the distribution parameters.

In this section we introduce a method to estimate reliability uncertainty (credible interval) from a Bayesian perspective. In a frequentist/classical framework, it is often difficult or impossible to propagate confidence intervals through complex system models. In a Bayesian framework, posterior distributions are true probability statements about unknown parameters, so they are easily propagated through such models. To demonstrate this in detail, the two examples in Section 6.2 are reanalyzed in this section.

R scripts are provided for two-level nested Monte Carlo simulations to assess the distribution of reliability. Distribution parameters are varied per the Bayesian posterior distributions in the Monte Carlo outer loop, thus uncertainty due to sample size limitation is considered. Based on reliability samples from the Monte Carlo simulations, reliability 95% credible intervals are estimated.

Here is how the two-level nested Monte Carlo simulations work: while the outer loop varies parameter values according to their Bayesian posterior distributions, the inner loop varies physical properties (e.g. dimension, force). Using R script **`6.2.1_Weibull_MLE.R`** as an example, to extend it to two-level nested Monte Carlo simulations, here are the steps:

1) Estimate posterior distributions for μ and σ in the lognormal distribution, and for scale and shape parameters in the Weibull distribution.
2) From the posterior distributions of the parameter estimates in step 1, select a pair of values for μ and σ for the lognormal distribution, and a pair of values of scale and shape for the Weibull distribution.

3) With the selected distribution parameter values in step 2, run Monte Carlo simulations in script **6.2.1_Weibull_MLE.R.** Iterations at this step are the inner loop. Output from this step is a single value of reliability.
4) Repeat step 2 by selecting a different set of distribution parameter values from the posterior distributions.
5) With the selected distribution parameter values in step 4, rerun the Monte Carlo simulations in script **6.2.1_Weibull_MLE.R.** Repeat steps 4 and 5 for a specified number of iterations. These iterations are the outer loop.

Note that the output from an inner loop is a single value of reliability estimate (probability of a response meeting a requirement). Output from the outer loop is a series of estimated reliability values, thus the reliability 95% credible interval can be estimated from these values. The flowchart in Figure 6.6 shows the steps of two-level nested Monte Carlo simulations.

6.3.1 Stress-strength Interference Example

Now let us revisit Example 6.1 using Bayesian methods. The JAGS models and R script are documented in **6.3.1_Lognormal.JAGS**, **4.3.3_Weibull.JAGS** (introduced in Chapter 4), and **6.3.1_StressStrengNestedMC.R**, respectively.

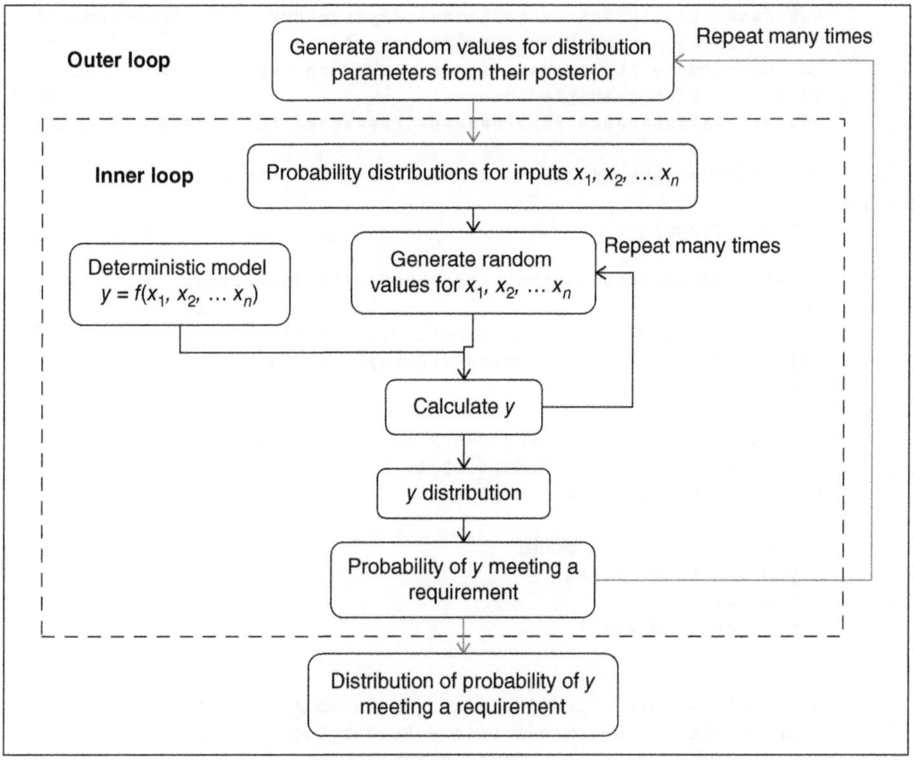

Figure 6.6 Flow chart of nested Monte Carlo simulations.

First, parameters in the lognormal and Weibull distributions are estimated using Bayesian methods with their posterior samples recorded. Then two-level nested Monte Carlo simulations are run to estimate reliability and its 95% credible interval. In this section, we divide the R script into a few sections and explain the function of each section.

```
 (6.3.1_Lognormal.JAGS)
model {

for( i in 1:n) {

# likelihood is Normal distribution of log(y) with mean = mu and
sd = sigma,
# or a Lognormal distribution of y
y[i] ~ dlnorm( mu, 1/sigma^2 )  # mu and sigma are parameters of
log(Y)
}

# non-informative priors
mu ~ dnorm(0,0.000001)  # mean=1, std=1000
sigma ~ dunif(0.01,100)

}

(6.3.1_StressStrengNestedMC.R)
#################################################################
## Two-level nested Monte Carlo analysis                       ##
## to estimate the probability of strength being greater than ##
## stress with credible interval                               ##
#################################################################

### Load package rjags (to connect to JAGS from R for Bayesian
analysis)
library(rjags)

####################################################
## 1st part                                       ##
## Fit stress data to a Lognormal distribution    ##
####################################################

# Data
stress = c(2.53,2.76,1.89,3.85,3.62,3.89,3.06,2.16,2.20,1.90,1.96,
2.09,1.70,5.77,4.35,5.30,3.61,2.63,4.53,4.77,1.68,1.85,2.32,2.11,1.94,1.81,
1.53,1.60,0.47,1.06,1.30,2.84,3.85,3.32)

# Data for the jags model
BayesianData <- list(
  y = stress,
  n = length(stress)
)

# Create (& initialize & adapt) a JAGS model object
ModelObject <- jags.model(file = "6.3.1_Lognormal.JAGS",
                          data= BayesianData, n.chains = 2, n.adapt
= 1000
)

# Burn-in stage
```

```
        update(ModelObject, n.iter=2000)

        # Run MCMC and collect posterior samples in coda format for
selected variables
        codaList_stress <- coda.samples(model= ModelObject,
                                   variable.names = c("mu", "sigma"),
                                   n.iter = 30000, thin = 1)

        codaMatrix_stress <- as.matrix( codaList_stress )

        mu_samples <- codaMatrix_stress[,"mu"]
        sigma_samples <- codaMatrix_stress[,"sigma"]

        # Show summary statistics for collected posterior samples
        summary(codaList_stress, quantiles = c(0.025, 0.05, 0.5, 0.95,
0.975))
```

The R script above provides Bayesian posterior distribution samples of lognormal distribution parameters for the stress data. The summary statistics are shown below. Non-informative priors are used for the unknown parameters μ and σ. Note that the Bayesian point estimates (mu = 0.8846, sigma = 0.5307) are very close to the maximum likelihood estimates (mu = 0.8846, sigma = 0.5023) in Section 6.2.1.

```
     > summary(codaList_stress, quantiles = c(0.025, 0.05, 0.5, 0.95,
0.975))

     Iterations = 3001:33000
     Thinning interval = 1
     Number of chains = 2
     Sample size per chain = 30000

     1. Empirical mean and standard deviation for each variable,
        plus standard error of the mean:

              Mean      SD  Naive SE Time-series SE
     mu     0.8846 0.09175 0.0003746      0.0004784
     sigma  0.5307 0.06915 0.0002823      0.0003965

     2. Quantiles for each variable:

              2.5%     5%    50%    95%  97.5%
     mu     0.7038 0.7347 0.8846 1.0349 1.0651
     sigma  0.4159 0.4308 0.5237 0.6553 0.6863
```

Now we fit the strength data to a Weibull distribution and sample the parameters from their posterior distributions. Vague priors are used for the shape and scale parameters. The R script is shown below.

```
        ####################################################
        ## 2nd part                                       ##
        ## Fit strength data to a Weibull distribution    ##
        ####################################################

        # Strength data (strength to failure; NA indicates right
censored data)
        stf <- c(7.52,NA,8.44,6.67,11.48,11.09,NA,5.85,13.27,13.09,12.73,
11.08,NA,8.41,12.34,8.77,6.47,10.51,7.05,10.90,12.38,7.78,14.61,NA,10.99,
11.35,4.72,6.72,11.74,8.45,13.26,13.89,12.83,6.49)
```

```
        # Censoring limit
        CenLimit <- rep(15, length(stf))
        Censor <- c(0,1,0,0,0,0,1,0,0,0,0,0,1,0,0,0,0,0,0,0,0,0,0,1,0,0,0,
0,0,0,0,0,0,0)

        # Data for the jags model
        BayesianData <- list(ttf = stf,
                             CenLimit = CenLimit, Censor = Censor,
n = length(stf)
        )

        # Initial values of censored data:
        stfInit <- rep(NA,length(stf))
        stfInit[as.logical(Censor)] = CenLimit[as.logical(Censor)]+1

        # Create a JAGS model object
        ModelObject <- jags.model(file = "4.3.3_Weibull.JAGS",
                                  data=BayesianData, inits=list
(ttf = stfInit),
                                  n.chains = 2, n.adapt = 1000
        )

        # Burn-in stage
        update(ModelObject, n.iter=2000)

        # Run MCMC and collect posterior samples in coda format for
selected variables
        codaList_strength <- coda.samples(model=ModelObject,
variable.names = c("shape", "scale"), n.iter = 30000, thin = 1)

        codaMatrix_strength <- as.matrix( codaList_strength )

        shape_samples <- codaMatrix_strength[,"shape"]
        scale_samples <- codaMatrix_strength[,"scale"]

        summary(codaList_strength, quantiles = c(0.025, 0.05, 0.5, 0.95,
0.975))
```

The Weibull distribution parameter posterior distribution summary statistics obtained after running the R script are shown below. The Bayesian point estimates of the parameters (scale = 12.02, shape = 3.278) are close to the maximum likelihood estimates (scale 12.00, shape 3.566) in Section 6.2.1.

```
     > summary(codaList_strength, quantiles = c(0.025, 0.05, 0.5, 0.95,
0.975))

     Iterations = 3001:33000
     Thinning interval = 1
     Number of chains = 2
     Sample size per chain = 30000

     1. Empirical mean and standard deviation for each variable,
        plus standard error of the mean:

              Mean     SD Naive SE Time-series SE
     scale 12.019 0.7107 0.002901       0.004227
```

```
shape  3.278 0.5056 0.002064        0.003484

2. Quantiles for each variable:

        2.5%     5%     50%     95%  97.5%
scale 10.71 10.907 11.989 13.222 13.508
shape  2.35  2.486  3.256  4.148  4.333
```

The following R script generates a histogram of stress vs. strength (lb) raw data with a series of curves based on 50 sets of parameter values sampled from the posterior distributions. The histogram is shown in Figure 6.7. There are 50 pairs of stress and strength curves plotted for demonstration purposes. Different pairs of stress and strength curves result in different stress-strength interference. For each pair of stress and strength curves, a reliability single value is calculated.

```
#################################################################
## Generate a histogram of stress vs. strength with fitted  ##
## curves                                                   ##
#################################################################

# plot stress raw data
# and predicted stress curves based on 50 sets of parameters
hist( stress , xlab="lb" , main = "Stress and Strength", breaks=30,
      col="pink" , border="white" , prob=TRUE , cex.lab=1.5,
xlim=c(0,15))
pltIdx = floor(seq(1,length(mu_samples),length=50))
x = seq( min(stress) , max(stress) , length=501 )
for ( i in pltIdx ) {
  lines( x ,
         dlnorm( x, mu_samples[i], sigma_samples[i] ),
         col="skyblue" )
}

# plot strength raw data
# and predicted strength curves based on 50 sets of parameters
hist( stf , xlab="lb" , breaks=30,
```

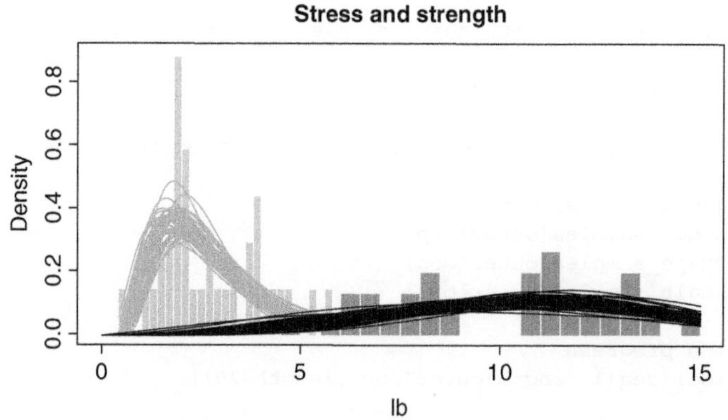

Figure 6.7 Stress (narrow bars on the left side) vs. strength (lb) (wide bars on the right side) raw data (right-censored data not shown), with a series of curves based on 50 sets of parameter values sampled from the posterior distributions.

```
            col="green" , border="white" , prob=TRUE , cex.lab=1.5,
add=T)

        # floor(x): returns the largest integer not greater than x.
        # seq(1,100,length=10): generate 10 numbers that are evenly
distributed
        # between 0 and 100
        Idx = floor(seq(1,length(shape_samples),length=50))

        x = seq( 0 , 15 , length=501 )

        for ( i in Idx ) {
          lines( x ,
                dweibull( x, shape=shape_samples[i], scale=scale_samples
[i] ),
                col="black" )
        }
        box()
```

Note that in this R script, the 50 values of a parameter selected to generate curves in Figure 6.7 are evenly distributed among the 30 000 values in its posterior sample. `seq` is used to generate a sequence of numbers. For example, the following code generates 10 numbers evenly distributed between 0 and 100:

```
> seq(1,100,length=10)
 [1]   1  12  23  34  45  56  67  78  89 100
```

The following R script runs two-level nested Monte Carlo simulations. There are 5000 iterations in the outer loop, and 10 000 iterations in the inner loop. Imagine in Figure 6.7 there are 5000 pairs of stress and strength curves, so that 5000 reliability values are calculated in the following R script.:

```
        ####################################################
        ## 3nd part                                       ##
        ## calculate reliability with nested Monte Carlo  ##
        ####################################################

        # sample 5000
        outerloop <- floor(seq(1,length(shape_samples),length=5000))
        innerloop <- 10000
        reliability <- rep(NA, length(outerloop))

        # get outerloop # of values for each parameter from posterior
sampling
        mu_pre <- mu_samples[outerloop]
        sigma_pre <- sigma_samples[outerloop]
        shape_pre <- shape_samples[outerloop]
        scale_pre <- scale_samples[outerloop]

        # for displaying progress
        progress <- floor(seq(1,length(outerloop),length=20))
        k <- 1

        cat( "running nested loop...\n" )

        for(i in 1:length(outerloop)) {
```

```
          # create stress and strength distributions
          stress_pre <- rlnorm(innerloop, meanlog = mu_pre[i], sdlog =
sigma_pre[i])
          strength_pre <- rweibull(innerloop, shape = shape_pre[i],
scale = scale_pre[i])
          # reliability is the percentage of strength elements >
stress elements
          reliability[i] <- mean((strength_pre>stress_pre)*1)
          # This may take some time. Display progress
          if(i>=progress[k]){
            cat(paste(round(i/length(outerloop)*100), "%", "...\n"))
            k <- k+1
          }
        }

        ## show results of reliability mean, median, and 95% credi-
ble interval
        print(paste("mean of the reliability is:", mean(reliability)))
        print(paste("median of the reliability is:", quantile(reliability,
0.50)))
        print(paste("95% Credible interval for the reliability is:",
quantile(reliability, 0.025), ",", quantile(reliability, 0.975)))

        # Histogram of reliability
        hist(reliability, main = " Histogram of reliability",
xlab="Reliability", breaks=50)
        box()
```

After running the R script above, the simulation results shown below are obtained. Reliability has a mean of 0.975 and a 95% credible interval of (0.937, 0.994). Figure 6.8 shows the histogram of estimated reliability.

```
[1] "mean of the reliability is: 0.9748368"
[1] "median of the reliability is: 0.978"
[1] "95% Credible interval for the reliability is: 0.9371 , 0.9936"
```

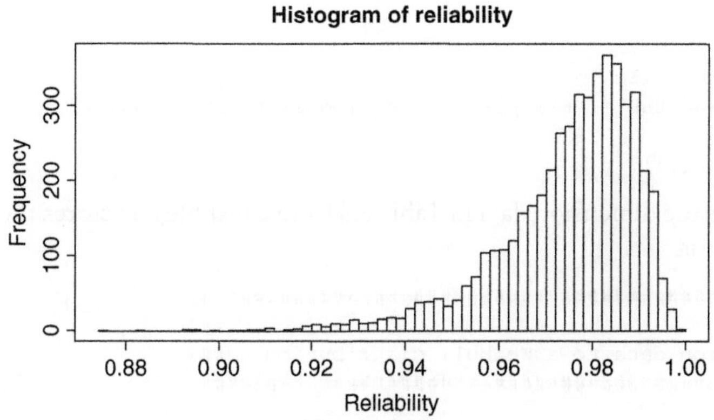

Figure 6.8 Histogram of estimated reliability.

6.3.2 Tolerance Stack-up Example

Now let us revisit Example 6.2 to include the reliability uncertainty information. In this section, we demonstrate design capability analysis with uncertainty estimated from Bayesian perspective. The JAGS model (to fit data to a Weibull distribution) and R script are shown in **`6.3.2_Weibull.JAGS`** and **`6.3.2_StressStrengNestedMC.R`**, respectively. First, parameters of the Weibull distribution are estimated using the Bayesian sampling method. Then two-level nested Monte Carlo simulations are run to estimate reliability and its 95% credible interval. In this section, we divide the R script into few sections and explain the function of each section.

```
(6.3.2_Weibull.JAGS)
  model {
  # Likelihood:
  lambda <- 1/pow(scale, shape)

  for( i in 1:n) {
  dim[i] ~ dweib(shape, lambda)
  }

  # shape MLE:54.5315 (standard error = 4.08639)
  # scale MLE: 0.0185069 (standard error = 0.0000341)
  # set priors to cover MLE +/- 6 standard error:
  # shape: [30.0, 79.0]; scale: [0.0183, 0.0187]
  shape ~ dunif(30.0,79.0)
  scale ~ dunif(0.0183,0.0187)

  }
```

```
(6.3.2_ToleranceStackupNestedMC.R)
  ###################################################################
  ## Nested Monte Carlo analysis                                   ##
  ## to estimate probability of proper electrical conductivity ##
  ## when there is tolerance stack-up                              ##
  ###################################################################

  ### Load package rjags (to connect to JAGS from R for Bayesian
analysis)
  library(rjags)

  # read data from file
  dim.data <- read.table("Example6.2Data_Dimension.txt",header=F,
sep="")
  dim <- dim.data[,1]
```

The R code above reads the dimension data in Table 6.3 from a text file and stores the data in a vector named `dim`.

```
  ####################################################
  ## 1st part                                       ##
  ## Fit dimension data to a Weibull distribution   ##
  ####################################################

  # Data for the jags model
  BayesianData <- list(dim = dim,
```

```
                    n = length(dim)
        )

        # Create, initialize and adapt a JAGS model object
        ModelObject <- jags.model(file = "6.3.2_Weibull.JAGS",
                                  data= BayesianData,
                                  n.chains = 2, n.adapt = 1000
        )

        # Burn-in stage
        update(ModelObject, n.iter=2000)

        # Run MCMC and collect posterior samples in coda format for
selected variables
        codaList_dim <- coda.samples(model= ModelObject, variable.names
= c("shape", "scale"), n.iter = 30000, thin = 1)

        codaMatrix_dim <- as.matrix( codaList_dim )

        shape_samples <- codaMatrix_dim[,"shape"]
        scale_samples <- codaMatrix_dim[,"scale"]

        summary(codaList_dim, quantiles = c(0.025, 0.05, 0.5, 0.95, 0.975))
```

The R script above provides Bayesian posterior distributions of Weibull distribution parameters for the dimension data. The summary statistics are shown below. The Bayesian point estimates of the parameters are close to the maximum likelihood estimates in Section 6.2.2.

```
> summary(codaList_dim, quantiles = c(0.025, 0.05, 0.5, 0.95, 0.975))

  Iterations = 3001:33000
  Thinning interval = 1
  Number of chains = 2
  Sample size per chain = 30000

  1. Empirical mean and standard deviation for each variable,
     plus standard error of the mean:

             Mean        SD  Naive SE Time-series SE
  scale  0.01851 3.445e-05 1.406e-07      1.962e-07
  shape 54.46500 4.096e+00 1.672e-02      2.345e-02

  2. Quantiles for each variable:

             2.5%        5%       50%       95%     97.5%
  scale  0.01844   0.01845   0.01851   0.01856   0.01858
shape 46.71530 47.92621 54.35589 61.35930 62.76763
```

The following R script generates a histogram of dimension data with a series of curves based on 50 sets of parameter values sampled from the posterior distributions. The plot is shown in Figure 6.9. There are 50 curves plotted for demonstration purposes. For a given curve, a single value of reliability is calculated. Different curves result in different reliability estimations.

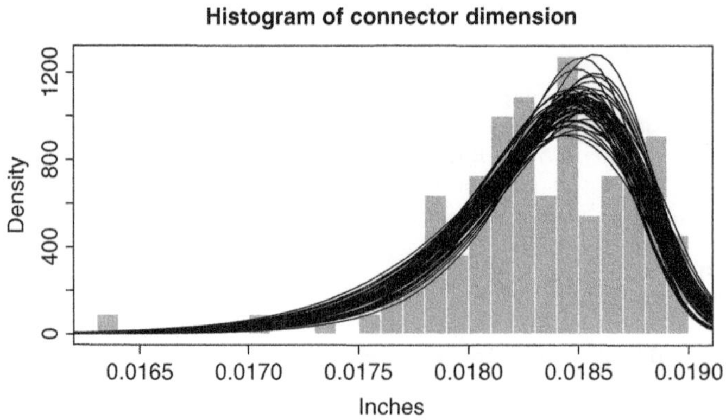

Figure 6.9 Histogram of the connector dimension with predicted curves based on 50 sets of parameters.

```
##########################################################
## Generate a histogram of dimension with fitted curves  ##
##########################################################

# plot stress raw data
# and predicted stress curves based on 50 sets of parameters
hist( dim , main = "Histogram of connector dimension", xlab="Inch"
, breaks=30,
      col="gray" , border="white" , prob=TRUE , cex.lab=1.5)
Idx = floor(seq(1,length(shape_samples),length=50))
x = seq( 0 , 0.02 , length=501 )
for ( i in Idx ) {
  lines( x ,
         dweibull( x, shape=shape_samples[i], scale=scale_samples
[i] ),
         col="black" )
}
box()
```

The following R script runs two-level nested Monte Carlo simulations. There are 5000 iterations in the outer loop, and 10 000 iterations in the inner loop. Imagine in Figure 6.9 there are 5000 curves, so that the 5000 reliability values are calculated in the following R script. Warning: it takes about 10 minutes to run this portion of R script in this case, so now it is a good time to take a break while having your computer run the code!

```
####################################################
## 2nd part                                       ##
## calculate reliability with nested Monte Carlo  ##
####################################################

outerloop <- 5000
innerloop <- 10000
scale <- rep(0,outerloop)
shape <- rep(0,outerloop)
connector_3_OOS_prob <- rep(0,outerloop)
```

```
        parameter_Idx = floor(seq(1,length(shape_samples),
length=outerloop))

        for (j in 1:outerloop) {

          scale[j] <- scale_samples[parameter_Idx[j]]
          shape[j] <- shape_samples[parameter_Idx[j]]
          connector_3_pos <- rep(0,innerloop)

          for (i in 1:innerloop){
            connector_1_dim <- rweibull(1, shape[j], scale[j])
            connector_2_dim <- rweibull(1, shape[j], scale[j])
            connector_3_dim <- rweibull(1, shape[j], scale[j])
            connector_3_pos[i] <- connector_1_dim + connector_2_dim +
connector_3_dim + 0.0540
          }

          connector_3_OOS_prob[j] <- length(which(connector_3_pos>0.1105))/
innerloop
        }

        connector_3_reliability <- 1 - connector_3_OOS_prob

        # Histogram of connector 3 position
        hist(connector_3_reliability, main = "Histogram of connector 3
reliability", xlab="Probability", breaks=30)
        box()

        summary(connector_3_reliability)
        sd(connector_3_reliability)
        quantile(connector_3_reliability, c(.025, .975))
        hist(connector_3_reliability)
```

After running nested Monte Carlo simulations using the R codes above, the estimated leftmost connector (connector 3) reliability mean, median, standard deviation, and 95% credible intervals obtained are as summarized below.

```
> summary(connector_3_reliability)
     Min. 1st Qu.  Median    Mean 3rd Qu.    Max.
   0.9417  0.9927  0.9955  0.9944  0.9973  1.0000

> sd(connector_3_reliability)
   [1] 0.004123029

> quantile(connector_3_reliability, c(.025, .975))
       2.5%     97.5%
  0.9840975 0.9992000
```

The analysis concludes that the probability of proper electrical contact resulting from the mold wear-out condition has a mean of 99.4% and the 95% credible interval is (98.4%, 99.9%). This prediction information, along with other manufacturing controls, can be used to guide manufacturing engineers to make the right decision and helps boost understanding in product reliability. Figure 6.10 shows the histogram of connector 3 reliability based on nested Monte Carlo simulations.

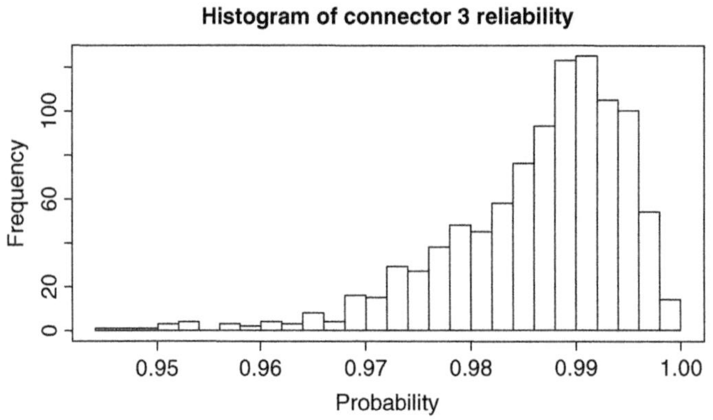

Figure 6.10 Histogram of connector 3 reliability.

6.4 Summary

We provided two examples of nested Monte Carlo simulations with Bayesian parameter estimations. These examples can be extended to any Monte Carlo simulations with a first principle model or an empirical transfer function. This method provides a way to estimate reliability with a 95% credible interval for design for reliability and design capability analysis, while the traditional single-level Monte Carlo simulations only provide the reliability point estimate. Other examples of Bayesian reliability estimation with two-level nested Monte Carlo simulations can be found in Haddad et al. (2014) and Hamada et al. (2010).

References

Haddad, T., Himes, A., and Campbell, M. (2014). Fracture prediction of cardiac lead medical devices using Bayesian networks. *Reliability Engineering and System Safety* 123: 145–157.

Hamada, M.S., Wilson, A.G., Shane Reese, C., and Martz, H.F. (2010). *Bayesian Reliability.* New York: Springer.

Maass, E. and McNair, P.D. (2010). *Applying Design for Six Sigma to Software and Hardware Systems.* Boston, MA: Pearson Education, Inc.

7

System Reliability Bayesian Model

This chapter introduces the basics of the use of Bayesian models for assessing system reliability. There are various methods to assess system reliability or failure rate. The focus of this chapter is to discusses Bayesian methods for system reliability estimation using reliability block diagrams, fault trees, and Bayesian networks.

7.1 Introduction

System reliability is the probability of the system operating without failures for a specified period under specified use conditions. Instead of obtaining data at the system level, which can be expensive or impractical during the development phase, reliability at the component or subsystem levels is often assessed individually. System level reliability is then estimated by aggregating reliability from component level data with the overall uncertainty quantified.

Commonly used methods for system reliability estimation include reliability block diagrams, fault trees, and Bayesian networks. In this chapter these methods are reviewed briefly, followed by the applications and examples in a Bayesian framework.

In a frequentist/classical framework, except in the simplest cases, it is difficult or impossible to propagate classical confidence intervals through complex system models, such as reliability block diagrams, fault trees, and other logic models. In a Bayesian framework, on the other hand, posterior distributions are true probability statements about unknown parameters, so they may be easily propagated through these system reliability models (Hamada et al. 2010). Using traditional methods, engineers can easily get the point estimate of system reliability, but not its confidence interval. Bayesian methods make it feasible and easy for engineers to estimate the confidence intervals of system reliability.

In some engineering practices, there could be no perfect source of information. For example, there could be underreporting or misclassification, or the sample size is too small. To get an accurate reliability estimation from these data sources can be very challenging. In these situations, the Bayesian network provides a flexible modeling method to account for such limitations and to allow various sources of imperfect information to be aggregated, so that the corrected reliability can be estimated.

Practical Applications of Bayesian Reliability, First Edition. Yan Liu and Athula I. Abeyratne.

Companion website: www.wiley.com/go/bayesian20

7.2 Reliability Block Diagram

A reliability block diagram is a method showing how successes or failures of components in a system contribute to system level reliability. A typical reliability block diagram includes several blocks connected in series or in parallel. In this section, we discuss a few commonly used reliability block diagrams, including a series system, a parallel system, and a 2-out-of-3 system, followed by an example to estimate the reliability of a series system from a Bayesian perspective.

A series system consists of n components or subsystems, where all components must work for the system to function successfully. The reliability block diagram for a series system is shown in Figure 7.1a. The reliability of a series system consisting of n independent components or subsystems is

$$R_{\text{system}} = \prod_{i=1}^{n} R_i \tag{7.1}$$

where R_i is the reliability of the ith component under the use conditions for a specified period of time.

A parallel system consists of n components or subsystems, and the system works when at least one component or subsystem works. The reliability block diagram for a parallel system is shown in Figure 7.1b. The reliability of a parallel system consisting of n independent components or subsystems is

$$R_{\text{system}} = 1 - \prod_{i=1}^{n} (1 - R_i) \tag{7.2}$$

A two-out-of-three system consists of three components or subsystems, and the system works when at least two components or subsystems work (Figure 7.2). The reliability of a two-out-of-three system is

$$R_{\text{system}} = \prod_{i=1}^{3} R_i + R_1R_2(1 - R_3) + R_2R_3(1 - R_1) + R_3R_1(1 - R_2) \tag{7.3}$$

To estimate system reliability from a Bayesian perspective, the posterior reliability of each component can be assessed individually from various types of test data. As the

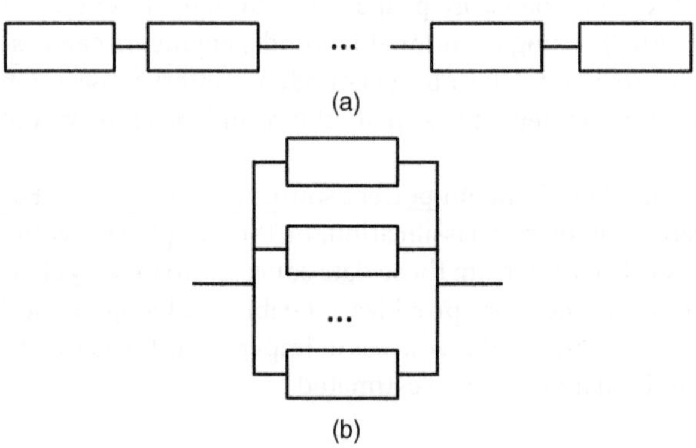

Figure 7.1 Reliability block diagram of (a) a series system and (b) a parallel system.

Figure 7.2 Reliability block diagram of a 2-out-of-3 system.

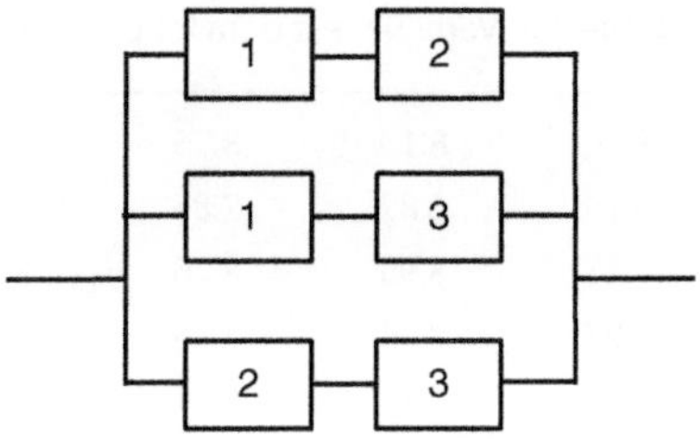

posterior distribution R_i is a random variable, the uncertainty of component level reliabilities can be propagated to the system level via Monte Carlo simulations, to estimate system reliability with uncertainty quantified. The same methodology can be extended to assess subsystem reliability, or to systems with more complex reliability block diagrams (Fenton and Neil 2012).

Bayesian estimation with Monte Carlo simulations provides a way to quantify the system reliability as a distribution. The reliability 95% confidence bounds (Bayesian credible interval) can be used to compare with a prespecified system reliability target. If the 95% lower bound of system reliability is below the target, sensitivity analysis can be performed to see which component has the largest impact on the variability of system reliability. When the source is identified, additional testing can be planned and performed, resulting in additional data that will reduce the variability of the component's reliability.

Hamada et al. (2010) discussed methods to select a prior distribution for the system reliability and used that information to generate prior distributions for the reliabilities of the components or subsystems. In this chapter, for simplicity, we use a beta distribution *Beta* (1, 1) as a non-informative prior of the component's reliability. Monte Carlo simulation can be used to check the prior distribution of the system reliability when the prior distributions of the component level reliability are known.

Example 7.1 A reliability engineer plans to estimate the reliability of a series system for a new product design during the development phase. Design verification testing data are available for the key components/design requirements. The testing results are shown in Tables 7.1 and 7.2, which indicate the performance of various components under

Table 7.1 Data type and test results for eight components in a series system.

Component #	Data type	Sample size	Results
1	Attribute	299	Zero failure
2	Attribute	299	Zero failure
3	Attribute	299	Zero failure
4	Attribute	299	Zero failure
5	Attribute	299	Zero failure
6	Attribute	59	Zero failure
7	Attribute	89	Zero failure
8	Continuous variable	30	See Table 7.2

Table 7.2 Variable test data for component #8.

8.16	8.10	8.75	6.12	9.56
8.19	6.87	7.85	7.24	8.53
7.11	8.95	8.26	8.19	7.94
8.27	8.24	7.33	6.31	8.04
7.36	9.13	6.44	8.79	8.99
7.05	7.60	7.46	6.42	7.92

specified use conditions at a specified lifetime. Note that some of the data are attribute data (pass/fail) and some are continuous variable data. For component #8, reliability is defined as the probability of meeting a specification (upper specification limit: 11). The objective is to get the system reliability point estimate and 95% credible interval.

The R script is shown in **7.2_Series_System.R**. The JAGS model **3.4_Normal.JAGS** introduced in Chapter 3 is reused in this case. In the R script, each component's reliability is estimated individually as a posterior distribution. For component #1 to #7, each reliability prior is represented by a non-informative prior *Beta* (1, 1) (also a uniform distribution), and each reliability posterior is also a beta distribution. For component #8, the data in Table 7.2 are fitted to a normal distribution using vague prior distributions, and then the reliability is computed from posterior samples based on the upper specification limit (the reliability of component #8 is the probability of meeting this specification). After posterior distributions of component reliabilities are estimated via closed-form solutions or Markov chain Monte Carlo (MCMC) sampling, the system reliability is estimated based on the series system structure by Eq. (7.1), via Monte Carlo sampling.

```
(7.2_Series_System.R)
###########################################
## Series system reliability estimation  ##
###########################################

##################################################
## reliability of component 1-7: attribute data ##
##################################################

  # set.seed(123)

  # Assume each prior reliability of component 1-7 has a flat
  distribution Beta(1,1).
  # Reliability posterior is Beta(x+1, n-x+1)

  # Sample 5000 iterations from each reliability posterior
  distribution
  R1 <- rbeta(50000,300,1)
  R2 <- rbeta(50000,300,1)
  R3 <- rbeta(50000,300,1)
  R4 <- rbeta(50000,300,1)
  R5 <- rbeta(50000,300,1)
  R6 <- rbeta(50000,60,1)
  R7 <- rbeta(50000,90,1)
```

```
####################################################################
## Estimate reliability of component #8                           ##
## Fit a Normal distribution using non-informative mean and       ##
## variance                                                       ##
####################################################################

  ### Load package rjags (to connect to JAGS from R for Bayesian
  analysis)
  library(rjags)

  # read data
  y <- c(8.16,8.19,7.11,8.27,7.36,7.05,8.10,6.87,8.95,8.24,9.13,7.60,
  8.75,7.85,8.26,7.33,6.44,7.46,6.12,7.24,8.19,6.31,8.79,6.42,9.56,
  8.53,7.94,8.04,8.99,7.92)

  # Data for Bayesian analysis
  BayesianData <- list(y = y, n=30)

  # Create, initialize, and adapt a JAGS model object
  ModelObject <- jags.model(file = "3.4_Normal.JAGS",
                            data= BayesianData,
                            n.chains = 2, n.adapt = 1000
  )

  # Burn-in stage
  update(ModelObject, n.iter=2000)

  # Run MCMC and collect posterior samples in coda format for selected
  variables
  codaList <- coda.samples(model= ModelObject,
  variable.names = c("mu", "sigma", "tau"), n.iter = 50000, thin = 1)

  codaMatrix <- as.matrix( codaList )

  mu_samples <- codaMatrix[,"mu"]
  sigma_samples <- codaMatrix[,"sigma"]

  # To estimate reliability of component #8
  loop <- floor(seq(1,length(mu_samples),length=50000))
  R8 <- rep(NA, length(loop))

  # get loop # of values for each parameter from posterior sampling
  mu <- mu_samples[loop]
  sigma <- sigma_samples[loop]

  for (i in 1:length(loop)) {
    R8[i] <- pnorm(11, mean = mu[i], sd = sigma[i])
  }

##################################
## Calculate system reliability ##
##################################

R_system <- R1*R2*R3*R4*R5*R6*R7*R8

## show results of reliability mean, median, and 95% credible
interval
```

```
    ## round results to 3 significant digits using signif()
    print(paste("mean of the system reliability is:",
    signif(mean(R_system),3)))
    print(paste("median of the system reliability is:",
signif(quantile(R_system, 0.50),3)))
    print(paste("95% credible interval for the system reliability is:",
signif(quantile(R_system, 0.025),3), ",", signif(quantile(R_system,
0.975),3)))

    # Histogram of reliability
    # jpeg("Example7.1_R_series_system_hist.jpeg", width = 6, height = 4,
units = 'in', res = 1800)  # save the plot as jpeg format
    hist(R_system, main = " Histogram of system reliability",
xlab="Reliability")
    box()
    # dev.off()
```

The results of system reliability estimation are shown below with the histogram of system reliability shown in Figure 7.3. The best point estimate of the system reliability is 0.96, represented by the mean of the distribution. The distribution also provides a probability statement on system reliability. The 95% credible interval for the system reliability is (0.91, 0.98), so we could say that there is a 95% chance that the system reliability is between 0.91 and 0.98 at a specified lifetime given existing test data. After system reliability is estimated, sensitivity analysis can be performed to see which component has the largest impact on the variability of system reliability. The average system reliability is influenced by the weakest link, which is the component with the least average reliability. The variability of the system reliability is influenced by the component level reliability distributions. Since sample size has an impact on the estimation of the component reliability distributions, larger sample size would reduce the variability of the system reliability. An example can be found in Liu et al. (2015).

```
[1] "mean of the system reliability is: 0.956"
[1] "median of the system reliability is: 0.960"
[1] "95% credible interval for the system reliability is: 0.906 , 0.984"
```

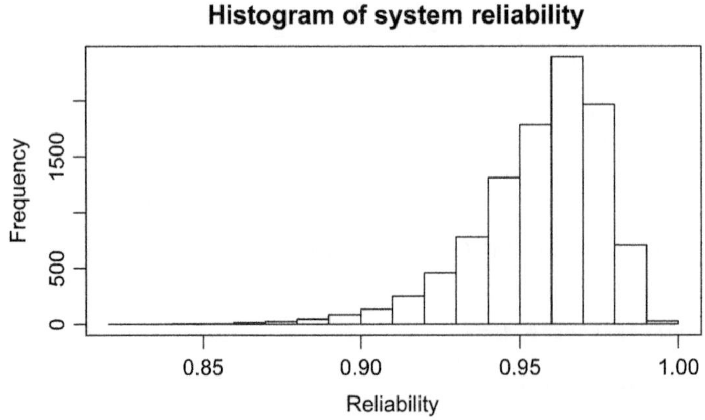

Figure 7.3 Predicted system reliability in Example 7.1 based on posterior distributions of component reliability.

Example 7.2 A reliability engineer plans to estimate the reliability and failure probability of a two-out-of-three system. Here a component level failure is defined as performance not meeting its design specification (lower specification limit = −51; upper specification limit = 84). Sample data on each individual component's performance are shown in Table 7.3.

Earlier in Section 7.1 we briefly introduced a two-out-of-three system. Such a system consists of three components or subsystems. The system works when at least two components work (see Eq. (7.3)). In addition, such a system fails when at least two components fail (see Eq. (7.4)). The failure probability of a two-out-of-three system is

$$\begin{aligned}F_{\text{system}} &= 1 - R_{\text{system}} \\ &= 1 - \left(\prod_{i=1}^{3} R_i + R_1R_2(1-R_3) + R_2R_3(1-R_1) + R_3R_1(1-R_2)\right) \\ &= 1 - R_1R_2 - R_2R_3 - R_1R_3 + 2R_1R_2R_3 \\ &= 1 - (1-F_1)(1-F_2) - (1-F_2)(1-F_3) - (1-F_1)(1-F_3) \\ &\quad + 2(1-F_1)(1-F_2)(1-F_3) \\ &= F_1F_2 + F_2F_3 + F_3F_1 - 2F_1F_2F_3 \\ &= \prod_{i=1}^{3} F_i + F_1F_2(1-F_3) + F_2F_3(1-F_1) + F_3F_1(1-F_2)\end{aligned} \tag{7.4}$$

Assuming the components are manufactured with the same process using the same design and materials, they are from the same population. Based on this assumption, the failure probabilities of the three components are identical, i.e. $F_i = F,\ i = 1,\ 2,\ 3$, where F is the component level failure probability.

The system level failure probability is

$$F_{\text{system}} = F^3 + 3F^2(1-F) = F^2(3-2F). \tag{7.5}$$

The R script is shown in **`7.2_2-out-of-3_system.R`**. The JAGS model **`3.4_Normal.JAGS`** introduced in Chapter 3 is reused in this case. In R script, the

Table 7.3 Observed performance of individual components.

37	19	−10	24	−30
41	7	−1	14	63
0	−3	1	−39	21
19	46	21	36	10
37	4	−28	−24	38
15	2	−3	−18	54
34	−17	10	43	43
7	47	32	−6	14
68	−17	37	52	23
40	12	−8	24	24

data in Table 7.3 are fitted to a normal distribution using vague prior distributions, and then the component level failure rate is computed from posterior samples based on the specification limits (component failure rate is the probability that it does not meet the specification). After the posterior distributions of the component failure rate are estimated via closed-form solutions or MCMC sampling, the system level failure rate is estimated by Eq. (7.5), via Monte Carlo sampling.

(7.2_2-out-of-3_system.R)

```
##############################################################
## To estimate system reliability of a 2-out-of-3 system   ##
##############################################################

###############################
## read data from a txt file ##
###############################

perf.data <- read.table("Example7.2Data.txt",header=F,sep="")
perf <- perf.data[,1]

### Load package rjags (to connect to JAGS from R for Bayesian
analysis)
library(rjags)

# Data for the jags model
BayesianData <- list(y=perf, n=length(perf) )


ModelObject <- jags.model(file = "3.4_Normal.JAGS",
                          data=BayesianData,
                          n.chains = 2, n.adapt = 1000
)

# Burn-in stage
update(ModelObject, n.iter=2000)

        # Run MCMC and collect posterior samples in coda format for
selected variables
        codaList <- coda.samples(model=ModelObject, variable.names =
c("mu", "sigma"), n.iter = 50000, thin = 1)

codaMatrix <- as.matrix( codaList )

mu_samples <- codaMatrix[,"mu"]
sigma_samples <- codaMatrix[,"sigma"]

# summary(codaList, quantiles = c(0.025, 0.05, 0.5, 0.95, 0.975))

####################################################################
## Calculate component failure rate                               ##
## component failure is defined as out of spec                    ##
## (LSL = -51; USL = 84)                                          ##
####################################################################
# sample 50000 data
loop <- floor(seq(1,length(mu_samples),length=50000))
CFR <- rep(NA, length(loop))
```

```
        # get loop # of values for each parameter from posterior sampling
        mu <- mu_samples[loop]
        sigma <- sigma_samples[loop]
        for (i in 1: length(loop)) {
          CFR[i] <- pnorm(-51, mean=mu[i], sd=sigma[i]) + pnorm(84,
mean=mu[i], sd=sigma[i],lower.tail=FALSE)
        }

        ###################################################################
        ## Calculate system level reliability and failure rate of       ##
        ## 2-out-of-3 system                                             ##
        ###################################################################
system_failure_rate <- CFR*CFR*(3-2*CFR)
        system_reliability <- 1-system_failure_rate

        ## show results of system reliability mean, median, and 95%
credible interval
        ## round results to 3 significant digits using signif()
        print(paste("mean of the system reliability is:",
signif(mean(system_reliability),3)))
        print(paste("median of the system reliability is:",
signif(quantile(system_reliability, 0.50),3)))
        print(paste("95% credible interval for the system
reliability is:", signif(quantile(system_reliability, 0.025),3), ",",
signif(quantile(system_reliability, 0.975),3)))

        # Histogram of system failure rate
        #jpeg("Example7.2_2-out-of-3_system_failure_rate_hist.jpeg",
width = 6, height = 4, units = 'in', res = 1800)  # save the plot as
jpeg format
        hist(system_failure_rate, main = " Histogram of system fail-
ure probability", xlab="Failure probability", xlim=c(0,0.01),breaks=100)
        box()
        #dev.off()
```

After running the R script, the simulation results are shown below. The system failure probability histogram is shown in Figure 7.4.

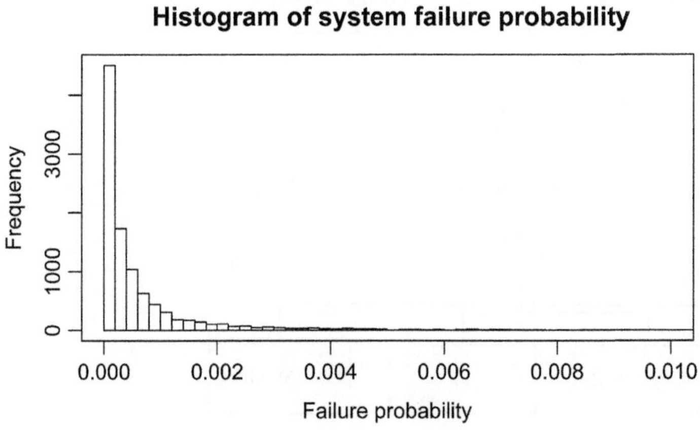

Figure 7.4 System failure probability histogram for Example 7.2.

```
[1] "mean of the system reliability is: 0.999"
[1] "median of the system reliability is: 1"
[1] "95% credible interval for the system reliability is: 0.997 , 1"
```

7.3 Fault Tree

A fault tree analysis is an approach to estimate the system level failure rate using logic relationships among higher-level systems/subsystems and lower-level components. The most commonly used logic relationships include AND gates and OR gates. When there is an AND gate between the higher-level system/subsystem and the lower-level components, the system/subsystem fails when all the components under the AND gate fail. When there is an OR gate between the higher-level system/subsystem and the lower-level components, the system/subsystem fails when at least one of the components under the OR gate fails. Figure 7.5 shows examples of an AND gate and an OR gate.

An OR gate in the fault tree is equivalent to a parallel system in a reliability block diagram. The system failure probability with an OR gate consisting of n independent failures is

$$F_{\text{system}} = 1 - \prod_{i=1}^{n}(1 - F_i), \tag{7.6}$$

where F_i is the failure probability of the ith component.

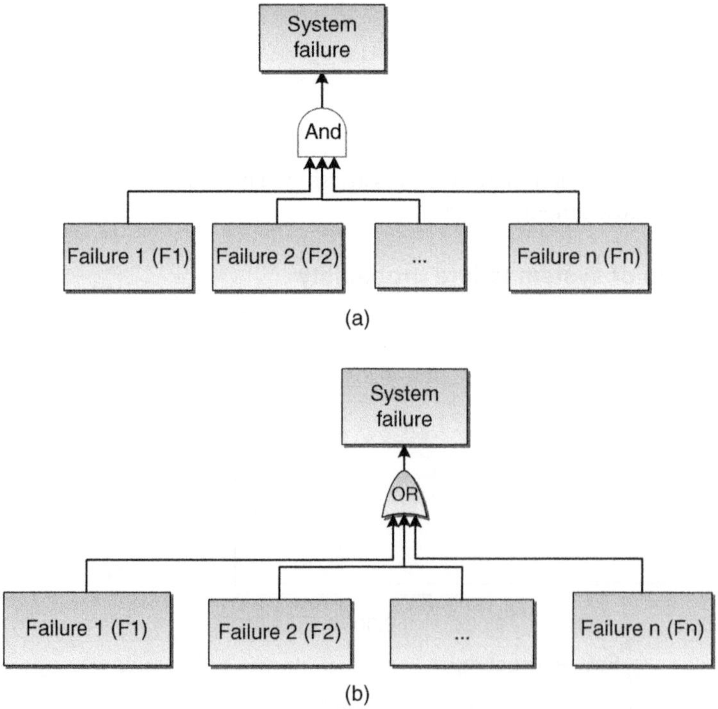

Figure 7.5 Fault tree for (a) an AND gate and (b) an OR gate.

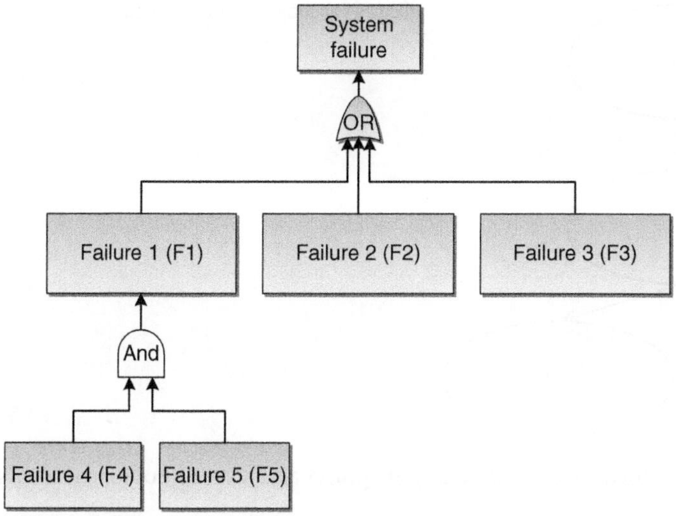

Figure 7.6 A fault tree containing various gates.

An AND gate in the fault tree is equivalent to a series system in a reliability block diagram. The system failure probability with an AND gate consisting of n independent component failures is

$$F_{\text{system}} = \prod_{i=1}^{n} F_i, \tag{7.7}$$

where F_i is the failure probability of the ith component. As an example, Figure 7.6 shows a fault tree consisting of both an OR gate and an AND gate. Based on the fault tree in Figure 7.6, the system failure probability is

$$\begin{aligned} F_{\text{system}} &= 1-(1-F1)(1-F2)(1-F3) \\ &= F1+F2+F3-F1F2-F1F3-F2F3+F1F2F3. \end{aligned}$$

Since

$$F1 = F4F5,$$

then

$$F_{\text{system}} = F4F5+F2+F3-F4F5F2-F4F5F3-F2F3+F4F5F2F3,$$

where F1–F5 are failure probabilities of events in the fault tree.

The Bayesian method to estimate the system level failure probability is very similar to the methods introduced in Section 7.2. First, component level failure rates are estimated individually. Second, the system level failure rate is estimated by equations such as (7.6) or (7.7), depending on the fault tree structure, via Monte Carlo sampling using the posterior distributions of component level failure rates.

7.4 Bayesian Network

A Bayesian network (aka Bayesian belief network) is a probabilistic graphical model that represents an explicit description of the direct dependencies among a set of variables.

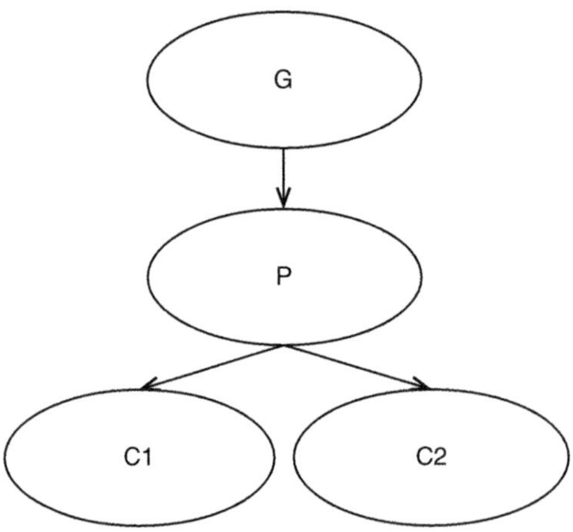

Figure 7.7 Bayesian network example.

This description is in the form of a directed acyclic graph and a set of node probability tables.

To understand how a Bayesian network works, basic probability theories need to be understood first. Some commonly used probability theory concepts were introduced in Chapter 2, including conditional probability, joint probability, independence, probability of the union, probability of the complement, and marginal probability.

Now we introduce a few more concepts and rules to illustrate dependency in a Bayesian network. Figure 7.7 shows a Bayesian network example. In this Bayesian network, there are four nodes, namely G, P, C1, and C2.

- There is an arc from P to C1, and another arc from P to C2. These arcs indicate that C1 and C2 directly depend on P. Node P is called a parent of nodes C1 and C2. Nodes C1 and C2 are children of node P.
- There is an arc from G to P, which indicates that P directly depends on G. Again, G is a parent of P and P is a child of G.
- There is no arc from node G to nodes C1 or C2. Since G is not linked to C1 or C2, C1 and C2 are not directly dependent on G.
- Cycles are not allowed in a Bayesian network to avoid circular reasoning. For example, since there is one arc from node G to node P, and another arc from node P to node C1, then there will be no arc from node C1 to node G.
- *Chain rule*
  For a set of n events $A_1, A_2, \ldots, A_n$, the full joint probability distribution of these n events $A_1, A_2, \ldots, A_n$ is

$$P(A_1, A_2, \ldots, A_n) = P(A_1 \mid A_2, \ldots, A_n) \times P(A_2 \mid A_3, \ldots, A_n) \times \ldots \times P(A_{n-1} \mid A_n) \times P(A_n). \quad (7.8)$$

In Figure 7.8, the joint probability of event G, P, C1, and C2 is

$$P(C_1, C_2, P, G) = P(C_1 \mid C_2, P, G) \times P(C_2 \mid P, G) \times P(P \mid G) \times P(G).$$

Since C_1 and C_2 are only directly dependent on P,

$$P(C_1 \mid C_2, P, G) = P(C_1 \mid P);$$

$$P(C_2 \mid P, G) = P(C_2 \mid P).$$

Figure 7.8 Bayesian network of a system consisting of two independent sensors to detect defective products.

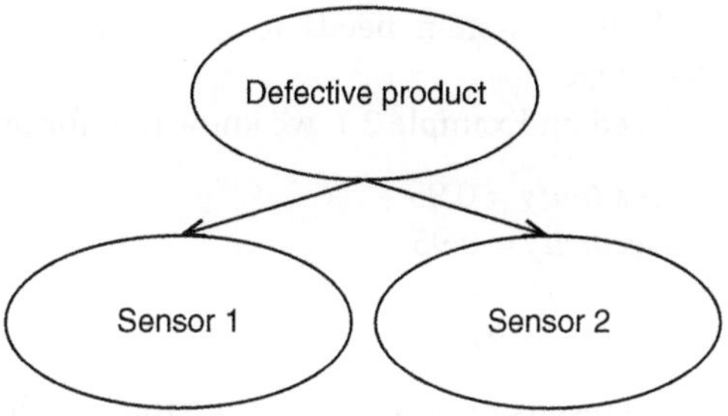

So the joint probability of G, P, C1, and C2 is

$$P(C_1, C_2, P, G) = P(C_1 \mid P) \times P(C_2 \mid P) \times P(P \mid G) \times P(G).$$

See Fenton and Neil (2012) for more theories on Bayesian networks.

In the following sections we will use a few Bayesian network examples to help readers understand these concepts. In Example 7.3, simple Bayesian network models are used to evaluate a multiple-sensor system. Example 7.4 is about using a Bayesian network to estimate system level reliability where there are dependent failure modes. These two examples are used to exercise conditional probability. They both have closed-form solutions. In Example 7.5 various sources of imperfect data are used to estimate reliability. Readers can use this example to exercise the chain rule. Since there is no closed-form solution, it is solved via MCMC sampling using R and JAGS script.

7.4.1 A Multiple-sensor System

Example 7.3 Let us revisit Example 2.1 and expand it to a system consisting of two or three independent sensors. The sensors are designed to identify whether or not the product is defective. The probability of fault is 1%. The sensitivity and specificity of each sensor are both 95%. We will consider the following different design options:

(1) There is only one sensor in the system. The product is claimed to be defective when a fault is detected by the sensor.
(2) There are two independent sensors in the system. The product is claimed to be defective when a fault is detected by both sensors.
(3) There are three independent sensors in the system. The product is claimed to be defective when a fault is detected by all three sensors.
(4) There are three independent sensors in the system. The product is claimed to be defective when a fault is detected by two out of the three sensors.

In order to choose the best design, we would like to calculate the sensitivity, specificity, positive predicted value (PPV), and negative predictive value (NPV) for each of the four design systems described above. The definitions of these terms in this specific example are the same as those defined in Example 2.1, i.e.

- $sensitivity = P(identified\ fault \mid fault)$
- $specificity = P(identified\ success \mid success)$
- $PPV = P(fault \mid identified\ fault)$
- $NPV = P(success \mid identified\ success)$.

Ideally a system needs to have sensitivity, specificity, PPV, and NPV all greater than 90%.

Based on Example 2.1, we know that for option 1 the results are:

- $sensitivity = 0.95$
- $specificity = 0.95$
- $PPV = 16.1\%$
- $NPV = 99.9\%$.

The results of option 1 indicate that the PPV is too low (only 16.1%). This means that option 1 will result in a large proportion of false alarms. As a matter of fact, majority of the alarms given by option 1 are false alarms.

Now let us examine option 2. The Bayesian network of a system consisting of two independent sensors is shown in Figure 7.8. Note that there is one parent node, the "defective product," and two child nodes, "sensor 1" and "sensor 2," in this Bayesian network. The parent node "defective product" has two states, with the probability of each state shown below:

$$P(product\ \ is\ \ defective) = 0.01$$
$$P(product\ \ is\ \ not\ \ defective) = 0.99.$$

Node "sensor 1" has two states: "Fault detected by sensor 1" and "No fault defected by sensor 1." Child node "sensor 1" is dependent on parent node "defective product," thus there are four total conditional probabilities, shown below. These four conditional probabilities can also be shown in a node probability table (see Table 7.4).

- $P(fault \text{ det } ected\ by\ sensor\ 1 \mid product\ is\ defective) = 0.95$ (i.e. sensitivity)
- $P(no\ fault \text{ det } ected\ by\ sensor\ 1 \mid product\ is\ defective) = 0.05$ (i.e. 1 − sensitivity)
- $P(fault \text{ det } ected\ by\ sensor\ 1 \mid product\ is\ not\ defective) = 0.05$ (i.e. 1 − specificity)
- $P(no\ fault \text{ det } ected\ by\ sensor\ 1 \mid product\ is\ not\ defective) = 0.95$ (i.e. specificity).

Node "sensor 2" is dependent on node "defective product." The node probability table of node "sensor 2" is the same as is shown in Table 7.4, indicating the conditional probabilities.

Now let us calculate this two-sensor system's sensitivity, specificity, PPV, and NPV. Note that in option 2 a product is claimed to be defective when a fault is detected by both sensors.

This two-sensor system's sensitivity, specificity, PPV, and NPV are calculated using Bayes' theorem. Since, sensors 1 and 2 are independent,

Table 7.4 Node probability table of node "sensor 1".

	Product is defective	Product is not defective
Fault detected by sensor 1	0.95	0.05
No fault detected by sensor 1	0.05	0.95

$$
\begin{aligned}
&\textit{system sensitivity}\\
&\quad = P(\textit{fault detected by sensor 1 and sensor 2} \mid \textit{product is defective})\\
&\quad = P(\textit{fault detected by sensor 1} \mid \textit{product is defective})\\
&\qquad \times P(\textit{fault detected by sensor 2} \mid \textit{product is defective})\\
&\quad = 0.95 \times 0.95 = 0.9025
\end{aligned}
$$

$$
\begin{aligned}
&\textit{system specificity}\\
&\quad = P(\textit{no fault detected by sensor 1 or sensor 2} \mid \textit{product is not defective})\\
&\quad = P(\textit{no fault detected by sensor 1} \mid \textit{product is not defective})\\
&\qquad + P(\textit{no fault detected by sensor 2} \mid \textit{product is not defective})\\
&\qquad - P(\textit{no fault detected by sensor 1} \mid \textit{product is not defective})\\
&\qquad \times P(\textit{no fault detected by sensor 2} \mid \textit{product is not defective})\\
&\quad = 0.95 + 0.95 - 0.95 \times 0.95 = 0.9975
\end{aligned}
$$

$$
\begin{aligned}
PPV &= P(\textit{product is defective} \mid \textit{fault detected by sensor 1 and sensor 2})\\
&= \frac{P(\textit{fault detected by sensor 1 and sensor 2} \mid \textit{product is defective}) \times P(\textit{product is defective})}{P(\textit{fault detected by sensor 1 and sensor 2})}\\
&= \frac{\textit{system sensitivity} * P(\textit{product is defective})}{\text{system sensitivity} * P(\textit{product is defective}) + (1 - \textit{system specificity}) * (1 - P(\textit{product is defective}))}\\
&= \frac{0.9025 * 0.01}{0.9025^{*}0.01 + (1 - 0.9975) * (1 - 0.01)}\\
&= 0.7848
\end{aligned}
$$

$$
\begin{aligned}
NPV &= P(\textit{product is not defective} \mid \textit{no fault detected by sensor 1 or sensor 2})\\
&= \frac{P(\textit{no fault detected by sensor 1 or sensor 2} \mid \textit{product is not defective}) \times P(\textit{product is not defective})}{P(\textit{no fault detected by sensor 1 or sensor 2})}\\
&= \frac{\textit{system specificity} * P(\textit{product is not defective})}{\textit{system specificity} * P(\textit{product is not defective}) + (1 - \textit{system sensitivity}) * P(\textit{product is defective})}\\
&= \frac{0.9975 * 0.99}{0.9975 * 0.99 + (1 - 0.95) * 0.01}\\
&= 0.9995
\end{aligned}
$$

Similarly, we calculated the system's sensitivity, specificity, PPV and NPV for the remaining design options. The results are shown in Table 7.5. Note that when including more sensors in the system to detect fault, PPV is greatly increased but sensitivity decreases as a cost. For example, compared to option 1, option 3 has a much better

Table 7.5 Sensitivity, specificity, PPV, and NPV of the four design options.

Option	Fault detection triggered by	Sensitivity	Specificity	PPV	NPV
1	1 sensor	95%	95%	16.1%	99.9%
2	2 sensors	90.3%	99.8%	78.5%	99.9%
3	3 of 3 sensors	85.7%	99.99%	98.6%	99.9%
4	2 of 3 sensors	99.3%	99.3%	58.0%	99.99%

Table 7.6 Sensitivity, specificity, PPV, and NPV of the fifth design option: each sensor has 99% sensitivity.

Option	Fault detection triggered by	Sensitivity	Specificity	PPV	NPV
5)	3 of 3 sensors	97.0%	99.99%	98.7%	99.97%

PPV, but a smaller sensitivity. This means that when all three sensors detect fault, there is a 98.6% chance that it is truly a defective part. On the other hand, when there is truly a defective part, there is only an 85.7% chance that this defect can be identified by all the three sensors when using option 3.

None of the four proposed design options provides a system level sensitivity, specificity, PPV, and NPV all greater than 90%. But if we add option 5, which is modified from option 3 by increasing each sensor's sensitivity to 99%, it is possible to have the system level sensitivity, specificity, PPV, and NPV all greater than 90% (see Table 7.6).

7.4.2 Dependent Failure Modes

In Sections 7.2 and 7.3 we discussed scenarios where components in a system fail independently. There are other cases where one failure mode is dependent on another failure mode. If a failure of one component causes another component to be more likely to fail, the dependence is positive. If a failure of one component causes another component to be less likely to fail, the dependence is negative.

Other types of dependent failures include common cause failures, where two or more failures are a result of a common cause, and cascading failures, where failure of one component results in increased load on the remaining components and thus increases the likelihood of failure of other components. Rausand and Hoyland (2004) introduced a few methods to model dependent failures. Here we provide an example using a Bayesian network to model dependent failures (see Example 7.4).

Example 7.4 A series system consists of four components, namely C1, C2, C3, and C4. The Bayesian network model of this system is shown in Figure 7.9. In this Bayesian network there is a parent node called "System" and four children nodes called "C1", "C2", "C3", and "C4". Each of the five nodes has two states: failure (indicated by F) and survival (indicated by S). The system fails if one of the four components fails. Failures of C1, C2, and C4 are independent of one another, with each of their failure probabilities shown

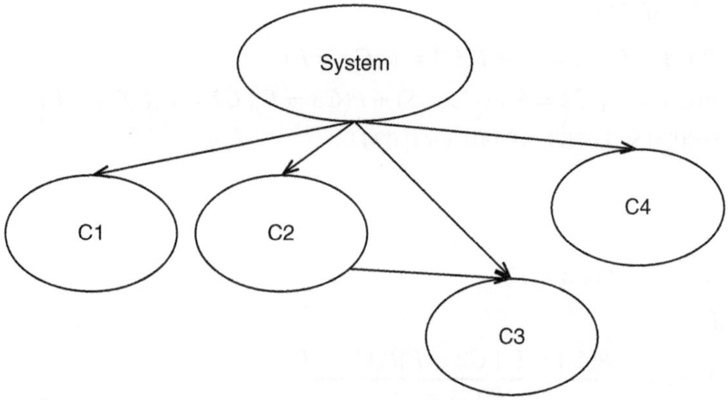

Figure 7.9 Bayesian network of a series system where the failure probability of C3 depends on the status of C2.

Table 7.7 Components C1, C2, and C4 failure and survival probabilities.

Component	Status(F, failure; S, survival)	Probability
C1	F	0.0020
	S	0.9980
C2	F	0.0241
	S	0.9759
C4	F	0.0054
	S	0.9946

Table 7.8 Component C3 failure and survival probabilities given whether C2 = F or C2 = S.

	C2 = F	C2 = S
F	0.0015	0.0030
S	0.9985	0.9970

in Table 7.7. C2 is a parent of C3, which indicates that failure probability of component C3 is dependent upon the status of component C2. Specifically, the failure of C2 causes the failure of C3 to become less likely. Table 7.8 shows the node probability table of C3 given the status of C2. Due to the existence of dependent failures, this Bayesian network model cannot be represented by a standard fault tree. The objective is to estimate the system level reliability in this example.

The probability that C3 fails given that C2 fails is

$$P(C3 = F \mid C2 = F) = 0.0015$$

The failure probability of C3 is

$$\begin{aligned} P(C3=F) &= P(C3=F, C2=S) + P(C3=F, C2=F) \\ &= P(C3=F \mid C2=S)P(C2=S) + P(C3=F \mid C2=F)P(C2=F) \\ &= 0.0030 \times 0.9759 + 0.0015 \times 0.0241 \\ &= 0.00296. \end{aligned}$$

So

$$P(C3=F \mid C2=F) < P(C3=F).$$

On the other hand,

$$\begin{aligned} P(C2=F \mid C3=F) &= \frac{P(C3=F \mid C2=F)P(C2=F)}{P(C3=F)} \\ &= \frac{0.0015 \times 0.0241}{0.00296} = 0.0122 \\ &< P(C2=F) = 0.0241. \end{aligned}$$

Since the failure of component C2 causes the failure of component C3 to become less likely and vice versa, the dependence between C3 and C2 is negative.

The survival probability of component C3 is calculated as follows,

$$\begin{aligned} P(C3=S) &= P(C3=S, C2=F) + P(C3=S, C2=S) \\ &= P(C3=S \mid C2=F)P(C2=F) + P(C3=S \mid C2=S)P(C2=S) \\ &= 0.9985 \times 0.0241 + 0.9970 \times 0.9759 \\ &= 0.99704. \end{aligned}$$

When there are dependent failures, based on the chain rule and Eq. (7.8), the system reliability is

$$R_{\text{system}} = P(R_4 \mid R_1, R_2, R_3) \times P(R_3 \mid R_1, R_2) \times P(R_2 \mid R_1) \times P(R_1), \tag{7.9}$$

where

R_{system} is the system reliability
R_i is the reliability of the ith component in the system.

Since C4 does not depend on C1, C2, or C3, and C3 only depends on C2, Eq. (7.9) becomes

$$\begin{aligned} R_{\text{system}} &= P(R_4) \times P(R_3 \mid R_2) \times P(R_2) \times P(R_1) \\ &= 0.9946 \times 0.9970 \times 0.9759 \times 0.9980 \\ &= 0.96578. \end{aligned}$$

7.4.3 Case Study: Aggregating Different Sources of Imperfect Data

Example 7.5 An engineer plans to estimate the reliability of a product released to the market based on three sources of data. The first data source is a database of field performance based on customer reporting. Though there exists a large amount of information about the product performance in this data source, there is a concern that there might be underreporting as not all issues are reported by customers. The second data source is a registry, where performances of some products in the field are monitored. The second data source, though having no issues of underreporting, has a limitation on its sample size. The third data source is bench testing data in the laboratory, where field use conditions are simulated as test conditions. Due to differences between test conditions and actual field usage conditions, reliability indicated by the bench testing results alone could be optimistic or conservative, compared to the actual field performance. In other words, there could be potential misclassification. Besides, there is also a sample size limitation in the bench testing data.

Since no single source of data is perfect, the engineer would like to utilize all three sources of data to estimate the product's actual reliability performance in the field. The data from different sources and their limitations are summarized in Table 7.9.

A Bayesian network model is used to estimate the actual reliability of this released product. The Bayesian network model is shown in Figure 7.10. The relationships among the failure probabilities of different data sources and the actual failure probability will be discussed later.

First let us look at nodes in this Bayesian network and the logical relationships among the nodes. To address underreporting, we introduce an additional parameter to represent the reporting rate of data source 1. To account for misclassification in data source 3, two other parameters, sensitivity and specificity, are introduced in the model. In addition, the actual failure rate is unknown. In the Bayesian network shown in Figure 7.11, based on the arcs, there are four parent nodes, named as follows:

Table 7.9 Different sources of data their limitations.

Data source	No. of failures	Sample size	Uncertainty
1. Database 1	33	70351	Underreporting
2. Database 2	1	1151	Smaller sample size
3. Bench test	0	81	Possible misclassification; small sample size

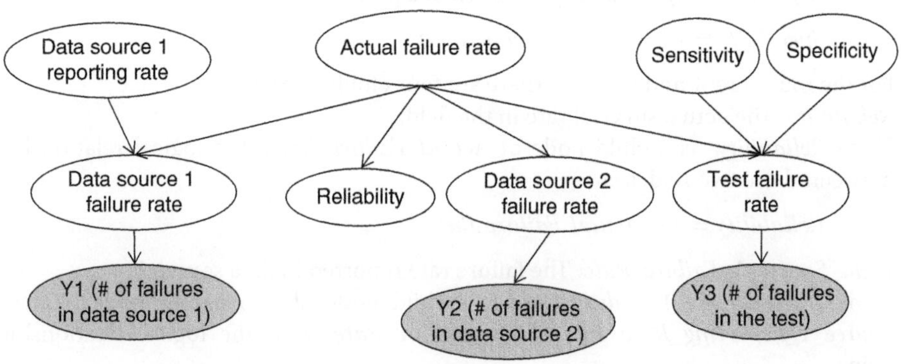

Figure 7.10 Bayesian network model to aggregate multiple sources of data. Data are available for the gray nodes (*Y1*, *Y2*, *Y3*). Posterior distributions of unknown parameters are conditional probabilities given the data.

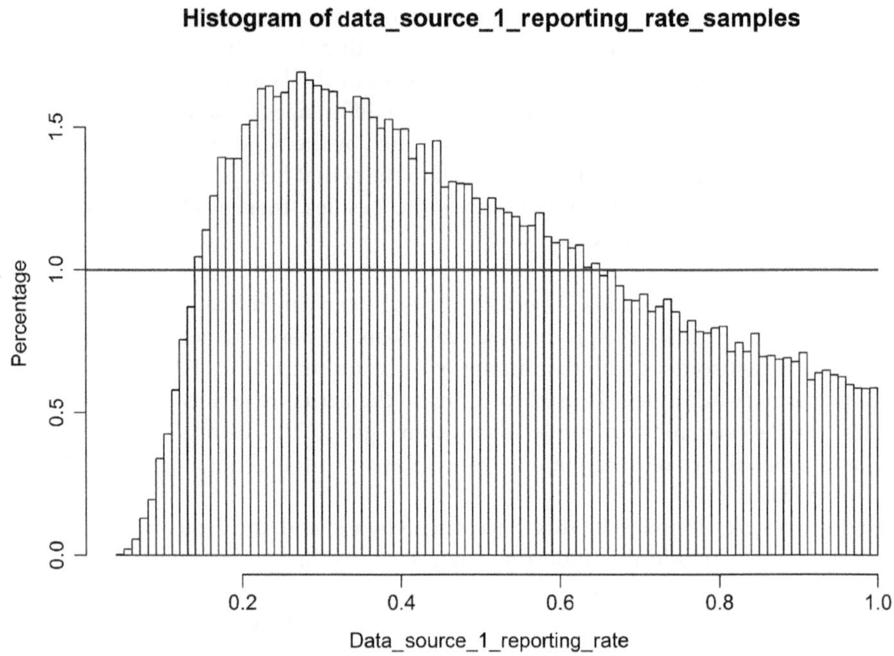

Figure 7.11 Posterior samples histogram vs. prior distribution *Uniform*(0, 1) (the line plot) of *Data_Source_1_Reporting_Rate*.

- *Data_Source_1_Reporting_Rate* : The probability of reporting a failure to database 1 when a part fails, i.e.

$$Data\_Source\_1\_Reporting\_Rate = P(report\ a\ failure\ to\ database\ 1 \mid part\ is\ bad).$$

- *Actual_Failure_Rate* The actual failure rate in the field.
- *Sensitivity* Probability of failing a test when a part is bad,

$$Sensitivity = P(fail\ test \mid part\ is\ bad).$$

- *Specificity* Probability of passing a test when a part is good,

$$Specificity = P(pass\ test \mid part\ is\ good).$$

For the four parent nodes above, there are four children nodes:

- *Reliability*: The actual survival rate in the field.
  Node *Reliability* is a child node of *Actual_Failure_Rate*. The logical relationship between these two nodes is

$$Reliability = 1 - Actual\_Failure\_Rate.$$

- *Data_Source_1_Failure_Rate*: The failure rate reported in data source 1.
  Node *Data_Source_1_Failure_Rate* is a child node of two parent nodes, *Data_Source_1_Reporting_Rate* and *Actual_Failure_Rate*, with the logical relationship being

$$Data\_Source\_1\_Failure\_Rate$$
$$= Data\_Source\_1\_Reporting\_Rate \times Actual\_Failure\_Rate.$$

- *Data_Source_2_Failure_Rate*: The failure rate reported in data source 2.
  Node *Data_Source_2_Failure_Rate* is a child node of the parent node *Actual_Failure_Rate*. Since there is no underreporting in database 2, the logical relationship is

$$Data\_Source\_2\_Failure\_Rate = Actual\_Failure\_Rate.$$

- *Test_Failure_Rate*: The failure rate reported in the test.
  Node *Test_Failure_Rate* is a child node of three parent nodes, *Actual_Failure_Rate*, *Sensitivity*, and *Specificity*, with the logical relationship being

$$\begin{aligned}&Test\_Failure\_Rate\\&\quad = Sensitivity \times Actual\_Failure\_Rate\\&\qquad + (1 - Specificity) \times (1 - Actual\_Failure\_Rate).\end{aligned}$$

The following equation explains why node *Test_Failure_Rate* has such a relationship with its three parent nodes:

$$\begin{aligned}&Sensitivity \times Actual\_Failure\_Rate\\&\qquad + (1 - Specificity) \times (1 - Actual\_Failure\_Rate)\\&\quad = P(fail\ test \mid part\ is\ bad) \times P(part\ is\ bad)\\&\qquad + (1 - P(pass\ test \mid part\ is\ good)) \times P(part\ is\ good)\\&\quad = P(fail\ test \mid part\ is\ bad) \times P(part\ is\ bad)\\&\qquad + P(fail\ test \mid part\ is\ good) \times P(part\ is\ good)\\&\quad = P(fail\ test, part\ is\ bad) + P(fail\ test, part\ is\ good)\\&\quad = P(fail\ test)\\&\quad = Test\_Failure\_Rate\end{aligned}$$

There are three "grandchildren" nodes.

- *Y1*: Number of failures reported in data source 1
  Node *Y1* is a child node of *Data_Source_1_Failure_Rate*, with the logical relationship being

$$Y1 \sim Binomial(Data\_Source\_1\_Failure\_Rate, 70351).$$

This means that *Y1* follows a binomial distribution with a trail size of 70 351 and a probability of *Data_Source_1_Failure_Rate*.

- *Y2*: Number of failures reported in data source 2
  Node *Y2* is a child node of *Data_Source_2_Failure_Rate*, with the logical relationship being

$$Y2 \sim Binomial(Data\_Source\_2\_Failure\_Rate, 1151).$$

This means that *Y1* follows a binomial distribution with a trail size of 1151 and a probability of *Data_Source_2_Failure_Rate*.

- *Y3*: Number of failures reported in the test
  Node *Y3* is a child node of *Test_Failure_Rate*, with the logical relationship being

$$Y3 \sim Binomial(Test\_Failure\_Rate, 81).$$

This means that *Y3* follows a binomial distribution with a trail size of 81 and a probability of *Test_Failure_Rate*.

Now let us define the prior distributions for all the unknown nodes. For unknown parameters *Data_Source_1_Reporting_Rate* and *Actual_Failure_Rate*, we do not have any prior knowledge, so we use uniform distributions *Uniform* (0, 1) as non-informative prior distributions. *Beta* (9, 1) is used as the prior distribution for nodes *Sensitivity* and *Specificity*, based on historical information, which indicates that on average the test has a 90% chance of detecting a good part when it is truly good, and a 90% chance of detecting a bad part when it is truly bad. Posterior distributions of these unknown parameters will be inferred based on data and this Bayesian network model. In this case the data is

$$Y1 = 33;\ \ Y2 = 1;\ \ Y3 = 0.$$

All the data are utilized in this Bayesian model for the inference of unknown nodes/parameters so that the information in different data sources are leveraged for the estimation of true reliability (node *Reliability*). There is no closed-form solution. The JAGS and R scripts to estimate the unknown parameters are shown in **7.4_Aggregating_Data.JAGS** and **7.4_Aggregating_Data.R.**

(7.4_Aggregating_Data.JAGS)

```
model {

     # prior
     Actual_Failure_Rate ~ dunif(0,1)
     Data_Source_1_Reporting_Rate ~ dunif(0,1)
     Sensitivity ~ dbeta(9,1)
     Specificity ~ dbeta(9,1)

     # likelihood
     Data_Source_1_Failure_Rate <- Data_Source_1_Reporting_Rate*Actual_
Failure_Rate
     Data_Source_2_Failure_Rate <- Actual_Failure_Rate
     TEST_Failure_Rate <- Sensitivity*Actual_Failure_Rate+(1-Specificity)*
(1-Actual_Failure_Rate)
     Reliability <- 1 - Actual_Failure_Rate

     Data_Source_1_Failures ~ dbin(Data_Source_1_Failure_Rate, 70351)
     Data_Source_2_Failures ~ dbin(Data_Source_2_Failure_Rate, 1151)
     TEST_Failures ~ dbin(TEST_Failure_Rate, 81)
     }
```

(7.4_Aggregating_Data.R)

```
     ####################################################################
     ## Aggregating multiple sources of data,                          ##
     ## taking into account underreporting and misclassification       ##
     ####################################################################

     ### Load package rjags (to connect to JAGS from R for Bayesian
analysis)
     library(rjags)
```

```
##########
## Data ##
##########

BayesianData <- list(
  Data_Source_1_Failures = 33,
  Data_Source_2_Failures = 1,
  TEST_Failures = 0
)

#############################################################
## Create the model, burnin, and collect posterior samples ##
#############################################################

# Create (& initialize & adapt) a JAGS model object
ModelObject <- jags.model(file = "7.4_Aggregating_Data.JAGS",
                         data=BayesianData,
                         n.chains = 3, n.adapt = 1000
)

# Burn-in stage
update(ModelObject, n.iter=3000)

# select variables to collect posterior samples
variable_names <- c("Reliability", "Data_Source_1_Reporting_Rate",
"Sensitivity", "Specificity")

# Run MCMC and collect posterior samples in coda format for selected
variables
codaList <- coda.samples(model=ModelObject, variable.names = vari-
able_names,
                            n.iter = 50000, thin = 1)

codaMatrix <- as.matrix( codaList )
Reliability_samples <- codaMatrix[,"Reliability"]
Data_Source_1_Reporting_Rate_samples <- codaMatrix[,"Data_Source_1_
Reporting_Rate"]
Sensitivity_samples <- codaMatrix[,"Sensitivity"]
Specificity_samples <- codaMatrix[,"Specificity"]

# Summary plots (trace plots and density plots)
# jpeg("Fig 7.4_summary_plots.jpeg", width = 8, height = 8, units =
'in', res = 1800)
# plot(codaList)
# dev.off()

# Summary statistics for Markov Chain Monte Carlo chains
# Quantiles of the sample distribution can be modified in the
quantiles argument
summary(codaList, quantiles = c(0.025, 0.05, 0.5, 0.95, 0.975))
```

Four nodes are selected to have posterior samples collected: *Data_Source_1_Reporting_Rate, Sensitivity, Specificity,* and *Reliability.* The summary statistics obtained after running the R script above are shown below. We also discuss the posterior samples vs. the prior distribution for each of the four parameters.

```
> summary(codaList, quantiles = c(0.025, 0.05, 0.5, 0.95, 0.975))

Iterations = 4001:54000
Thinning interval = 1
Number of chains = 3
Sample size per chain = 50000

1. Empirical mean and standard deviation for each variable,
   plus standard error of the mean:

                               Mean        SD  Naive SE Time-series SE
Data_Source_1_Reporting_Rate 0.4855 0.2380841 6.147e-04      3.517e-03
Reliability                  0.9987 0.0008211 2.120e-06      1.325e-05
Sensitivity                  0.8992 0.0910078 2.350e-04      4.752e-04
Specificity                  0.9890 0.0109141 2.818e-05      6.737e-05

2. Quantiles for each variable:

                               2.5%     5%    50%    95%  97.5%
Data_Source_1_Reporting_Rate 0.1300 0.1552 0.4504 0.9181 0.9576
Reliability                  0.9965 0.9971 0.9989 0.9995 0.9995
Sensitivity                  0.6616 0.7144 0.9251 0.9942 0.9971
   Specificity                  0.9597 0.9670 0.9923 0.9994 0.9997
```

Data_Source_1_Reporting_Rate has a uniform distribution *Uniform*(0, 1) as its prior, which has a mean of 0.50 and a standard deviation of 0.29 (variance = 1/12). This prior belief indicates a belief that when there is a failure in the field, the chance that customers choose to report it is equal to not reporting it. Its posterior distribution has a mean of 0.49 and a median of 0.45. This suggests that the reporting rate of data source 1 is less optimistic compared to the previous belief. In other words, customers are more likely to choose not reporting a failure when there is truly a failure in the field. The posterior distribution has a standard deviation of 0.24, which is smaller than the standard deviation of the prior distribution (0.29). This indicates that certainty is improved in the posterior distribution compared to the prior. Summary statistics of *Data_Source_1_Reporting_Rate* prior and posterior distributions are shown in Table 7.10. Figure 7.11 shows the posterior samples (histogram) vs. prior distribution (line plot).

Sensitivity has a beta distribution *Beta* (9, 1) as its prior distribution, which has a mean of 0.90 and a standard deviation of 0.09. Its posterior distribution has a mean of 0.90, which is the same as its prior distribution mean. Figure 7.12 shows the posterior samples (histogram) vs. prior distribution (line plot). Note that this posterior distribution overlaps with the prior distribution. In other words, the data has negligible impact to change this prior belief.

Specificity has a beta distribution *Beta* (9, 1) as its prior distribution, which has a mean of 0.90 and a standard deviation of 0.09. The posterior distribution has a mean of 0.99, which is larger than the prior distribution mean. The posterior distribution has a standard deviation of 0.01, which is smaller than the prior distribution standard

Table 7.10 *Data_Source_1_Reporting_Rate* prior and posterior distribution statistics.

Statistics	Prior	Posterior
Mean	0.50	0.49
Standard deviation	0.29	0.24
2.5%	0.025	0.13
5%	0.050	0.16
50%	0.50	0.45
95%	0.95	0.92
97.5%	0.975	0.96

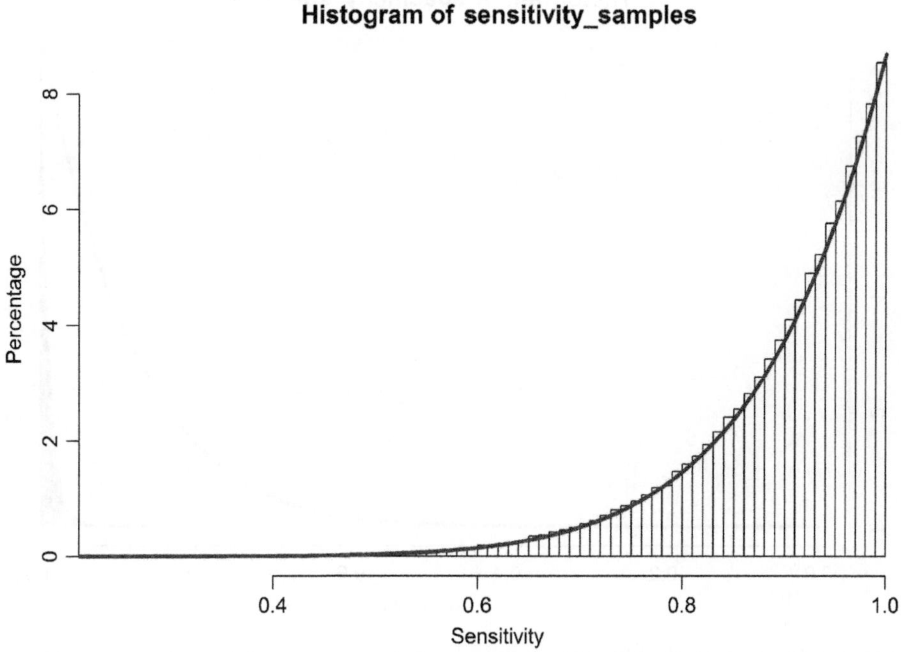

Figure 7.12 Posterior samples histogram vs. prior distribution *Beta* (9, 1) (the line plot) of *Sensitivity*.

deviation. Summary statistics of *Specificity* prior and posterior distributions are shown in Table 7.11. Figure 7.13 shows the posterior samples (histogram) vs. prior distribution (line plot). Note that compared to the prior distribution, the posterior distribution shifts to the right and has a smaller spread. The results show that, conditional on the data, on average 99% of the good parts can be successfully identified in the test, which is more optimistic compared to the prior belief.

Reliability has a uniform distribution *Uniform*(0, 1) as its prior, which has a mean of 0.500 and a standard deviation of 0.289 (variance = 1/12). This prior belief indicates that we are completely ignorant about the reliability in the field, so we assume equal chance over the entire range of reliability (from 0 to 1). Summary statistics of *Reliability*

Table 7.11 *Specificity* prior and posterior distribution statistics.

Statistics	Prior *Beta* (9, 1)	Posterior
Mean	0.90	0.99
Standard deviation	0.09	0.01
2.5%	0.66	0.96
5%	0.72	0.97
50%	0.93	0.99
95%	0.99	1.00
97.5%	1.00	1.00

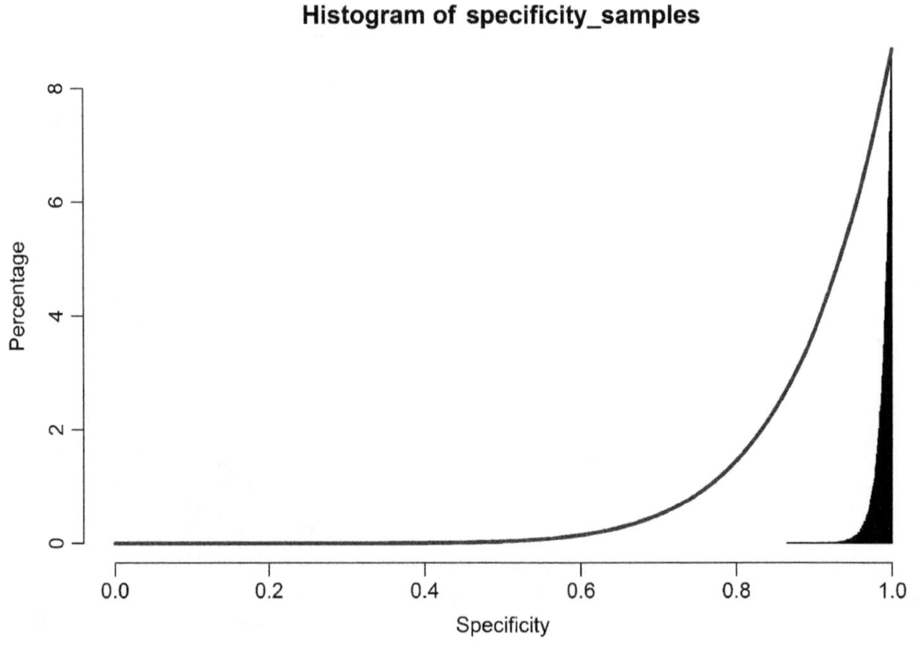

Figure 7.13 Posterior samples histogram vs. prior distribution *Beta* (9, 1) (the line plot) of *Specificity*.

prior and posterior distributions are shown in Table 7.12. The posterior distribution has a mean of 0.999, a standard deviation of 0.001, and a 95% credible interval of (0.997, 1.000). Compared to the prior belief, *Reliability* posterior has a much larger mean and a much smaller standard deviation. Figure 7.14 shows the posterior samples (histogram) vs. prior distribution (line plot).

In this Bayesian model, *Reliability* posterior can be viewed as a weighted average of estimations from the three data sources. Readers are free to run sensitivity analysis to see the impacts of different prior beliefs and different data sources on the reliability posterior distribution. For example, readers can rerun the analysis by changing the prior distributions of some nodes, or by deleting one data source in the Bayesian analysis to see their impact on the final reliability prediction.

Table 7.12 *Reliability* prior and posterior distribution statistics.

Statistics	Prior	Posterior
Mean	0.500	0.999
Standard deviation	0.289	0.001
2.5%	0.025	0.997
5%	0.050	0.997
50%	0.500	0.999
95%	0.950	1.000
97.5%	0.975	1.000

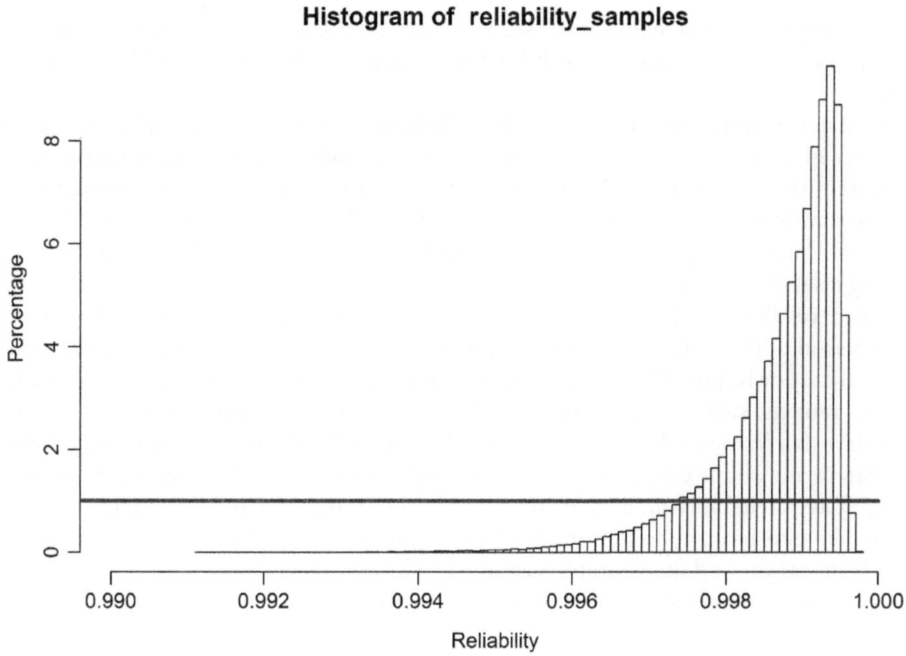

Figure 7.14 Posterior samples histogram vs. prior distribution (the line plot) of *Reliability*.

In this example, the Bayesian network model is used to predict the reliability by combining various sources of field/test data. There is a concern that some data sources may have been biased because of potential misclassification and underreporting. In addition, some data sources have a limitation of small sample size. The Bayesian network model allows these to be considered and predicts the corrected reliability. In this example, the actual failure rate is inferred as a weighted average of the failure rate from different sources. Note that if using the frequentist method with data source 2 alone, to achieve a 95% confidence interval lower bound of 0.9970, one would need a sample size of 1855 when there is one failure in data source 2 (1590 sample size if lower bound is 0.9965). In other words, even if data source 2 does not have underreporting or misclassification issues, its current sample size is not large enough to demonstrate the reliability 95% lower confidence bound to be as high as 0.997. The other two sources of data, although they have limitations, made contributions to decrease the uncertainty in the reliability estimation. Their contribution is equivalent to increasing the sample size in data source 2. More examples of correcting misclassification can be referred to MacLehose et al. (2009).

For other Bayesian models on reliability estimation using complex system models or leveraging multiple sources of information, readers can be referred to Guo and Wilson (2013), Reese et al. (2011), Wilson and Huzurbazar (2007), Wilson et al. (2006), Hamada et al. (2004), etc.

7.5 Summary

In this chapter various methods to estimate system reliability in a Bayesian framework are introduced. These include reliability block diagrams, fault trees, and Bayesian networks.

A reliability block diagram is a method showing how successes or failures of components in a system contribute to system level reliability. A fault tree analysis is an approach to estimate system level failure rate using logical relationships among higher level system/subsystem and lower level components. A Bayesian network is a probabilistic graphical model that represents an explicit description of the direct dependencies among a set of variables.

System reliability is the probability of the system to operate without failures for a specified period of time. In a frequentist/classical framework, except in the simplest cases, it is difficult or impossible to propagate classical confidence intervals though complex system models, such as reliability block diagrams, fault trees, and other logic models. In a Bayesian framework, on the other hand, posterior distributions are true probability statements about unknown parameters, so they may be easily propagated through these system reliability models. In addition, Bayesian network models provide flexibility to take into account limitations in actual engineering practices, e.g. underreporting, misclassification, and small sample size, and to allow different sources of imperfect information to be aggregated, so that the corrected reliability can be estimated.

References

Fenton, N. and Neil, M. (2012). *Risk Assessment and Decision Analysis with Bayesian Networks*. Boca Raton, FL: CRC Press.

Guo, J. and Wilson, A.G. (2013). Bayesian methods for estimating system reliability using heterogeneous multilevel information. *Technometrics* 55 (4): 461–472.

Hamada, M., Martz, H.F., Reese, C.S. et al. (2004). A fully Bayesian approach for combining multilevel failure information in fault tree quantification and optimal follow-on resource allocation. *Reliabiity Engineering & Sytstem Safety* 86 (3): 297–305.

Hamada, M.S., Wilson, A.G., Shane Reese, C., and Martz, H.F. (2010). *Bayesian Reliability*. New York: Springer.

Liu, Y., Berg, R., Chen, X., Abeyratne, A., Wang, X., and Haddad, T. (2015) Bayesian reliability prediction of a medical device system. The First International Conference on Reliability Systems Engineering (ICRSE), Beijing, China.

MacLehose, R.F., Olshan, A.F., Herring, A.H. et al. (2009). Bayesian methods for correcting misclassification – an example from birth defects epidemiology. *Epidemiology* 20 (1): 27–35.

Rausand, M. and Hoyland, A. (2004). *System Reliability Theory – Models, Statistical Methods, and Applications*. Hoboken, NJ: John Wiley & Sons.

Reese, C.S., Wilson, A.G., Guo, J. et al. (2011). A Bayesian model for integrating multiple sources of lifetime information in system-reliability assessments. *Journal of Quality Technology* 42 (2): 127–141.

Wilson, A.G., Graves, T.L., Hamada, M.S., and Reese, C.S. (2006). Advances in data combination, analysis and collection for system reliability assessment. *Statistical Science* 21 (4): 514–531.

Wilson, A.G. and Huzurbazar, A.V. (2007). Bayesian networks for multilevel system reliability. *Reliability Engineering and System Safety* 92 (10): 1413–1420.

8

Bayesian Hierarchical Model

This chapter introduces the basics of Bayesian hierarchical models. For a family of products whose designs are similar in nature, under the assumption of exchangeability, Bayesian hierarchical models can be applied to allow partial information to be leveraged among different products. The potential benefits of applying Bayesian hierarchical models include reducing sample size and/or reducing the uncertainty of predicted reliability for a new product when there is no other prior information of its reliability performance. Specifically, in this chapter a Bayesian hierarchical binomial model and a Bayesian hierarchical Weibull model are discussed with examples.

8.1 Introduction

In Chapter 2 we demonstrated that if an informative prior distribution is available and if it indicates high reliability, to demonstrate the same level of confidence/reliability sample size could be reduced by using an informative prior distribution, compared to using a vague/noninformative prior. This is one benefit of Bayesian analysis. Let us briefly refresh here.

There are a few methods to convert this reliability prior information into a distribution. For example, the reliability of a product is believed to have a mean of 0.95 and a standard deviation of 0.01 based on other sources of information. Let us convert this information to a distribution. A standard beta distribution can be a good choice to model probability, since it is a continuous distribution defined on the interval [0, 1] (for convenience we call standard beta just beta distribution in this book). Equations (8.1) and (8.2) calculate the mean and variance of a beta distribution, respectively.

$$E(X) = \frac{\alpha}{\alpha + \beta}; \tag{8.1}$$

$$Var(X) = \frac{\alpha\beta}{(\alpha + \beta)^2(\alpha + \beta + 1)}. \tag{8.2}$$

Mean and variance can also be converted to parameters in a beta distribution, based on Eqs. (8.3) and (8.4).

$$\alpha = E(X)\left(\frac{E(X)\ (1 - E(X))}{Var(X)} - 1\right); \tag{8.3}$$

$$\beta = (1 - E(X))\left(\frac{E(X)\ (1 - E(X))}{Var(X)} - 1\right). \tag{8.4}$$

Practical Applications of Bayesian Reliability, First Edition. Yan Liu and Athula I. Abeyratne.

Companion website: www.wiley.com/go/bayesian20

For a distribution with a mean of 0.95 and a standard deviation of 0.01 (variance = 0.0001), the corresponding α and β in a beta distribution are 450.3 and 23.7 per Eqs. (8.3) and (8.4). This beta distribution can be used as an informative prior distribution *Beta* (450.3, 23.7) for reliability. This prior is equivalent to having a virtual test with a sample size of $450.3 + 23.7 = 474$, where 450.3 out of 474 parts have passed.

In addition to the prior belief, to further assess the reliability engineers conducted a test to collect data. Assume 12 parts have been tested and all passed. Using the above informative prior distribution, the reliability posterior distribution is *Beta* (462.3, 23.7). Based on Eqs. (8.1) and (8.2), the posterior reliability has a mean of 0.951 and a standard deviation of 0.00976. Note that the mean and the standard deviation of this posterior reliability distribution are very close to those of the prior. This indicates that the chosen informative prior may have a relatively higher weight compared to the data.

How is this posterior reliability compared to the reliability estimation when using a non-informative prior? If using a uniform prior *Beta* (1, 1), the reliability posterior distribution is *Beta* (13, 1), which has a mean of 0.929 and a standard deviation of 0.0665. Compared to the reliability posterior *Beta* (462.3, 23.7), *Beta* (13, 1) has a smaller mean and a larger standard deviation. In this case, when the sample size is small in the test, there is a larger uncertainty in the posterior reliability estimation.

R script **`(8.1_BetaDensityPlots.R)`** is used to plot the informative prior distribution, *Beta* (450.3, 23.7), and the posterior reliability distribution, *Beta* (462.3, 23.7), based on this informative prior, and posterior reliability, *Beta* (13, 1), based on a uniform prior. The three density plots are shown in Figure 8.1.

Note that when using the informative prior, the posterior is very similar to the prior. For *Beta* (13, 1), the one-sided 95% credible interval lower bound (5th percentile) is 0.7942. The one-sided 95% credible interval lower bound for posterior *Beta* (462.3, 23.7) is 0.9342. This lower bound is equivalent to the reliability posterior lower bound when the prior is a uniform distribution and when the data shows 43 parts all passed the test. When using a strong informative prior, to achieve the target confidence/reliability the sample size in the test might be smaller compared to using a non-informative prior or using the frequentist method.

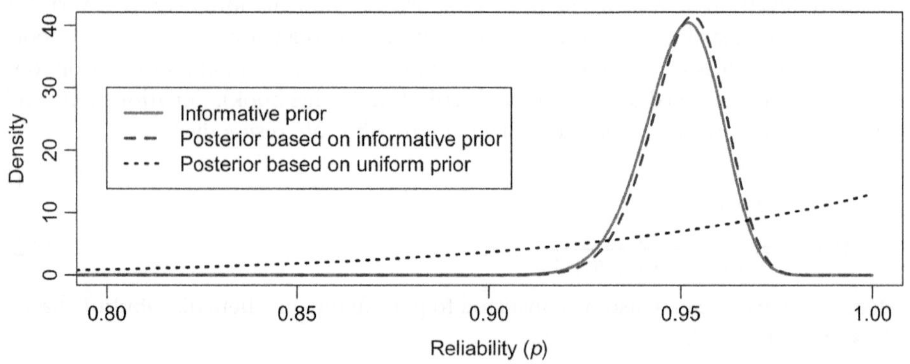

Figure 8.1 Reliability density plots: reliability informative prior *Beta* (450.3, 23.7), reliability posterior *Beta* (462.3, 23.7) based on the informative prior, and reliability posterior *Beta* (13, 1) based on a uniform prior.

(8.1_BetaDensityPlots.R)

```
# Create density plots of reliability Beta prior and posteriors

p <- seq(0,1,length=10001)
plot(p, dbeta(p,450.3,23.7),type="l",ylab="Density",xlab="Reliability
(p)",lty=1,lwd=2,col="red", xlim=c(0.8,1)) # plot reliability
informative prior

lines(p,dbeta(p,462.3,23.7),lty=2,lwd=2,col="blue") # add reliability
posterior based on informative prior
lines(p,dbeta(p,13,1),lty=3,lwd=2,col="black") # reliability posterior
based on uniform prior

# add legend
legend(.8,30,c("Informative prior","Posterior based on informative
prior", "Posterior based on uniform prior"),lty=c(1,2,3),lwd=c(2,2,2),
col=c("red","blue", "black"))
```

Other ways to build a reliability prior distribution include acquiring information on two quantiles of reliability belief (Albert 2009), or specifying the prior by center and spread (BayesWeb n.d.), and converting the information to a beta distribution.

However, often in engineering practices none of the prior belief information mentioned above is available. Without published information or data, typically it is not feasible to build an informative reliability prior distribution. Instead, engineers may find out that reliability testing data of previous product generations are available. Different generations of products that belong to the same family could have slight differences in design and/or manufacturing process. The design changes could be driven by business needs or to improve a product's user experience, etc. In these cases, products from the same family may have similar reliability performance. Their reliability performances are close, but not identical. Since there could be design/manufacturing changes, testing data on the previous generation of products cannot be directly leveraged to build a prior distribution for the current product. Let us look at an example.

Example 8.1 A reliability engineer is characterizing the performance of a product by testing several parts and recording the number of successes and failures against a key performance requirement. Eight generations of products have already been released to the market. For each of the eight generations of product, historical design verification test (DVT) data are available. For the first four generations, the sample size in each DVT is 59. For generations #5 to #8, the sample size in each DVT is 299. For the ninth generation of product, 12 parts have been tested. The objectives are to estimate the reliability of the ninth generation of product based on the existing data, as well as the reliability of any future product (called tenth generation here) when there is no testing data at all (see Table 8.1).

For product generation #9, only 12 parts have been tested. This sample size is too small to derive any useful information about the product's reliability performance. In this case, it might be desirable to pool the data of other generations in some way and leverage some information from those data.

Table 8.1 Design verification test data.

Product generation #	Sample size	# of successes
1	59	59
2	59	59
3	59	59
4	59	59
5	299	299
6	299	299
7	299	299
8	299	299
9	12	12
10	NA	NA

Different from the frequentist method where no historical information can be leveraged, or a one-level Bayesian model where historical information can be fully utilized in the prior distribution, a Bayesian hierarchical model allows partial strength to be borrowed from previous data. Thus, the estimated reliability for the new product may have reduced uncertainty compared to not leveraging historical information at all.

First, let us introduce the assumption of exchangeability. Parameters are exchangeable in their joint distribution if the joint distribution is invariant to permutations of the parameter indexes. Under the exchangeability assumption, reliabilities of different product generations are not expected to depend on the order in which the test results are observed or any other re-ordering. In this case, assume products have similar but not identical designs. If we consider the reliability of one product generation is no more likely to be higher or lower compared to other products (i.e. we are ignorant about the reliability performance of each product generation), exchangeability can be assumed.

Similar to physical measurement data (e.g. dimension, weight), which are usually assumed as a sample drawn from a common population (and thus can be modeled using a distribution), exchangeable models here consider reliabilities of various product generations (or other parameters to be modeled) a random sample drawn from a common distribution. More discussions on exchangeability can be found in Gelman et al. (2014).

For data structure that is hierarchical in nature, the number of levels in a hierarchical model depends on the data structure. Typically, the data can be partitioned into several subgroups which share some common characteristics. If the subgroups can be further divided, more levels can be added to the hierarchical model structure.

Examples and applications of hierarchical Bayesian models include the design and analysis of clinical trials for medical devices (US Department of Human Services & Food and Drug Administration et al. 2010), hospital mortality rates (Albert 2009), etc. Bayesian hierarchical models applied in reliability practices include Andrade and Teixeira (2015), Johnson et al. (2005), and Mishra et al. (2018). This chapter introduces two conceptual models with the distributions most commonly used in reliability practices. Hopefully these examples can inspire more modeling ideas and applications.

8.2 Bayesian Hierarchical Binomial Model

To understand how partial strength is leveraged in Bayesian hierarchical models, let us look at Example 8.1 using a Bayesian hierarchical binomial model vs. other methods.

8.2.1 Separate One-level Bayesian Models

First, we use basic one-level separate Bayesian models and the traditional statistical method (frequentist method) to predict the reliability of each generation of product, then we apply a Bayesian hierarchical model in Section 8.2.2 and compare the results from these methods.

When estimating the reliability of each product separately using a one-level Bayesian method, there is only one unknown parameter (reliability θ_i of the ith product). Assuming the DVT observations follow a binomial distribution, the model structure is

$$y_i \sim Binomial\ (n_i,\ \theta_i),$$

where

y_i is the number of successes in the ith test
n_i is the sample size of the ith test
θ_i is the reliability of the ith generation product.

If we assume each reliability prior is a uniform distribution or *Beta* (1, 1), then the posterior reliability is also a beta distribution due to the fact that the beta distribution is a conjugate prior of the binomial success rate (or reliability). The resulting posterior distributions are given by;

- for product generation #1 to #4: *Beta* (60, 1)
- for product generation #5 to #8: *Beta* (300, 1)
- and for product generation #9: *Beta* (13, 1).

For each generation of product, the corresponding lower bound of the 95% credible interval of reliability is shown in Table 8.2. Table 8.2 also lists the lower bound of the 95% confidence interval (CI) for each generation of product estimated using the frequentist method (computed using Minitab® version 17 One-Sample Proportion with exact method). The results show that reliability estimation based on the one-level Bayesian method is close to the frequentist method, which is expected since we used a non-informative prior on the reliability of each generation of product.

Note that in Table 8.2 the purpose is not to compare the reliability of different products. Instead, our intention is to compare the frequentist solution to the Bayesian solution. Even though generation #9 product has a relatively smaller lower bound for reliability, this only means that its reliability uncertainty is larger than that of other products. There is no indication that its reliability is worse compared to other products. If the designs of different product generations are not identical, the data may not be directly pooled together, thus information in the previous DVT cannot be leveraged to reduce sample size in the new DVT. For product generation #9, in order to demonstrate

Table 8.2 Estimated reliability using one-level separate Bayesian models (assuming uniform reliability prior) and the frequentist method.

Product generation #	Sample size	# of successes	Reliability 95% credible interval lower bound using the one-level Bayesian method	Reliability 95% lower confidence bound using the frequentist method (using Minitab version 17 One-Sample Proportion with exact method)
1 to 4	59	59	0.95	0.95
5 to 8	299	299	0.99	0.99
9	12	12	0.79	0.78
10	NA	NA	NA	NA

95%/99% confidence/reliability, a sample size of 299 with 0 failures is required in a DVT, using the frequentist method.

Another way to apply a one-level Bayesian model is to assume

$$\theta_i \sim Beta\ (\alpha, \beta),$$

where α and β are constants.

To estimate θ_9, first α and β are estimated from $\theta_1, \ldots, \theta_i, \ldots, \theta_8$ point estimates. This is achieved by calculating the mean and standard deviation of the sample point estimates $\theta_1, \ldots, \theta_i, \ldots, \theta_8$, and then converting the mean and standard deviation to corresponding α and β based on Eqs. (8.1) and (8.2), as discussed in Section 8.1. In this case, since there is no failure in any historical test, $\theta_1, \ldots, \theta_i, \ldots, \theta_8$ point estimates are identical, i.e. $\theta_1, \ldots, \theta_i, \ldots, \theta_8 = 1$. The sample mean of historical reliability is 1, with standard deviation 0. Thus constants α and β cannot be calculated per Eqs. (8.1) and (8.2), since the denominator cannot be 0.

Furthermore, Gelman et al. (2014) questioned the logic of directly estimating a prior distribution from existing data. Disadvantages of this approach include double counting the data when the objective is to estimate each θ_i using the estimated prior distribution and underestimating posterior uncertainty.

8.2.2 Bayesian Hierarchical Model

The data in Example 8.1 are hierarchical in nature. Now let us apply a Bayesian hierarchical binomial model to the dataset in Table 8.1. The structure of the Bayesian binomial hierarchical model is shown in Figure 8.2.

Compared to the one-level Bayesian model, in a Bayesian hierarchical model two more unknown parameters, α and β, are added at the second stage. They are called hyperparameters (parameters of a prior distribution). Instead of being constants, hyperparameters have their own distributions in a Bayesian hierarchical model.

In this case, hyperparameters α and β define the reliability distribution of each product, θ_i,

$$\theta_i \sim Beta\ (\alpha + 1,\ \ \beta + 1).$$

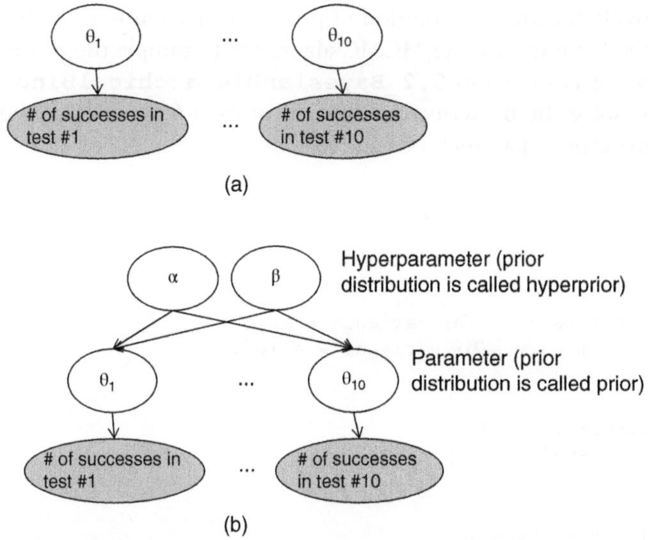

Figure 8.2 (a) Separate one-level Bayesian model and (b) Bayesian hierarchical binomial model (data are available for the gray nodes).

Note that it is more common to see $\theta_i \sim Beta\ (\alpha, \beta)$ in literature. We modified the distribution of θ_i in this specific case (there are no failures in Example 8.1) to avoid sampling challenges when results are concentrated to a single value.

Hyperparameters are unknown and will be inferred given the data. In the hierarchical model, all the unknown reliabilities of various product generations are considered to be a random sample drawn from a common distribution.

This hierarchical model structure allows test results of each generation of product to affect the inference of α and β, which in turn affects the inference of each reliability θ_i. With this hierarchical model, the reliability of each generation of product is not only influenced by its own test results, but is also partially influenced by the test results of other generations of products.

We assigned vague/weakly informative prior distributions to the second-stage parameters (hyperparameters). Specifically, here we assign the following vague distributions:

$$\alpha \sim Gamma\ (1,\ 1)$$

$$\beta \sim Gamma\ (1,\ 1).$$

There is no standard way to specify non-informative or vague distributions. Generally, the ranges in a non-informative distribution should be wide enough to cover all possible values for the parameter of interest. It is always good practice to change a non-informative prior distribution and run sensitivity analysis (e.g. change the mean or the standard deviation, etc.) to see if the specified prior distribution truly has negligible impact on the posterior distribution results.

In this case, we use *Gamma* (1, 1) as a prior for both α and β. This distribution has a mean of 1, a standard deviation of 1, and a range of $(0, \infty)$. With these hyperpriors, the prior distribution of θ_i has a mean of 0.50, a median of 0.50, and a standard deviation of 0.26 (this can be estimated via Monte Carlo simulations). This prior distribution of θ_i is considered a vague prior.

Gelman et al. (2014) provide the analytic solution of posterior distributions for the Bayesian hierarchical model. Here we use the MCMC algorithm to sample the posterior distributions. The R script is shown in **8.2_BayesianHierarchicalBinomial.R.** JAGS is used to solve the posterior distribution of each θ_i, as well as the posterior distributions of hyperprior parameters α and β.

(8.2_BayesianHierarchicalBinomial.JAGS)

```
model {

#alpha ~ dunif(0.1, 10)
alpha ~ dgamma(1, 1) # mean = a/b; variance = a/(b^2)
beta ~ dgamma(1, 1) # mean = a/b; variance = a/(b^2)

for( i in 1:9) {
theta[i] ~ dbeta(alpha+1,beta+1)
Successes[i] ~ dbin(theta[i], SampleSize[i])
}

theta[10] ~ dbeta(alpha+1,beta+1)
}
```

(8.2_BayesianHierarchicalBinomial.R)

```
##################################################################
## Bayesian hierarchical Binomial model                         ##
## to estimate reliability of various product generations       ##
##################################################################

### Load package rjags (to connect to JAGS from R for Bayesian
analysis)
library(rjags)

##########
## Data ##
##########

BayesianData <- list(
  Successes = c(59,59,59,59,299,299,299,299,12),
  SampleSize = c(59,59,59,59,299,299,299,299,12)
)


##############################################################
## Create the model, burnin, and collect posterior samples ##
##############################################################

# Create (& initialize & adapt) a JAGS model object
ModelObject <- jags.model(file = "8.2_BayesianHierarchicalBino-
mial.JAGS",
                          data=BayesianData,
                          inits=list(alpha = 1, beta = 1),
                          n.chains = 3, n.adapt = 1000
)
```

```
# Burn-in stage
update(ModelObject, n.iter=1000)

# Select variables to collect posterior samples
variable_names <- c("theta[1]", "theta[5]", "theta[9]", "theta[10]",
"alpha", "beta")

# Run MCMC and collect posterior samples in coda format for selected
variables
codaList <- coda.samples(model=ModelObject, variable.names =
variable_names,
                              n.iter = 10000, thin = 1)

# Summary statistics for Markov Chain Monte Carlo chains
# Quantiles of the sample distribution can be modified in the
quantiles argument
summary(codaList, quantiles = c(0.025, 0.05, 0.5, 0.95, 0.975))
```

After running the R and JAGS script, the summary statistics are:

```
> summary(codaList, quantiles = c(0.025, 0.05, 0.5, 0.95, 0.975))

Iterations = 2001:12000
Thinning interval = 1
Number of chains = 3
Sample size per chain = 10000

1. Empirical mean and standard deviation for each variable,
   plus standard error of the mean:

             Mean       SD Naive SE Time-series SE
alpha     8.24176 2.897922 1.673e-02      2.351e-02
beta      0.04287 0.042583 2.459e-04      5.844e-04
theta[10] 0.89082 0.102901 5.941e-04      6.163e-04
theta[1]  0.98501 0.014499 8.371e-05      8.455e-05
theta[5]  0.99662 0.003357 1.938e-05      1.988e-05
theta[9]  0.95204 0.045373 2.620e-04      2.687e-04

2. Quantiles for each variable:

              2.5%       5%     50%     95%   97.5%
alpha     3.512766 4.085226 7.92252 13.4611 14.7768
beta      0.001107 0.002186 0.02981  0.1290  0.1601
theta[10] 0.612335 0.678809 0.92098  0.9937  0.9968
theta[1]  0.946470 0.955839 0.98926  0.9991  0.9995
theta[5]  0.987732 0.990006 0.99763  0.9998  0.9999
theta[9]  0.831330 0.861506 0.96547  0.9972  0.9985
```

Table 8.3 list lower bounds of 95% confidence intervals of reliability using the traditional method vs. lower bounds of 95% credible intervals of reliability using the Bayesian hierarchical binomial model for different products. The intention is to compare the frequentist solution to the solution from the Bayesian hierarchical model. For product #9,

Table 8.3 Reliability 95% lower bound using traditional method vs. Bayesian hierarchical binomial model.

Test #	Sample size	# of successes	Reliability 95% lower confidence bound using frequentist method	Reliability 95% lower confidence bound using Bayesian hierarchical binomial model
1	59	59	0.95	0.96
5	299	299	0.99	0.99
9	12	12	0.78	0.86
10	NA	NA	N.A.	0.68

NA, not available.

its Bayesian solution has smaller uncertainty compared to the frequentist solution of reliability.

Note that for product #9, the sample size is only 12. If using the frequentist method without leveraging other sources of information, the reliability 95% lower bound for product #9 is 0.78. The reliability 95% lower bound using the Bayesian hierarchical binomial model is 0.86, which is higher compared to the results based on the frequentist method. This shows that partial information in previous tests is leveraged in the reliability estimation for product #9 using the Bayesian hierarchical model. This is why even though the sample size is small, the Bayesian reliability estimation of product #9 is higher.

Similarly, for product #10 or any other future product, which has no testing data at all, its reliability cannot be estimated using the frequentist method. However, its reliability can be predicted using the Bayesian hierarchical model (the 95% lower bound is 0.68). This means that for any future product, without observing any testing data, we have a prediction that its reliability has a lower bound of 0.68. This reliability estimation for product #10 (or any other future product) is a predictive analysis result based on the Bayesian hierarchical model by partially leveraging information from earlier products.

The following script can be used to show a histogram of p_9 posterior distributions.

```
codaMatrix <- as.matrix( codaList )
theta9_samples <- codaMatrix[,"theta[9]"]

hist(theta9_samples, breaks=100)
```

For convergence diagnosis, for example, the following script can be used to show the trace plot of p_9 posterior samples.

```
traceplot(codaList[,"theta[9]"])
```

Posterior distributions of p_1, p_5, p_9, p_{10}, *alpha*, and *beta* are shown in Figure 8.3. Figure 8.4 shows two trace plots to examine the convergence. As the two chains are mixed well, and there are no patterns observed, we can assume that convergence is achieved. Other diagnostics plots or results are not shown here. It is recommended that users check those diagnostics results themselves.

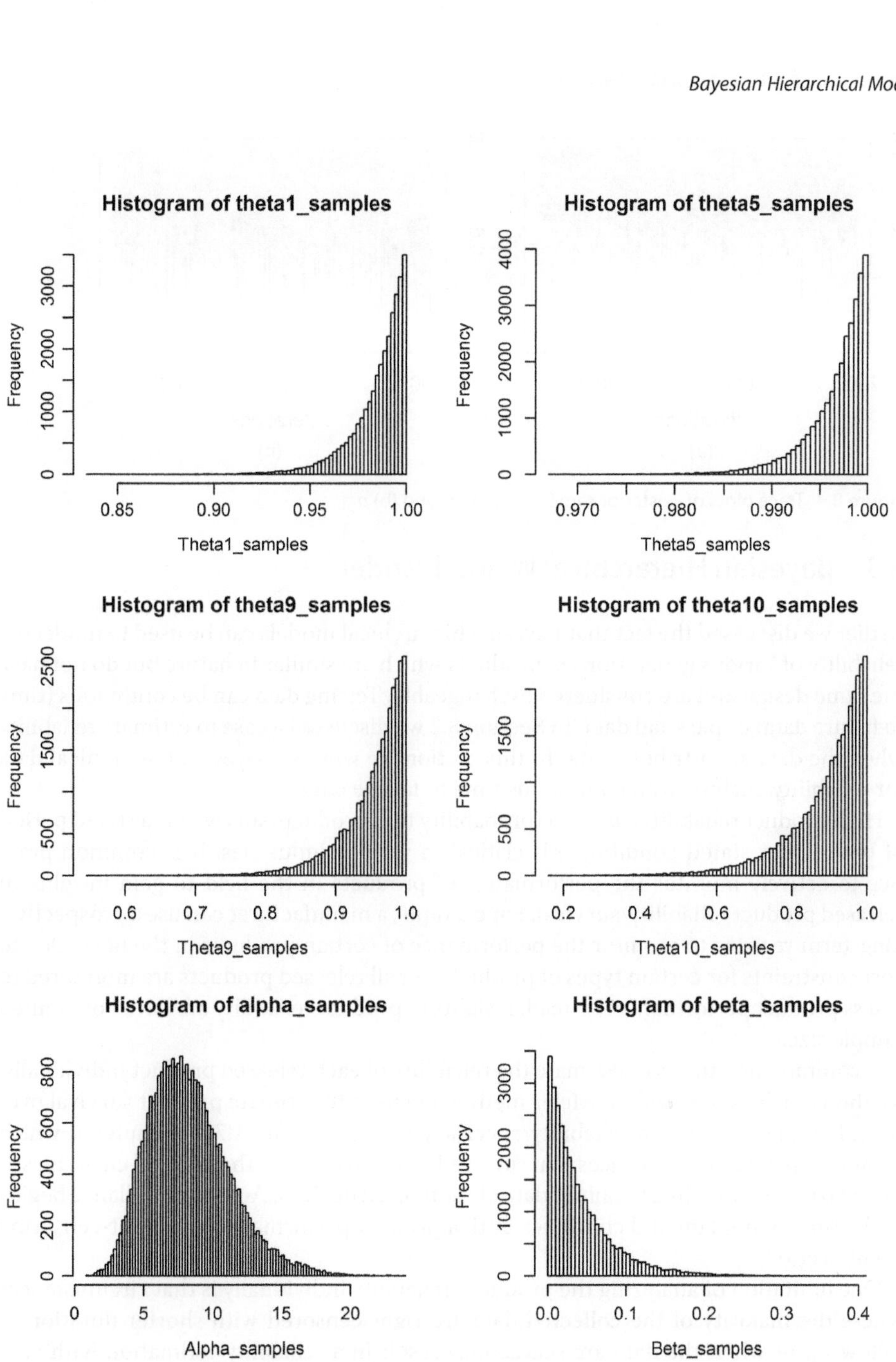

Figure 8.3 Posterior distributions of selected parameters.

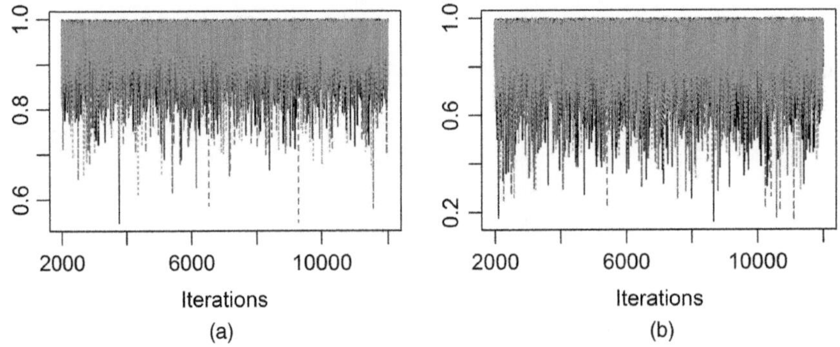

Figure 8.4 Trace plots of posterior samples for (a) p_9 and (b) p_{10}.

8.3 Bayesian Hierarchical Weibull Model

Earlier we discussed the fact that Bayesian hierarchical models can be used to model the reliability of various generations of products which are similar in nature but do not have the same design and are considered exchangeable. Testing data can be continuous (time to failure data) or pass/fail data. In Section 8.2 we discussed a case to estimate reliability when the data are attribute data. In this section we will use Bayesian hierarchical data for reliability analysis with continuous time to failure data.

High product reliability/survival (probability that products survive for a stated period of time under stated conditions) is critical in various industries. It is common practice to actively monitor the performance of products in the field to gain insights of released product reliability/survival. For example, a manufacturer can use a prospective, long-term registry to monitor the performance of certain products in the field. Due to cost constraints for certain types of products not all released products are monitored to assess product performance. Instead, insights of product reliability are based on limited sample size.

A common practice is to estimate the reliability of each released product individually. Methods such as the Kaplan–Meier method are used to estimate product survival over time. For an example of a reliability/survival curve, see Figure 1.7 in Chapter 1, which shows the percentage of devices that survived a period of time. The survival curve is estimated based on the time to failure data of monitored devices. Active surveillance begins at the time of implant and continues until a product performance or a right-censoring event occurs.

One limitation of analyzing the product's reliability individually is that circumstances where the majority of the collected data are right-censored with shorter duration of follow-up or when the data are scarce may result in a reliability estimation with very large uncertainty (i.e. the confidence intervals are too wide to be useful). On the other hand, it is not uncommon to see released products from the same family/category have very similar reliability performance, while products from different families have dramatically different reliability estimations. When using traditional methods to analyze the product's reliability individually, reliability information on previous products in the same family cannot be leveraged to predict the reliability of the newly released product.

Since products from the same family have very close reliability estimation, it is reasonable to build a Bayesian hierarchical model that can partially leverage the reliabilities of previous products as prior information. Thus, when estimating the reliability of a newly released product, partial strength can be borrowed from previous similar products to reduce the uncertainty of estimated reliability. In Example 8.2, a Bayesian hierarchical Weibull model is discussed.

Example 8.2 Assume time to failure data ttf (unit: month) is recorded for ten released products from the same family (simulated data are used in this example). For each released product, 100 devices were sampled and monitored. The ten released product generations are named Gen_1, Gen_2, …, Gen_10. These products are released to the market at different times and their market experience varies from 2 to 30 years.

- Sample data of *ttf[i, j]* are shown in Table 8.4, where *ttf[i, j]* is observed months to failure of the *i*th device of product Gen_j (NA indicates the data point is right-censored).
- Sample data of *CenLimit[, j]* are shown in Table 8.5, where *CenLimit[, j]* is the censoring limit (unit: month) for product Gen_j. For example, the censoring limit of product Gen_9 is 60 months. This means that Gen_9 has been released to the market for 60 months, so the right-censored time to failure data for Gen_9 are censored at 60 months.
- Sample data of *CensorLG[i, j]* are shown in Table 8.6. *CensorLG[i, j]* is an indicator of whether the data ttf[i, j] is a censored data (TRUE means censored data).

The objective in this example is to estimate the reliability of product Gen_10 at seven years (84 months). Note product Gen_10 has only been released to the market for 24 months, and all the time to failure data for product Gen_10 are right-censored data (there has not been observed failure yet).

Table 8.4 Sample data of ttf (months-to-failure) for each of the ten released products.

	"Gen_1"	"Gen_2"	"Gen_3"	"Gen_4"	"Gen_5"	"Gen_6"	"Gen_7"	"Gen_8"	"Gen_9"	"Gen_10"
1	235.68	NA	NA	NA	NA	NA	NA	NA	NA	NA
2	99.95	NA	86.75	NA	NA	NA	NA	NA	NA	NA
3	198.5	37.24	NA	NA	NA	NA	NA	NA	NA	NA
4	91.52	NA	NA	NA	NA	73.06	NA	NA	NA	NA
5	NA	NA	240.88	NA	NA	NA	NA	NA	NA	NA
6	NA	106.48	171.3	NA	NA	133.5	NA	NA	NA	NA
7	NA	NA	NA	NA	176.76	NA	NA	NA	NA	NA
8	NA	139.45	NA	NA	NA	NA	NA	NA	NA	NA
9	229.25	NA	NA	NA	NA	NA	NA	NA	NA	NA
10	8.83	NA	249.6	NA	NA	NA	NA	NA	NA	NA
…	…	…	…	…	…	…	…	…	…	…
100	NA	NA	NA	NA	163.05	NA	NA	NA	NA	NA

NA, not available.

Table 8.5 Sample data of CenLimit (censoring limit in months) for each of the ten released products.

"Gen_1"	"Gen_2"	"Gen_3"	"Gen_4"	"Gen_5"	"Gen_6"	"Gen_7"	"Gen_8"	"Gen_9"	"Gen_10"
360	300	252	216	180	144	120	96	60	24
…	…	…	…	…	…	…	…	…	…

Table 8.6 Sample data of CensorLG for each of the ten released products.

"Gen_1""	"Gen_2"	"Gen_3"	"Gen_4"	"Gen_5"	"Gen_6"	"Gen_7"	"Gen_8"	"Gen_9"	"Gen_10"
FALSE	TRUE	TRUE	TRUE	TRUE	TRUE	TRUE	TRUE	TRUE	TRUE
FALSE	TRUE	FALSE	TRUE	TRUE	TRUE	TRUE	TRUE	TRUE	TRUE
FALSE	FALSE	TRUE	TRUE	TRUE	TRUE	TRUE	TRUE	TRUE	TRUE
FALSE	TRUE	TRUE	TRUE	TRUE	FALSE	TRUE	TRUE	TRUE	TRUE
…	…	…	…	…	…	…	…	…	…

Note that the data format in Tables 8.4–8.6 is set up in a way that the data are easily readable by the R and JAGS scripts (**`8.3_BayesianHierarchicalWeibull.R`**). For example, in Table 8.4, time to failure information for right-censored data is not available, and is indicated by NA. Table 8.5 shows the censoring limit for each product. Each product has one censoring limit, so each column in Table 8.5 has identical censoring limit values. Table 8.6 is a table of logical values indicating whether a value in Table 8.4 is right-censored or not. For example, row 1, column 1 in Table 8.4 is not censored data, thus row 1, column 1 in Table 8.6 is marked as FALSE. Row 1, column 2 in Table 8.4 is right-censored data, thus row 1, column 2 in Table 8.6 is marked as TRUE. If readers would like to use this script for their own data, you can prepare your own data tables following the format shown in Tables 8.4–8.6.

Since for Gen_10 product there are only right-censored data, and the monitoring window is only 24 months, data from sampled Gen_10 devices alone are not sufficient to estimate the reliability of Gen_10 product at seven years (84 months) if treating the data as continuous.

Time to failure data in Example 8.2 are simulated from Weibull distributions with fixed scale and shape parameters, and then truncated at different censoring limits. Different products' time-to-failure data have different fixed scale and shape parameters, sampled from normal distributions. In reliability practices, time-to-failure data of different products from the same family are usually found to have close scale or shape parameters, especially when the designs are very similar from one generation to another.

True values of shape and scale parameters used for the simulated data are listed in Table 8.7. True reliability values of each product at 84 months are calculated per the true values of shape and scale parameters based on the following equation:

$$reliability = \exp\left(-\left(\frac{84}{scale}\right)^{shape}\right).$$

Let us apply a Bayesian hierarchical model in this example. The model structure is shown in Figure 8.5 and is discussed as follows.

Table 8.7 True values of Weibull shape and scale parameters, and corresponding reliability (shape and scale parameters are sampled from normal distributions).

Product name	Shape	Scale	Reliability
Gen_1	1.05271	681.011	0.895419
Gen_2	1.30417	610.300	0.927470
Gen_3	1.25654	489.721	0.896623
Gen_4	1.53451	439.992	0.924240
Gen_5	1.58226	570.550	0.952891
Gen_6	1.33131	549.467	0.921222
Gen_7	1.38997	674.867	0.946269
Gen_8	2.15500	510.555	0.979744
Gen_9	1.47402	610.915	0.947733
Gen_10	1.28420	667.784	0.932594

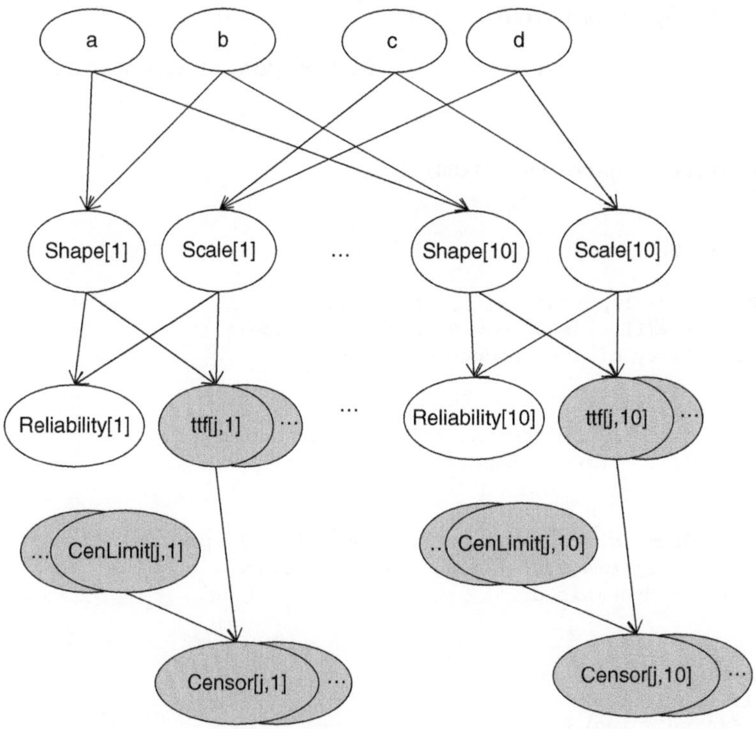

Figure 8.5 Bayesian hierarchical Weibull model (data are available for the gray nodes).

Let R_i be the reliability of the ith product at 84 months, and η_i and β_i are the scale and shape parameters of a Weibull distribution for the ith product, then R_i and ttf_{ij} are both dependent on η_i and β_i. i.e.

$$R_i = e^{-(84/\eta_i)^{\beta_i}}$$

$$ttf_{ij} \sim Weibull\ (\eta_i, \beta_i).$$

η_i and β_i are unknown. We assume they each have a gamma distribution (gamma distributions are convenient to model positive variables), i.e.

$$\beta_i \sim Gamma\ (a, b)$$

$$\eta_i \sim Gamma\ (c, d).$$

If parameters a, b, c, and d are constants, then all the β_i have the same prior distributions (the same case for η_i) and the model becomes a basic Bayesian model instead of a Bayesian hierarchical model. Here in the hierarchical model, we assume a, b, c, and d each has its own distribution. These four parameters are hyperparameters in the hierarchical model. Since there is no other information on these parameters, non-informative distributions are used to represent the prior distributions for hyperparameters a, b, c, and d. Specifically, the following hyperpriors are applied:

$$a, b, c, d \sim Uniform\ (0.001,\ 100).$$

The R and JAGS scripts are shown in **8.3_BayesianHierarchical-Weibull.R.**

```
(8.3_BayesianHierarchicalWeibull.JAGS)
model {
    # Likelihood:
    for( i in 1:10) {

        lambda[i] <- 1/pow(scale[i], shape[i])
        reliability[i] <- exp(-pow(84/scale[i],shape[i]))
        # reliability of ith product at 84 months

        for( j in 1:100){
            Censor[j,i] ~ dinterval(ttf[j,i], CenLimit[j,i])
            ttf[j,i] ~ dweib(shape[i], lambda[i])
        }

        # vague Gamma distribution for shape and scale priors
        shape[i] ~ dgamma(a,b) # mean = a/b; variance = a/(b^2)
        scale[i] ~ dgamma(c,d)
    }

    # flat hyperpriors
    a ~ dunif(0.001, 100)
    b ~ dunif(0.001, 100)
    c ~ dunif(0.001, 100)
    d ~ dunif(0.001, 100)
}
```

(8.3_BayesianHierarchicalWeibull.R)

```
##################################################################
## To estimate the reliability of product #10 at 84 months,     ##
## a Bayesian hierarchical model is used                        ##
##################################################################

### Load package rjags (to connect to JAGS from R for Bayesian analysis)
library(rjags)

# The Weibull distribution with shape parameter a and scale parameter b
has density given by
# f(x) = (a/b) (x/b)^(a-1) exp(- (x/b)^a)
# for x > 0. The cumulative distribution function is
F(x) = 1 - exp(- (x/b)^a) on x > 0


# read data from files
ttf <- read.table("Example8.2Data_ttf_Table8.4.txt",header=T,sep="",
na.strings = "NA")
CenLimit <- read.table("Example8.2Data_CenLimit_Table8.5.txt",
header=T,sep="")
CensorLG <- read.table("Example8.2Data_CensorLG_Table8.6.txt",
header=T,sep="")

# convert a data frame to a logical matrix
ttf <- data.matrix(ttf, rownames.force = NA)
CenLimit <- data.matrix(CenLimit, rownames.force = NA)

# convert a data frame to a numeric matrix
CensorLG <- sapply(as.data.frame(CensorLG), as.logical)

##########
## Data ##
##########

#JAGS dinterval needs 0,1 so convert logical CensorLG to numeric
BayesianData <- list(ttf = ttf,
                   CenLimit = CenLimit, Censor = CensorLG*1
)

##############################################################
## Create the model, burn-in, and collect posterior samples ##
##############################################################

# intial values of censored data:
ttfInit <- array(NA,c(100,10))
ttfInit[CensorLG] = CenLimit[CensorLG]+1

# Define different initial values for multiple chains
# Use .RNG.name and .RNG.seed to make results reproducible
Initial1 <- list(.RNG.name="base::Super-Duper", .RNG.seed=3456,
ttf=ttfInit, a=50, b=30, c=60, d=0.1)
Initial2 <- list(.RNG.name="base::Super-Duper", .RNG.seed=6543,
ttf=ttfInit, a=60, b=40, c=70, d=0.5)
InitialValues <- list(Initial1, Initial2)
```

```
ModelObject <- jags.model(file = "8.3_BayesianHierarchicalWeibull.JAGS",
                        data=BayesianData, inits=InitialValues,
                        n.chains = 2, n.adapt = 1000
)

# Burn-in stage
update(ModelObject, n.iter=2000)

# Select variables to collect posterior samples
variable_names <- c("a", "b", "c", "d", "shape[10]", "scale[10]",
"reliability")

# Run MCMC and collect posterior samples in coda format for selected
variables
codaList <- coda.samples(model=ModelObject, variable.names =
variable_names,
                            n.iter = 100000, thin = 2)

# Summary statistics for Markov Chain Monte Carlo chains
# Quantiles of the sample distribution can be modified in the quantiles
argument
summary(codaList, quantiles = c(0.025, 0.5, 0.975))
```

After running the R script, the results are shown below. The trace plot (left) and density plot (right) of posterior samples of R_{10} (reliability[10] in R script) are shown in Figure 8.6. The trace plot shows that the two chains are mixing well and there are no patterns observed, so we can assume that convergence is achieved. Figure 8.7 shows histograms of posterior samples of reliability vs. true reliability (indicated by dashed lines).

```
> summary(codaList, quantiles = c(0.025, 0.5, 0.975))

Iterations = 3002:103000
Thinning interval = 2
Number of chains = 2
Sample size per chain = 50000

1. Empirical mean and standard deviation for each variable,
   plus standard error of the mean:

                  Mean       SD  Naive SE Time-series SE
a              57.7779 24.14152 7.634e-02      1.122e+00
b              37.7017 16.48355 5.213e-02      7.748e-01
c              68.9947 21.72975 6.872e-02      8.538e-01
d               0.1381  0.04657 1.473e-04      1.827e-03
```

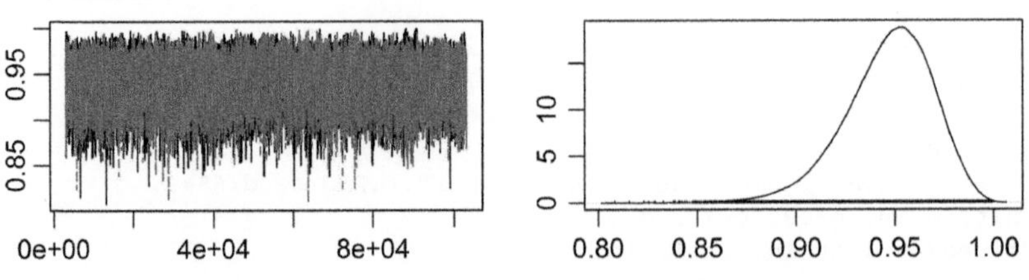

Figure 8.6 Trace plot (left) and density plot (right) of posterior samples of reliability [10].

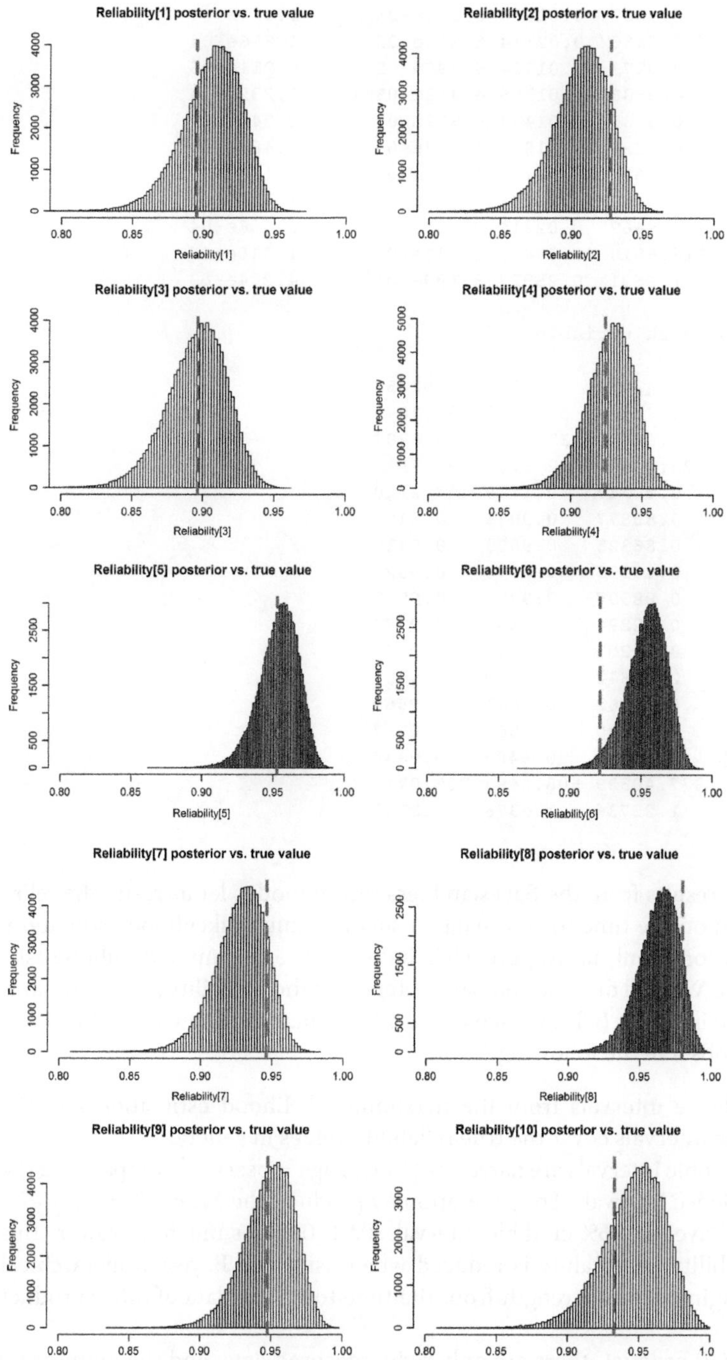

Figure 8.7 Histograms of reliability posterior samples vs. true reliability (dashed lines).

```
reliability[1]     0.9055  0.02106 6.660e-05         2.226e-04
reliability[2]     0.9067  0.01987 6.284e-05         1.412e-04
reliability[3]     0.8956  0.02114 6.686e-05         1.876e-04
reliability[4]     0.9273  0.01704 5.390e-05         9.211e-05
reliability[5]     0.9538  0.01395 4.412e-05         1.299e-04
reliability[6]     0.9536  0.01427 4.512e-05         1.246e-04
reliability[7]     0.9283  0.01813 5.732e-05         8.483e-05
reliability[8]     0.9628  0.01446 4.571e-05         2.121e-04
reliability[9]     0.9486  0.01796 5.680e-05         1.473e-04
reliability[10]    0.9467  0.02214 7.000e-05         2.186e-04
scale[10]        516.6810 83.38453 2.637e-01         1.346e+00
shape[10]          1.6691  0.25927 8.199e-04         4.204e-03

2. Quantiles for each variable:

                     2.5%      50%     97.5%
a                15.75141  57.2442   97.7875
b                 9.64506  37.0407   66.6787
c                22.93903  72.5703   98.7933
d                 0.04247   0.1439    0.2120
reliability[1]    0.85877   0.9075    0.9409
reliability[2]    0.86325   0.9083    0.9410
reliability[3]    0.84953   0.8973    0.9322
reliability[4]    0.88997   0.9287    0.9567
reliability[5]    0.92294   0.9551    0.9772
reliability[6]    0.92203   0.9549    0.9776
reliability[7]    0.88842   0.9299    0.9591
reliability[8]    0.93062   0.9641    0.9869
reliability[9]    0.90890   0.9503    0.9787
reliability[10]   0.89733   0.9489    0.9838
scale[10]       377.40529 508.6493  705.0712
shape[10]         1.25736   1.6376    2.2745
```

To compare with results from the Bayesian hierarchical model, let us rerun the reliability analysis based on the time to failure data using maximum likelihood estimation, which is currently a commonly used approach in industry. The maximum likelihood estimates (MLEs) of the Weibull distribution parameters, and the reliability point estimates and 95% confidence intervals (CI) for each of the 10 products are shown in Table 8.8.

In Figure 8.8 we observe the following:

- Both the confidence intervals from the maximum likelihood estimation and the Bayesian credible intervals cover the true reliability values in general.
- Bayesian 95% credible intervals are narrower ("shrinkage" observed) compared to the MLE 95% confidence intervals. For generation #9 product, the MLE 95% CI is (0.74, 0.99), while the Bayesian 95% credible interval, (0.91, 0.98), is much narrower. This shows that reliability uncertainty is reduced when using the Bayesian hierarchical model, by leveraging partial strength from the time to failure data of other product generations.
- For generation #10 product, there are only right-censored data, and the monitoring window is only 24 months, so the MLE solution does not exist. The Bayesian hierarchical Weibull model enables the reliability of generation #10 product to be assessed. The estimated reliability of Gen_10 product at 7 years (84 months) has a mean of 0.95 and a 95% Bayesian credible interval of (0.90, 0.98).

Table 8.8 MLE of the Weibull distribution parameters and reliability.

Product name	Product generation #	No. of uncensored value	No. of right censored value	Weibull shape parameter MLE	Weibull scale parameter MLE	Reliability at 84 months	Reliability 95% CI lower bound	Reliability 95% CI upper bound
Gen_1	1	45	55	1.08862	579.958	0.885109	0.819387	0.927953
Gen_2	2	45	55	1.31223	444.486	0.893748	0.829684	0.934651
Gen_3	3	40	60	1.22392	436.275	0.875342	0.807959	0.920235
Gen_4	4	30	70	1.75798	388.159	0.934420	0.878544	0.965093
Gen_5	5	13	87	1.73560	562.638	0.963817	0.914924	0.984840
Gen_6	6	9	91	1.89815	498.291	0.966506	0.918401	0.986457
Gen_7	7	12	88	1.42802	503.446	0.925399	0.865406	0.959271
Gen_8	8	2	98	2.71626	403.376	0.986004	0.940531	0.996765
Gen_9	9	2	98	2.58478	271.384	0.952889	0.738095	0.992361
Gen_10	10	0	100	N.A.[a)]	N.A. [a)]	N.A. [a)]	N.A. [a)]	N.A. [a)]

a) Maximum likelihood estimates could not be calculated when all the data are right-censored.

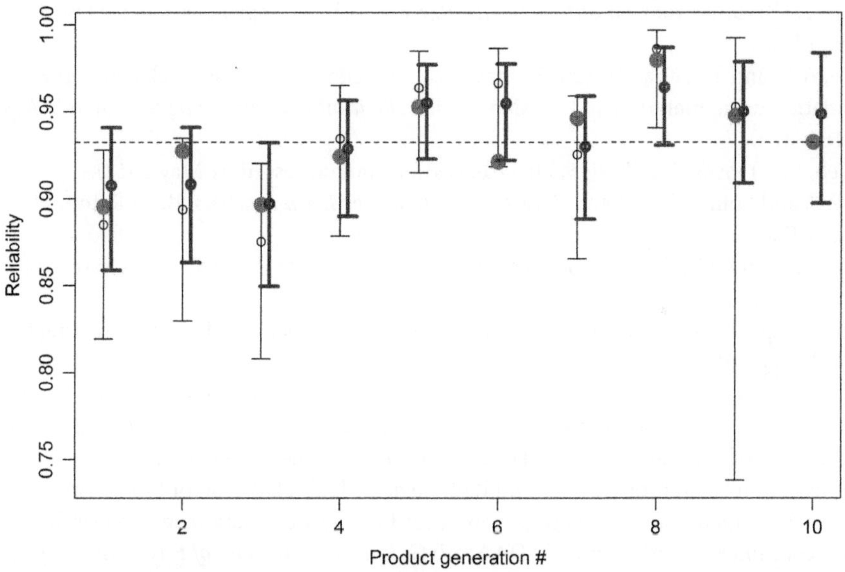

Figure 8.8 True reliability (large solid circles) vs. reliability MLE (thin error bars on the left, with open circles indicating point estimates from MLE and error bars indicating 95% confidence interval) and Bayesian reliability estimates (thick error bars on the right, with solid circles indicating posterior medians and error bars indicating 95% credible interval). The horizontal dash line shows the average of the ten products' true reliability.

For model assessment and selection discussions, readers may refer to Carlin and Louis (2009). Sensitivity analysis can be used to evaluate the impact of prior distributions and the robustness of the model. Methods such as the sum of residual squares can be used to compare and select various Bayesian models with different likelihood or model structures.

8.4 Summary

In this chapter Bayesian hierarchical models were introduced to estimate product reliability. In reliability practice, it is often challenging to estimate the reliability of highly reliable products as this usually requires lots of data and/or long testing/monitoring time. Bayesian hierarchical models allow partial strength to be borrowed from previous products to predict the reliability of a newly developed/released product. This is especially helpful when failure data are scarce for the new product. The potential benefits of applying Bayesian hierarchical models include reduced sample size in a reliability test and/or reduced uncertainty of predicted reliability for a new product.

References

Albert, J. (2009). *Bayesian Computation with R.* New York, NY: Springer Science+Business Media, LLC.

Andrade, A.R. and Teixeira, P.F. (2015). Statistical modelling of railway track geometry degradation using hierarchical Bayesian models. *Reliability Engineering & System Safety* 142: 169–183.

BayesWeb, (n.d.) *Prior distribution,* http://bayesweb.com (accessed 01 May 2018).

Carlin, P.C. and Louis, T.A. (2009). *Bayesian Methods for Data Analysis.* Boca Raton, FL: Taylor & Francis.

Gelman, A., Carlin, J.B., Stern, H.S. et al. (2014). *Bayesian Data Analysis.* Boca Raton, FL: Taylor & Francis.

Johnson, V.E., Moosman, A., and Cotter, P. (2005). A hierarchical model for estimating the early reliability of complex systems. *IEEE Transactions on Reliability* 54 (2): 224–231.

Mishra, M., Martinsson, J., Rantatalo, M. et al. (2018). Bayesian hierarchical model-based prognostics for lithium-ion batteries. *Reliability Engineering & System Safety* 172: 25–35.

US Department of Human Services & Food and Drug Administration and Drug Administration Center for Devices and Radiological Health Division of Biostatistics Office of Surveillance and Biometrics Center for Biologics Evaluation and Research (2010). *Guidance for Industry and FDA Staff: Guidance for the Use of Bayesian Statistics in Medical Device Clinical Trials.* FDA.

9

Regression Models

This chapter introduces how to use Bayesian methods to build linear regression and binary logistic regression models. Examples and R scripts are provided. We also show how to use a Bayesian regression model to make predictions.

9.1 Linear Regression

Linear regression builds a relationship between a continuous response variable (also called a dependent variable) and one or more predictors (also called independent variables) by fitting a linear equation to observed data.

A simple linear regression model builds a relationship between a continuous response variable and one predictor, and the relationship is a straight line, i.e.

$$y = \beta_0 + \beta_1 x + \varepsilon, \tag{9.1}$$

where

y is the response (or dependent variable, or predicted variable)
x is the predictor (or independent variable, or regressor variable)
β_0 and β_1 are unknown regression coefficients
β_0 is the intercept, which is the mean of the distribution of y when x equals to zero
β_1 is the slope coefficient, which indicates the change of the mean of the distribution of y when x changes by a unit
ε is the error term, which is assumed to follow a normal distribution with a mean of 0 and unknown variance σ^2.

Thus, at each given value of x, the distribution of y has a mean of $\beta_0 + \beta_1 x$ and variance σ^2. This makes a simple linear regression model easy to use in various fields.

For example, a linear regression equation can be used to establish the relationship between infant weight and height. In general, when height increases, the baby's weight will increase. If we are interested in predicting weight based on height information, then body weight is the response here and height is the predictor. When a baby's height is given, the baby's weight can be estimated based on a linear regression equation.

As another example, consider the price and size of a house in a community in a particular time period (e.g. summer 2017). In general, house prices rise when the size of the house (in square feet) increases. If our interest is to predict house prices based on the size of the house, then price is the response and size is the predictor. Using a known

Practical Applications of Bayesian Reliability, First Edition. Yan Liu and Athula I. Abeyratne.

Companion website: www.wiley.com/go/bayesian20

linear regression equation, when the size of a house is given, the price can be estimated in that community in the same time period (e.g. summer 2017).

Note that a predictor variable and a response in a linear regression equation do not necessarily have a causal relationship. Linear regression only indicates the correlation between predictors and responses. For example, people may find that the number of traffic accidents in Minnesota is highly correlated with the sales volume of snow pants in local stores. In this case, a linear regression model can be established in which the number of car accidents is the response and the sales volume of snow pants is the predictor. However, this does not mean that sales of snow pants lead to car accidents. The response and predictor in this case may be due to an actual reason: the snow season. In the snow season, both the number of car accidents and the sales of snow pants increase as compared to summer.

In a simple linear regression model, there is only one predictor. A multiple linear regression model builds a relationship between a continuous response variable and two or more independent predictors. As an example, assume a response y is dependent on multiple predictors, and the relationship can be described as

$$y = \beta_0 + \sum_{i=1}^{n} \beta_i x_i + \varepsilon, \tag{9.2}$$

where

y is the response variable
x_i $(i = 1, \ldots, n)$ are the predictors
n is the number of predictors
β_0 and β_i $(i = 1, \ldots, n)$ are the unknown coefficients
β_0 is the mean of the distribution of y when all the predictors (i.e. x_i $(i = 1,\ldots, n)$) are equal to zero
β_i $(i = 1,\ldots, n)$ indicates the change of the mean of the distribution of y when x_i changes by a unit while other predictors remain unchanged
ε is the error term.

Equation (9.2) is a multiple regression model with n predictors. Interaction effects and curvature effects (polynomial models) can also be analyzed by multiple regressions. Equations (9.3) and (9.4) are also multiple regression models.

$$y = \beta_0 + \beta_1 x_1 + \beta_2 x_2 + \beta_3 x_1 x_2 + \varepsilon, \tag{9.3}$$

$$y = \beta_0 + \beta_1 x_1 + \beta_2 {x_1}^2 + \varepsilon. \tag{9.4}$$

In Eq. (9.3), β_3 is an unknown coefficient of a second-order term $x_1 x_2$, which indicates interaction effects. When x_2 is held constant, for every unit change of x_1, the mean of the distribution of y changes by $\beta_3 x_2$, which not only depends on β_3, but also depends on the value of x_2. In Eq. (9.4), β_2 is an unknown coefficient of a quadratic term, which indicates curvature effects.

These types of multiple linear regression models are widely used in engineering practices via methods like design of experiments (Montgomery 2013) and response surface methodology (Myers et al. 2016). Multiple regression models represent engineering empirical transfer functions between responses and predictors for analysis and prediction purposes. They are particularly useful when first principle equations between responses and predictors are a challenge to establish.

A regression equation is called a linear regression when it is linear in the parameters. Equations (9.2)–(9.4) are all linear regressions since the relationships in these models are linear functions of the unknown coefficients $\beta_0, \beta_1, \ldots, \beta_n$. More discussions on multiple linear regressions can be found in Montgomery et al. (2012).

Example 9.1 provides a simple linear regression data analysis example. We first analyze the data using Minitab®. Then we provide the Bayesian solution with R script and results. We also demonstrate the Bayesian way to estimate the response when the predictor value is given.

As discussed in Chapter 2, compared to frequentist methods a key difference in Bayesian statistics is that unknown parameters are treated as random variables with probability distributions assigned, rather than unknown constants. The frequentist solution to a linear regression model provides point estimates and the associated confidence intervals of the regression coefficients. The confidence intervals quantify the uncertainty associated with the point estimates. In the Bayesian solution, posterior samples of the coefficients represent the uncertainty and therefore the probability distributions of these parameters. As an example, in a simple linear regression model the Bayesian solution comprises a series of slopes and intercepts values. In a Bayesian framework, uncertainty of unknown parameters can be easily propagated to the final outcome.

Example 9.1 Table 9.1 lists simulated data of predictor x and response y generated from a simple linear regression model. First, x is a sequence generated from a starting value of 0 to an ending value of 30, with the increment being 0.5. Next, the noise term, *error*, is sampled from a normal distribution with a mean of 0 and a standard deviation of 8. y is calculated from the following linear regression of y vs. x:

$$y = 3x + 10 + error.$$

This means that y has a mean of $3x + 10$ and a standard deviation of 8.

Assume the regression coefficients are unknown, and we would like to estimate unknown slope and intercept coefficients, and to estimate the value of y when x is 12.8.

It is always a good idea to explore the data graphically first before establishing a statistical model. First, the data are shown in a scatter plot (Figure 9.1). A scatter plot is a commonly used plot to display values for two variables as a collection of points. For each point, its position on the horizontal axis equals the value of one variable and its position on the vertical axis equals the value of the other variable. The scatter plot in Figure 9.1 shows y vs. x. It appears that the amount of increased y is proportional to the increase of x, so there could be a linear relationship between these two variables.

Since we would like to predict the value of y when x is given, in this case y is the response and x is the predictor.

First, we use Minitab to run linear regression analysis (to fit least squares models). Figure 9.2 shows the Minitab linear regression results. The linear regression model estimates y as a function of x. The analysis of variance table shows the p-value is 0.000 for the variable x. Since $p < 0.05$ for the variable x, this indicates that x significantly impacts y. The coefficient for the variable x is 3.026, which indicates that for every unit increase of x, the value of y increases by 3.026.

Table 9.1 *x* and *y*.

x	*y*	*x*	*y*	*x*	*y*
0.0	11.5995	10.5	31.3845	21.0	65.143
0.5	19.0216	11.0	37.0165	21.5	64.864
1.0	17.3595	11.5	37.0324	22.0	82.264
1.5	19.0145	12.0	41.1153	22.5	94.490
2.0	4.3508	12.5	55.9709	23.0	64.860
2.5	16.7596	13.0	49.1119	23.5	77.142
3.0	16.9405	13.5	52.3296	24.0	83.121
3.5	29.6552	14.0	51.4863	24.5	82.212
4.0	26.9851	14.5	50.5312	25.0	90.267
4.5	29.4851	15.0	65.6465	25.5	97.302
5.0	18.4533	15.5	62.0491	26.0	80.749
5.5	20.9916	16.0	55.0116	26.5	94.555
6.0	29.3405	16.5	54.1408	27.0	98.919
6.5	24.0566	17.0	63.4129	27.5	108.880
7.0	35.6257	17.5	55.8068	28.0	92.869
7.5	15.5164	18.0	73.7722	28.5	90.362
8.0	43.7983	18.5	76.7759	29.0	102.920
8.5	29.6881	19.0	67.7105	29.5	104.359
9.0	48.3340	19.5	67.3502	30.0	81.743
9.5	28.9686	20.0	68.1820		
10.0	48.9138	20.5	65.0644		

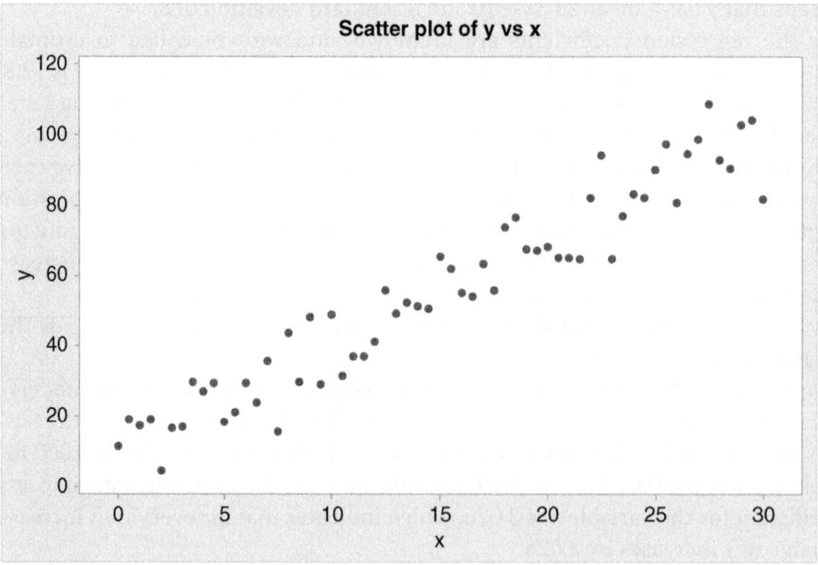

Figure 9.1 Scatterplot of ***y*** vs. ***x***.

Based on the linear regression output from Minitab, the prediction model for y as a function of x is given by

$$y = 9.91 + 3.026\ x + error,$$

where error follows a normal distribution with a mean of 0 and a standard deviation of 7.92 (the standard error, reported as S in Figure 9.2).

Based on this linear regression equation, when $x = 12.8$, y has a mean of $9.91 + 3.026 \times 12.8 = 48.6$ and a standard deviation of 7.92. The estimations of regression coefficients and the prediction of y are close to the true values. Note that the true value of slope and intercept are 3 and 10, respectively. Based on the true values of linear regression coefficients, the value of y has a mean of 48.4 and a standard deviation of 8.

Now we use Just Another Gibbs Sampler (JAGS) and R scripts to provide Bayesian estimation for this problem. The structure (directed acyclic graph) of the linear regression Bayesian model is shown in Figure 9.3.

In this Bayesian model there are three unknown variables, named *beta*0, *beta*1, and *sigma*. The logical relationship between $\boldsymbol{y}$, $\boldsymbol{x}$, and the unknown variables is

$$y[i] \sim Normal(beta0 + beta1 \times x[i],\ sigma),\ \ i = 1, \ldots n,$$

where n is the length of the vector $\boldsymbol{x}$ or $\boldsymbol{y}$

```
Regression analysis: y versus x

Analysis of variance

Source        DF   Adj SS    Adj MS   F-Value   P-Value
Regression     1    43274   43274.0    689.61     0.000
  x            1    43274   43274.0    689.61     0.000
Error         59     3702      62.8
Total         60    46976

Model summary

      S     R-sq   R-sq(adj)    R-sq(pred)
7.92161   92.12%      91.99%        91.54%

Coefficients

Term         Coef    SE Coef       T-Value      P-Value    VIF
Constant     9.91    2.00          4.95           0.000
x            3.026   0.115         26.26          0.000   1.00

Regression equation

y = 9.91 + 3.026 x
```

Figure 9.2 Linear regression results from Minitab.

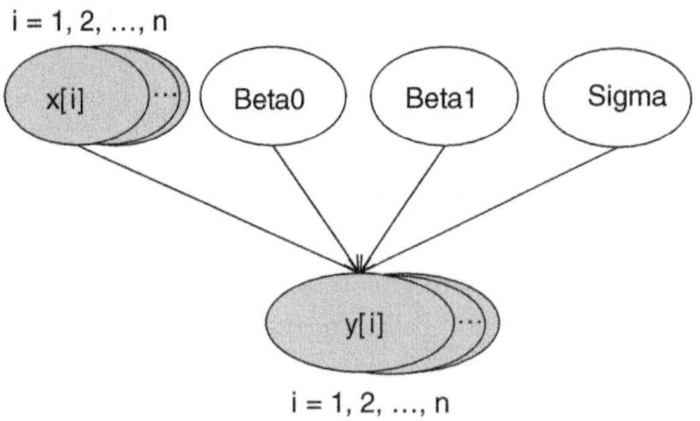

Figure 9.3 Linear regression Bayesian model (directed acyclic graph).

beta 0 is the unknown intercept
beta 1 is the unknown slope coefficient
sigma is the unknown standard deviation of the error term.

In other words, every data point in vector ***y***,$y[i]$, follows a normal distribution with a mean of $beta0 + beta1 \times x[i]$ and a standard deviation of *sigma*.

The JAGS model is shown in **9.1_Linear_Regression.JAGS**. The R script, shown in **9.1_Linear_Regression.R**, estimates the intercept and slope coefficient in the linear regression, and predicts the value of y for a given x. To estimate the value of y when $x = 12.8$, simply add 12.8 to the vector ***x*** and a missing value (NA) to the vector ***y***.

(9.1_Linear_Regression.JAGS)

```
model{

# likelihood model: linear regression
for( i in 1:n) {
      Y[i]  ~ dnorm(mu[i] , tau)                    #  regression model
        mu[i] <- beta0 + beta1*x[i]
} # end of  [i] loop

# diffuse priors
beta0 ~ dnorm(0,0.001)
beta1 ~ dnorm(0,0.001)

tau ~ dgamma(0.1, 0.1)   # gamma prior on precision tau
sigma <- 1/sqrt(tau)

} #  end of model
```

(9.1_Linear_Regression.R)

```
##################################################################
## Bayesian model for Linear regression                          ##
##################################################################

### Load package rjags (to connect to JAGS from R for Bayesian
analysis)
library(rjags)

# read data from a file
data <- read.table("Example9.1_LinearRegression.txt",header=T,sep="")
```

```
    x <- data[,1]
    y <- data[,2]

    # For prediction purposes, add 1 data row with x = 12.8
    # add NA to predict the y value for this x
    x <- c(x, 12.8)
    y <- c(y, NA)

    # Data for the JAGS model
    BayesianData <- list(x = x,
                         Y = y, n = length(x)
    )

    # Create (& initialize & adapt) a JAGS model object
    ModelObject <- jags.model(file = "9.1_Linear_Regression.JAGS",
                              data=BayesianData,
                              n.chains = 2, n.adapt = 1000
    )

    # Burn-in stage
    update(ModelObject, n.iter=10000)

    # Run MCMC and collect posterior samples in coda format for selected
variables
    codaList <- coda.samples(model=ModelObject, variable.names =
c("beta0", "beta1", "sigma", "Y[62]"),
                                  n.iter = 50000, thin = 1)

    # Summary statistics for Markov Chain Monte Carlo chains
    # Quantiles of the sample distribution can be modified in the
quantiles argument
    summary(codaList, quantiles = c(0.025, 0.05, 0.5, 0.95, 0.975))
```

Summary plots of unknown parameters, *beta*0, *beta*1, *sigma*, and $Y[62]$ (the unknown y value when $x = 12$) are shown in Figure 9.4. After Markov chain Monte Carlo (MCMC) sampling, summary statistics of unknown parameters are shown below. The mean of *beta*0, *beta*1, and *sigma* Bayesian posterior distributions are 9.9, 3.0, and 8.0, respectively, which are close to the corresponding point estimates from Minitab (9.9, 3.0, and 7.9, respectively). The 95% credible intervals of *beta*0, *beta*1, and *sigma* cover the corresponding parameter true values (10, 3, and 8, respectively). When $x = 12.8$, the y value ($Y[62]$) has a mean of 48.6 and standard deviation of 8.1, which is close to the true value (the true value of y has a mean of 48.4 and a standard deviation of 8 when $x = 12.8$).

```
> summary(codaList, quantiles = c(0.025, 0.05, 0.5, 0.95, 0.975))

Iterations = 10001:60000
Thinning interval = 1
Number of chains = 2
Sample size per chain = 50000

1. Empirical mean and standard deviation for each variable,
   plus standard error of the mean:

          Mean     SD  Naive SE Time-series SE
```

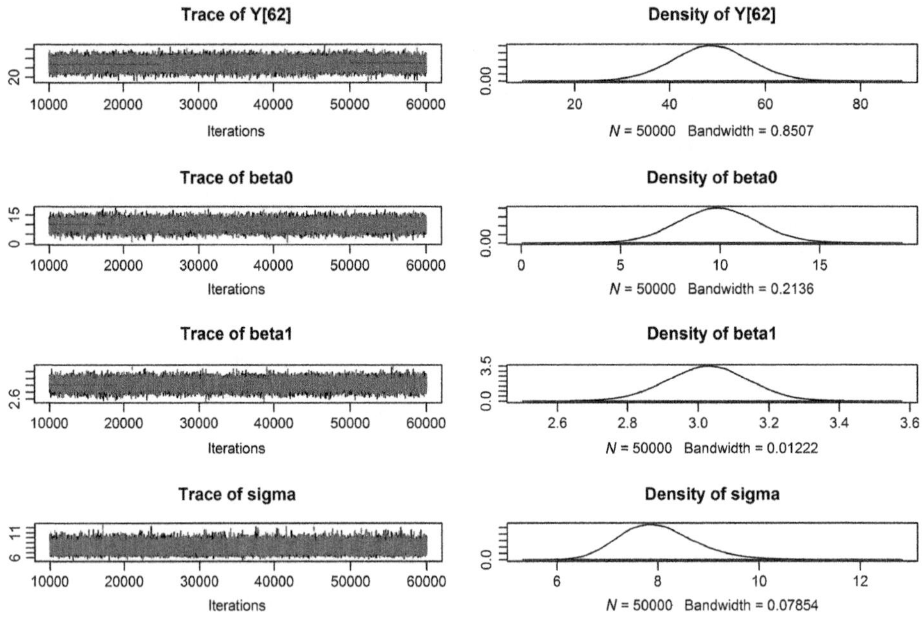

Figure 9.4 Summary plots of unknown parameters.

```
Y[62] 48.599 8.1203 0.0256786       0.0256786
beta0  9.873 2.0198 0.0063872       0.0166837
beta1  3.027 0.1163 0.0003678       0.0009552
sigma  8.007 0.7521 0.0023782       0.0024858

2. Quantiles for each variable:

        2.5%     5%    50%    95%  97.5%
Y[62] 32.564 35.241 48.598 61.939 64.543
beta0  5.912  6.572  9.870 13.211 13.868
beta1  2.797  2.835  3.028  3.217  3.254
sigma  6.696  6.875  7.951  9.330  9.638
```

9.2 Binary Logistic Regression

Logistic regression is used to build a relationship between a categorical response and one or more predictors. The predictors can be continuous or categorical. When the response is binary (i.e. the response has two categories), this type of logistic regression is called binary logistic regression. Examples of such responses include testing pass/fail, defective/non-defective parts, yes/no decision, etc. Other types of logistic regression include ordinal logistic regression and nominal logistic regression, both of which are for situations when the response has three or more categories. For ordinal logistic regression, the categories of response have a natural ordering, e.g. poor/fair/good. For nominal logistic regression, there is no natural ordering in the response categories, e.g. red/green/blue.

In this section, we discuss binary logistic regression and provide an example. Binary logistic regression is commonly used for pass/fail or event/no-event cases. The following summarizes basic concepts and equations behind binary logistic regression when there is only one continuous predictor.

Let x be a continuous predictor. Let a response Y be whether or not there is an event, i.e. $Y_i = 0$ or 1 (1 indicates an event and 0 indicates non-event; $i = 1, \ldots, n$; n is sample size).

Let P_i be the event probability of the ith part. Y_i is assumed to be Bernoulli distributed. The probability of Y_i being 1 is P_i. The likelihood of each observed data point Y_i is calculated as

$$L_i = P_i^{Y_i}(1 - P_i)^{1-Y_i}$$

when $Y_i = 0$ and $L_i = 1 - P_i$. When $Y_i = 1$, $L_i = P_i$.

The overall likelihood of the observed data is

$$L = \prod_{i=1}^{n} L_i = \prod_{i=1}^{n} P_i^{Y_i}(1 - P_i)^{1-Y_i}.$$

The *odds* of the event are a function of the probability of the event P_i, i.e.

$$\text{odds} = P_i/(1 - P_i).$$

For example, if the probability of winning a game is 20%, then the odds of winning the game is $0.2/0.8 = 0.25$.

Odds ratio is the odds of the event at a new level of predictor x divided by the odds at the original level of x. Let us revisit Example 2.1. It is known that the probability to win game A is 60%. The probability to win game B is 40%. The odds and odds ratio are calculated as follows:

Odds of winning game A = $0.6/0.4 = 3/2$
Odds of winning game B = $0.4/0.6 = 2/3$
Odds ratio = $(3/2)/(2/3) = 9/4 = 2.25$.

The *logit* of the probability P_i is the logarithm of the odds $P_i/(1 - P_i)$, i.e.

$$Y_i' = \log(P_i/(1 - P_i)). \tag{9.5}$$

Since P_i is probability, the values for P_i range from 0 to 1. With the logit link function shown in Eq. (9.5), values for Y_i' range from $-\infty$ to ∞. Thus Y' can be the response variable for a linear regression model, shown in Eq. (9.6),

$$Y_i' = \beta_0 + \beta_1 x_i, \tag{9.6}$$

where β_0 and β_1 are unknown coefficients in the linear regression equation between Y_i' and x_i.

Equations (9.5) and (9.6) show the form of a binary logistic regression model. The event probability P_i can be calculated via the inverse logit function, i.e.

$$P_i = \exp(Y_i')/(1 + \exp(Y_i')).$$

From the equations above,

$$\text{odds} = P_i/(1 - P_i) = \exp(Y_i');$$

$$P_i = \text{odds}/(1 + \text{odds}).$$

The logarithm of the odds ratio is the difference between the logits of two probabilities, i.e.

$$\log\left(\frac{P_1/(1-P_1)}{P_2/(1-P_2)}\right) = \log(P_1/(1-P_1)) - \log(P_2/(1-P_2)) = \text{logit}\,(P_1) - \text{logit}\,(P_2).$$

By building a logistic regression, the probability of a categorical response can be estimated for a given (set of) predictor value(s). This helps predict the response probability, or to increase or decrease the response probability by properly controlling the predictor(s).

Example 9.2 In a characterization study, mechanical designers found that the probability of the event of interest (called event in the following text, figures, and analysis) could be affected by the amount of a parameter named offset. Table 9.2 shows a random sample data of offset and event. The event data is coded with 1 for events and 0 for non-events. The objectives are:

1) to understand the relationship between the amount of offset (as a predictor) and the probability of the event (as a response)
2) to estimate the probability of event given offset = 10, 30, 50, respectively
3) to estimate the probability of event for an old design with offset mean = 50 and standard deviation = 5, and for a new design with offset mean = 30 and standard deviation = 5.

For objectives 1 and 2, we show both the results using Minitab and the Bayesian solution using R and JAGS. For objective 3, Minitab does not provide a solution, and we will only provide the Bayesian solution using R scripts.

The scatterplots of the event vs. offset are shown in Figure 9.5a. The x axis shows the amount of offset, which is the predictor (independent variable) in this case. The y axis indicates the actual events. From the scatterplot, it appears that the larger the offset, the more likely it is to have an event.

Figure 9.6 show the Minitab binary logistic regression results. The binary logistic regression model estimates the probability of event (response is 1 when an event occurs) vs. offset value. The response information summarizes that there are 78 events and 33 non-events. The deviance table shows the p value is 0.000 for the offset. Since $p < 0.05$ for the offset, this indicates that the amount of offset significantly impacts the event probability.

The odds ratio for offset is 1.08, indicating that the odds of event increases by 1.08 times with each increase of offset by 1 unit. The 95% confidence interval lower bound of the odds ratio is 1.0466, which is greater than 1. If the odds ratio is 1, this means that offset has no impact on the probability of events. Since the odds ratio 95% confidence interval does not include 1, it is concluded at 95% confidence that the amount of offset does impact the event probability.

Since the coefficient for offset (0.0804) is positive, and the odds ratio is more than 1 (1.0837), the model indicates that more offset is related to higher probability of event.

Based on the binary logistic regression output, the prediction model for probability of event is given by

$$\boldsymbol{P} = \exp(\boldsymbol{Y}')/(1+\exp(\boldsymbol{Y}')), \tag{9.7}$$

Table 9.2 Offset and event values.

Offset	Event	Offset	Event	Offset	Event	Offset	Event
14.167	0	23.500	0	30.167	1	47.167	1
18.167	0	57.333	0	130.500	1	43.167	1
28.000	0	27.667	0	52.167	1	40.833	1
30.167	0	17.500	0	62.667	1	70.167	1
33.167	0	70.000	0	33.000	1	76.000	1
27.833	0	94.833	1	89.500	1	45.333	1
34.833	0	101.833	1	111.000	1	51.500	1
44.667	0	88.167	1	40.167	1	35.833	1
25.167	0	65.500	1	66.667	1	49.667	1
48.833	0	77.667	1	26.000	1	45.833	1
27.667	0	75.667	1	53.000	1	24.500	1
58.500	0	91.500	1	71.667	1	37.833	1
34.833	0	59.667	1	44.833	1	54.333	1
14.167	0	27.333	1	62.833	1	31.667	1
28.167	0	67.667	1	63.667	1	51.000	1
16.833	0	70.333	1	25.000	1	17.500	1
33.333	0	33.167	1	38.833	1	71.333	1
26.500	0	48.833	1	26.500	1	24.333	1
22.000	0	81.333	1	85.333	1	33.167	1
17.167	0	75.000	1	57.667	1	30.000	1
18.667	0	61.500	1	37.167	1	29.000	1
12.667	0	45.167	1	48.167	1	47.500	1
14.667	0	7.667	1	59.667	1	54.833	1
29.667	0	69.500	1	48.000	1	16.833	1
22.667	0	75.333	1	42.667	1	45.333	1
23.000	0	50.833	1	44.167	1	52.000	1
26.833	0	76.333	1	95.500	1	27.333	1
3.333	0	98.333	1	94.667	1		

1, there is an event; 0, there is no event.

where

$$Y' = -2.289 + 0.0804 \times \text{offset}. \tag{9.8}$$

The scatterplots of the event and model fits vs. offset are shown in Figure 9.5b. The y axis indicates the actual events (shown as dots) and estimated event probabilities (shown as squares). Note that the raw data has only two categories, while the model fits are shown as an S curve.

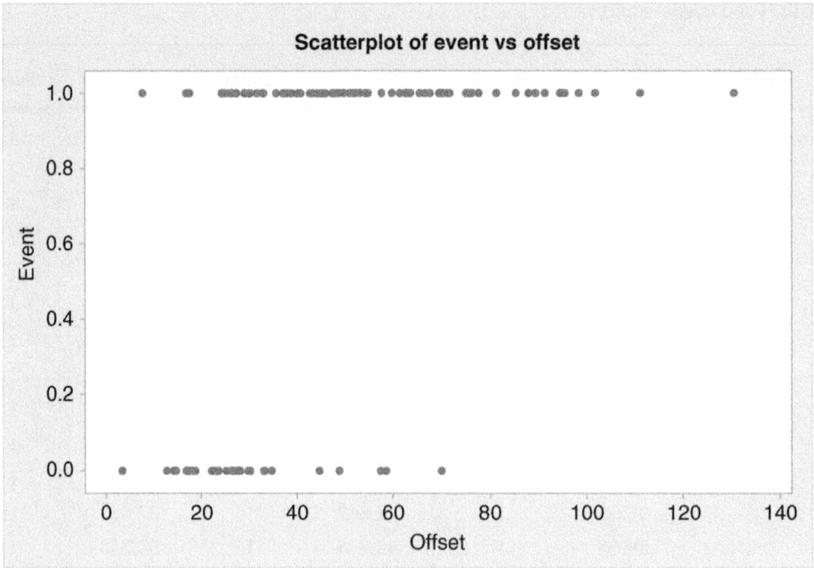

(a)

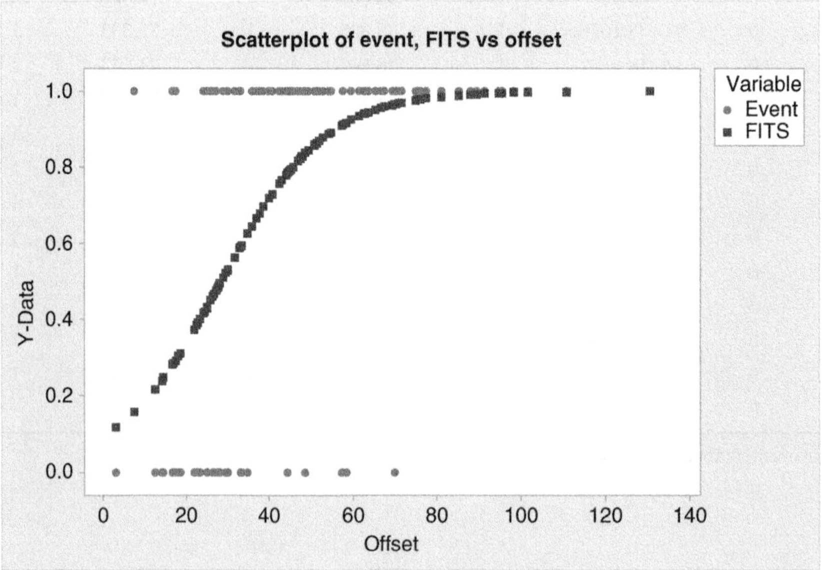

(b)

Figure 9.5 (a) Scatterplot of events vs offset. (b) Scatterplot of events, model fits vs offset.

Equations (9.7) and (9.8) provide the relationship between the amount of offset and the event probability, which shows the answers for objective 1. For objective 2, the probabilities of event given offset = 10, 30, 50 are estimated as follows based on the above model:

Binary logistic regression: event versus offset

```
* WARNING * When the data are in the Response/Frequency format, the Residuals versus fits
            plot is unavailable.

Method

Link function  Logit
Rows used      111

Response information

Variable  Value  Count
Event     1         78  (Event)
          0         33
          Total    111

Deviance table

Source       DF  Adj Dev  Adj Mean  Chi-Square  P-Value
Regression    1    38.21   38.2118       38.21    0.000
  Offset      1    38.21   38.2118       38.21    0.000
Error       109    96.89    0.8889
Total       110   135.10

Model summary

Deviance   Deviance
   R-Sq  R-Sq(adj)      AIC
 28.28%     27.54%   100.89

Coefficients

Term         Coef  SE Coef   VIF
Constant   -2.289    0.652
Offset     0.0804   0.0178  1.00

Odds ratios for continuous predictors

        Odds Ratio        95% CI
Offset      1.0837  (1.0466, 1.1222)

Regression equation

P(1)  =  exp(Y')/(1 + exp(Y'))

Y' = -2.289 + 0.0804 Offset
```

Figure 9.6 Binary logistic regression results from Minitab.

- when offset = 10, probability of event = 0.185
- when offset = 30, probability of event = 0.531
- when offset = 50, probability of event = 0.850.

Now we use JAGS and R scripts to provide the Bayesian estimation for this problem. The directed acyclic graph of the binary logistic regression Bayesian model is shown in Figure 9.7. Diffuse prior distributions are used for the two unknown parameters *beta*0 and *beta*1. The main logic relationships are shown in Eqs. (9.7) and (9.8):

$$Event[i] \sim Bernoulli(p[i]) \quad (9.9)$$

$$\log\text{it}(p[i]) = beta0 + beta1 \times \text{offset}[i] \quad (9.10)$$

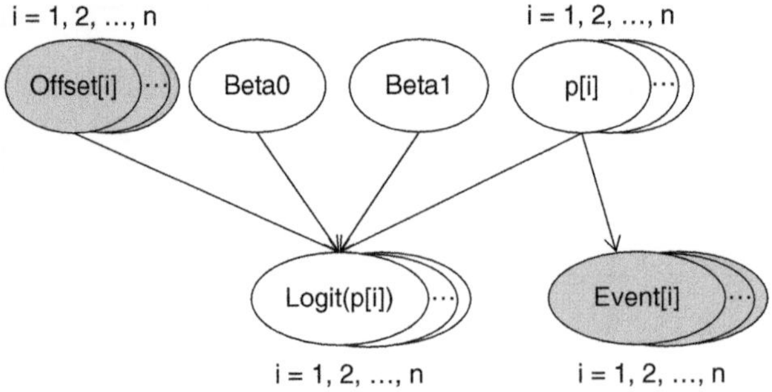

Figure 9.7 Binary logistic regression Bayesian model (directed acyclic graph).

The JAGS model is shown in **9.2_Logistic_Regression.JAGS**. The R script, shown in **9.2_Logistic_Regression.R**, is to estimate the unknown coefficients in the binary logistic regression. To estimate the probability of event when offset = 10, 30, 50, simply add these three offset values to the offset vector, and add three missing values (NA) to the event vector. The R script is shown below.

```
offset <- c(offset, 10, 30, 50)
Event <- c(Event, NA, NA, NA)
```

(9.2_Logistic_Regression.JAGS)

```
model{

  for( i in 1:n) {

    Event[i] ~ dbern( p[i] )

    # likelihood model: binary logistic regression
    logit(p[i]) <- beta0 + beta1*offset[i]
#  logistic model

  } # end of  [i] loop

  # diffuse priors
  beta0 ~ dnorm(0,0.001)
  beta1 ~ dnorm(0,0.001)

} #  end of model
```

(9.2_Logistic_Regression.R)

```
################################################################
## Bayesian model for Logistic regression                   ##
################################################################

### Load package rjags (to connect to JAGS from R for Bayesian
analysis)
library(rjags)

# read data from a file
data <-read.table("Example9.2_LogisticRegression.txt",header=T,sep="")
```

```
offset <- data[,1]
Event <- data[,2]

# For prediction purposes, add 3 data rows with offset = 10, 30, 50
# To predict Event probability for these offset values, simply add NA
offset <- c(offset, 10, 30, 50)
Event <- c(Event, NA, NA, NA)

# Data for the JAGS model
BayesianData <- list(offset = offset,
                 Event = Event, n = length(offset)
)

# Create (& initialize & adapt) a JAGS model object
ModelObject <- jags.model(file = "9.2_Logistic_Regression.JAGS",
                        data=BayesianData,
                        n.chains = 2, n.adapt = 1000
)

# Burn-in stage
update(ModelObject, n.iter=10000)

# Run MCMC and collect posterior samples in coda format for selected
variables
codaList <- coda.samples(model=ModelObject, variable.names =
c("beta0", "beta1", "p[112]", "p[113]", "p[114]"),
                           n.iter = 50000, thin = 1)

## Convergence diagnostics with 'CODA' package

# Trace plots:
# traceplot(codaList)  # displays a trace plot for each variable
traceplot(codaList[,'beta0'])    # trace plot for only one variable
traceplot(codaList[,'beta1'])
# Desntity plots: displays a plot of the density estimate for each
variable
densplot(codaList[,'beta0'])
densplot(codaList[,'beta1'])
```

After running the R script, the posterior sample trace plots and density plots of specified parameters are shown in Figure 9.8. Since the two chains are mixed well (not separated) with no patterns observed, we can assume that the convergence is achieved. Other diagnostic results are not shown here. Readers are encouraged to run diagnostics themselves.

The summary statistics are shown below.

```
> summary(codaList, quantiles = c(0.025, 0.05, 0.5, 0.95, 0.975))

Iterations = 61001:111000
Thinning interval = 1
Number of chains = 2
Sample size per chain = 50000
```

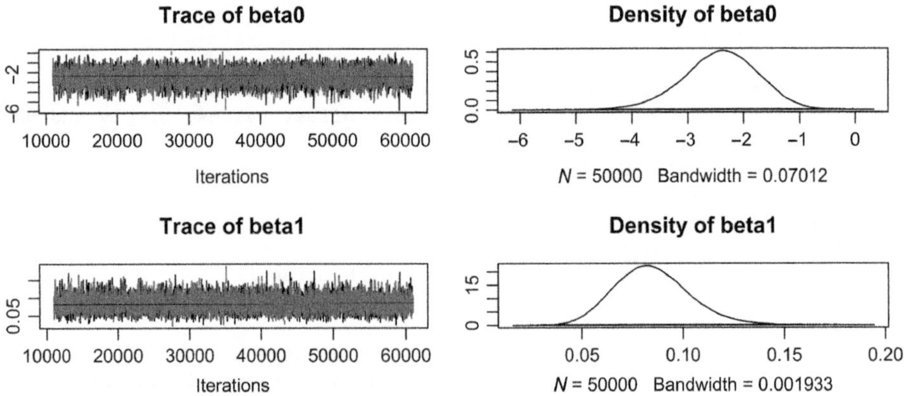

Figure 9.8 Summary plots of posterior samples.

```
1. Empirical mean and standard deviation for each variable,
   plus standard error of the mean:

              Mean      SD  Naive SE Time-series SE
beta0   -2.41640 0.66451 2.101e-03      0.0095154
beta1    0.08465 0.01825 5.771e-05      0.0002629
p[112]   0.18338 0.07229 2.286e-04      0.0009807
p[113]   0.53023 0.06449 2.039e-04      0.0004472
p[114]   0.85363 0.04681 1.480e-04      0.0004189

2. Quantiles for each variable:

              2.5%        5%       50%      95%    97.5%
beta0   -3.79500 -3.55062 -2.39419 -1.3579 -1.1722
beta1    0.05158  0.05640  0.08364  0.1165  0.1235
p[112]   0.06910  0.08191  0.17438  0.3169  0.3488
p[113]   0.40174  0.42274  0.53101  0.6349  0.6543
p[114]   0.75239  0.77028  0.85739  0.9239  0.9345
```

Parameter estimation from Minitab binary logistic regression output vs. Bayesian solution is shown in Table 9.3. The intercept and slope parameter estimations from Minitab are −2.3 and 0.080, respectively. The Bayesian point estimates are close to the

Table 9.3 Parameter estimation from Minitab binary logistic regression output vs. Bayesian solution.

Parameter	Results from Minitab binary logistic regression output	Bayesian solution	
		Mean	Median
Beta0	−2.3	−2.4	−2.4
Beta1	0.080	0.085	0.084
Event probability when offset = 10	0.18	0.18	0.17
Event probability when offset = 30	0.53	0.53	0.53
Event probability when offset = 50	0.85	0.85	0.86

Minitab solutions. When offset = 10, 30, 50, estimated probabilities of events from Minitab are also close to the corresponding event probability Bayesian posterior sample means and medians. This is no surprise, since non-informative prior distributions are used in the Bayesian estimation. In the Bayesian framework, the probability of event given a specific offset is a distribution, thus the 95% credible intervals of event probability when offset = 10, 30, 50 are also provided via quantiles in the summary statistics.

For the third objective, Minitab does not provide a solution. We use Bayesian solutions with Monte Carlo simulations to estimate the probability of event given a distribution of offset (old design vs. new design). Specifically, The binary logistic regression model for the event probability is

$$\boldsymbol{P} = \exp(beta0 + beta1 \times \boldsymbol{offset})/(1 + \exp(beta0 + beta1 \times \boldsymbol{offset})),$$

with *beta*0, *beta*1, and ***offset*** each being a variable with certain distribution. After Bayesian computations, the posterior samples of *beta*0 and *beta*1 from MCMC simulations are recorded. Monte Carlo simulations can be run using the posterior samples of these two parameters and a random sample of offset, based on the binary logistic regression model. When changing the offset distribution from the old design to the new design, the probability of event is updated and compared.

Assuming the old design has average offset of 50, the new design has an offset of 30 and the standard deviation is 5 for both designs. Normal distributions are assumed for the offset values. The additional R script to compare the event probability of the old design vs. the new design is shown below.

```
######################################################################
## 2nd part: to estimate the event probability for the old vs.     ##
## new design                                                       ##
## Old design: Normal(mean = 50,sd = 5);                            ##
## New design: Normal(mean = 30,sd = 5).                            ##
######################################################################

# change posterior samples from coda format to a matrix
codaMatrix <- as.matrix( codaList )

# collect posterior samples of beta0 and beta1 as numeric vectors
beta0_samples <- codaMatrix[,"beta0"]
beta1_samples <- codaMatrix[,"beta1"]

# generate random samples from Normal distributions
offset_old <- rnorm(100000, mean=50, sd=5)
offset_new <- rnorm(100000, mean=30, sd=5)

# calculate event probability for the old design
P_old <- exp(beta0_samples+beta1_samples*offset_old)/
(1+exp(beta0_samples+beta1_samples*offset_old))

# calculate event probability for the new design
P_new <- exp(beta0_samples+beta1_samples*offset_new)/
(1+exp(beta0_samples+beta1_samples*offset_new))

# overlapping the histograms of P_old and P_new
# Histogram Grey Color
```

```
    # Create plot
    # jpeg("P_old_new.jpeg", width = 6, height = 4, units = 'in',
res = 1800)  # save the plot as jpeg format
    hist(P_new, col=rgb(0.1,0.1,0.1,0.5), main = "Overlapping
Histogram: P_old and P_new", xlab="Event Probability", xlim=range(0,1),
ylim=range(0,8000), breaks=50 ) # dark grey
    hist(P_old, col=rgb(0.8,0.8,0.8,0.5), breaks=50, add=T) # light grey
    box()
    # dev.off()

    # summary statistics of P_old and P_new
    summary(P_old)
    summary(P_new)

    # show 95% credible intervals of P_old and P_new
    quantile(P_old, c(0.025,0.975))
    quantile(P_new, c(0.025,0.975))
```

Summary statistics of the event probability of the old design (P_old) and of the new design (P_new) are shown below. Compared to the old design, the mean of the event probably dropped from 0.85 to 0.53 with the new design. This is consistent with the point estimation shown in Table 9.3. Since P_old and P_new are distributions, their 95% credible intervals can be found from the quantile function. With the old design, the event probably has a 95% credible interval of (0.69, 0.96). With the new design, the event probably has a 95% credible interval of (0.29, 0.75). Figure 9.9 shows the histograms of event probability from the old design vs. the new design. This type of analysis enables designers to adjust design parameters so as to change the event probability to achieve a target.

```
> summary(P_old)
   Min. 1st Qu.  Median    Mean 3rd Qu.    Max.
 0.4405  0.8032  0.8535  0.8456  0.8971  0.9957
> summary(P_new)
   Min. 1st Qu.  Median    Mean 3rd Qu.    Max.
0.07656 0.44980 0.53340 0.52930 0.61280 0.95210
```

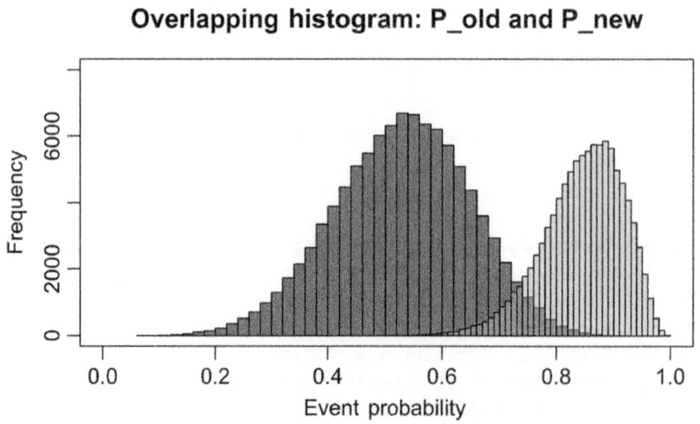

Figure 9.9 Histograms of event probability from the old design (light gray) vs. the new design (dark gray).

```
> quantile(P_old, c(0.025,0.975))
     2.5%     97.5%
0.6899364 0.9557236
> quantile(P_new, c(0.025,0.975))
     2.5%     97.5%
0.2880135 0.7497032
```

9.3 Case Study: Defibrillation Efficacy Analysis

In Sections 9.1 and 9.2 we introduced Bayesian solutions to linear regression and binary logistic regression with examples. These models can be further extended to Bayesian hierarchical regression models when applicable (Kruschke 2011; Gelman et al. 2014; Carlin and Louis 2009).

Next, we will discuss another application of a logistic regression model to analyze the performance of two different designs of an implantable cardioverter defibrillator (ICD). ICDs are designed to treat sustained ventricular tachycardia (VT) and ventricular fibrillation (VF), and therefore to prevent sudden cardiac death in patients who are at high risk of having such a condition. When VT or VF is detected in a patient, the device delivers a high-energy electrical shock to the heart to get it back to normal sinus rhythm. The amount of energy (measured in Joules) required to terminate the harmful arrhythmia varies from patient to patient and depends on the design of the device and whether the electrodes are placed inside or outside of the heart.

ICD efficacy testing is a common procedure performed at ICD device implant to ensure that each individual patient receives a device and accompanied lead (electrode) configuration with adequate shock efficacy. The shock efficacy is usually defined as having a safety margin (SM) of at least 10 J between the maximum energy output of the device and the energy needed for successful defibrillation. For example, if the device has 35 J maximum output, then a VT/VF episode in the patient should be able to successfully defibrillate with an energy less than or equal to 25 J. Several metrics have been used to quantify ICD efficacy testing in the literature. The most commonly used metric is the defibrillation threshold (DFT), which is defined as the lowest energy needed to achieve defibrillation. Another metric is the SM, which is characterized by the number of successes required in a specified number of attempts at a single shock energy, for example requiring two successful defibrillations out of two attempts at 20 J. This is abbreviated as SM (2/2@20 J). Various defibrillation efficacy testing protocols are discussed in Smits et al. (2013). During defibrillation efficacy testing, a series of VFs are induced in the patient and the device is used to deliver defibrillation shocks at specified energy levels. If the device fails to terminate the VF, an external rescue shock is delivered at a much higher energy level to terminate the VF.

In general, there does not exist an energy level above which success in defibrillation is guaranteed. The defibrillation success is probabilistic in nature and the outcome of each individual defibrillation attempt is uncertain. The probability of successful defibrillation at a given energy level is called the defibrillation shock efficacy (DSE) at that energy. The probability of successful defibrillation is characterized by a dose–response relationship as a function of shock energy (Smits and Virag 2010). The probability of successful defibrillation at very low energy is close to zero and for high energy it is close to one. The amount of energy required for successful defibrillation depends on the patient and the design of the ICD and its electrode configuration. To describe the relationship between DSE and the shock energy, let $p_i(E)$ be the probability of successful defibrillation at energy E for the ith patient. Then the relationship between $p_i(E)$ and E is given by the logistic regression model

$$\log\left(\frac{p_i(E)}{1-p_i(E)}\right) = \gamma(\log(E) - \log(E_{50(i)})), \tag{9.11}$$

where γ is the slope parameter and $E_{50(i)}$ is the energy for which the probability of successful defibrillation for the ith patient is 50%. A more commonly used version of the above formula is given by

$$p_i(E) = \frac{1}{1+\left(\frac{E_{50(i)}}{E}\right)^{\gamma}}. \tag{9.12}$$

From Eq. (9.11) it is easy to see that when $E = E_{50(i)}$, then $p_i(E) = 0.5$ (or 50%). The slope parameter γ describes how fast the probability of successful defibrillation, $p_i(E)$, increases as the energy increases for a given $E_{50(i)}$ value.

There are many test protocols that are used for DFT and SM testing. One of the frequently used protocols is the binary search protocol.

Example 9.3 Suppose we are interested in comparing two different designs of defibrillator/electrode configurations. Let us call the previous and new designs design-*A* and design-*B*, respectively. We already have a considerable amount of defibrillation testing data from a clinical study for the design-*A* device patients and some early stage defibrillation test data for the design-*B* patients from a feasibility study. In the clinical study, 325 patients went through defibrillation testing by inducing up to three VF episodes and then using 60 J shocks by the device to terminate the VF. For each induction, up to two test shocks were delivered. If both shocks failed, then a rescue shock was given by an external defibrillator. The implant criterion required two consecutive successes out of a maximum of four shocks given at 60 J for the device to be implanted in the patient. The third VF was not induced if either the implant criterion was already met or there was no possibility of meeting it. The clinical study outcomes are shown in Table 9.4. The feasibility study of design-*B* had 50 patients. A single VF was induced and a 30 J shock was delivered to terminate it. A maximum of two test shocks was used. No device was implanted in any of the patients. The results of the study are shown in Table 9.5. The last four rows of Table 9.5 provide data for the patients who were not able to complete the implant testing protocol due to either not being able to induce up to three VFs or other clinical reasons. By design it is expected that design-*B* requires much lower energy to successfully defibrillate. The maximum device outputs for design-*A* and design-*B* devices are 70 J and 40 J, respectively.

A new clinical study with 300 patients is designed to investigate the defibrillation performance of design-*B* devices. The defibrillation efficacies between two different device designs are believed to be related and therefore it is desired to combine design-*A* clinical study data and design-*B* feasibility study data to make a prediction for the planned 300-patient design-*B* clinical study.

Table 9.4 Defibrillation efficacy testing outcomes from the clinical study of design-*A* ICD devices.

First induced VF		Second induced VF		Third induced VF		Number of patients
First shock	Second shock	First shock	Second shock	First shock	Second shock	
S	NA	S	NA	NA	NA	275
S	NA	F	S	S	NA	11
F	S	S	NA	NA	NA	12
F	F	S	NA	S	NA	7
F	S	F	S	NA	NA	3
F	F	F	S	NA	NA	4
S	NA	F	F	NA	NA	2
S	NA	F	S	NA	NA	4
S	NA	NA	NA	NA	NA	4
F	S	NA	NA	NA	NA	2
F	F	NA	NA	NA	NA	1

NA, not applicable or not available.

Table 9.5 Defibrillation efficacy testing outcomes from the feasibility study of design-*B* ICD devices.

Induced VF		Number of patients
First shock	Second shock	
S	NA	44
F	S	4
F	F	2

S, Success; F, Failure; NA, not applicable or not available.

We would like to estimate the following defibrillation efficacies for the two ICD designs and compare model predictions of defibrillation efficacy performances for design-*A* devices with those observed in the clinical study:

1) Probability of first shock success for design-*A* devices at 60 J, 65 J, and the maximum output of 70 J and the same for design-*B* devices at 30 J, 35 J, and the maximum output of 40 J.
2) Probability of first and second shocks success for design-*A* devices at 60 J, 65 J, and 70 J and the same for and design-*B* devices at 30 J, 35 J, and 40 J.
3) Probability of implant success (2 consecutive successes in 4 attempts) at 60 J for design-*A* devices

4) Probability of meeting the same implant criterion as in item 3 above for design-*B* devices with 30 J shocks.
5) Probability of first shock success at 70 J for design-*A* devices given that 2 of 4 implant success criteria in item 3 above have met.
6) Probability of first shock success at 40 J for design-*A* devices given that 2 of 4 implant success criteria in item 4 above have met.

As shown in Eq. (9.8), the defibrillation efficacy of a patient at a given energy for a specific design of an ICD mainly depends on the E_{50}. The literature suggests that the slope parameter γ does not vary much between different designs of ICDs. Therefore, we consider a transfer function that relates only E_{50} values of design-*A* and design-*B* devices. To mathematically express this relationship, suppose $E_{50}^{i}(A)$ is that energy for which there is a 50% chance of defibrillation for the *i*th patient with a design-*A* device and $E_{50}^{i}(B)$ is the same for a design-*B* device patient. The relationship between these parameters is given by Eq. (9.13):

$$E_{50}^{i}(B) = (E_{50}^{i}(A))^{\delta^{I_D}} * (1 - \kappa I_D) \tag{9.13}$$

where $I_D = 1$ for design-*B* and $I_D = 0$ for design-*A*, κ $(0 < \kappa < 1)$ is a shrinkage parameter, and $\delta(>0)$ is a scaling parameter.

Bayesian analysis is performed using models (9.9) and (9.10) to obtain the posterior distributions of the parameters $E_{50}^{i}(A)$, δ, κ, and $E_{50}^{i}(B)$. These posterior distributions are subsequently used for estimating various defibrillation efficacy parameters for the two ICD designs. Several R scripts are used as described below.

a) **`9.3_Defib_Efficacy_Pred.R`**: Makes defibrillation efficacy predictions for design-*B* devices based on the clinical data gathered for design-*A* devices and the limited amount of feasibility study data collected on design-*B* devices. This script creates the text file **`9.3_JAGSModel.txt`** that contains the Bayesian model to be used with **`jags.model`**. Defibrillation efficacy estimates are computed for design-*B* devices based on a new 300-patient clinical study.
b) **`9.3_Plot_DSE_Curve_For_DesA_Clin_Pts.R`**: Creates the function, Plot-DesA_Clin_DSE_Curve to plot the average DSE curve for design-*A* device clinical study patients.
c) **`9.3_Plot_DSE_Curve_For_Given_Trial_Size_DesB.R`**: Creates the function PlotDesBDSECurves for generating a predicted average DSE curve for design-*B* device patients that is based on the posterior distributions of the model parameters obtained by fitting the combined data from 325 design-*A* clinical study patients and 50 design-*B* device feasibility study patients.
d) **`9.3_Comp_DesA_Obs_Predicted_Perf.R`**: Creates the function PrintDe-sAReslts to generate and print the comparison results of the observed and predicted defibrillation performance of design-*A* devices for the clinical study.
e) **`9.3_Print_DesB_New_Study_Predictions.R`**: This program creates the function PrintDesBReslts to generate and print the results for various defibrillation efficacy estimates for a new 300-patient study for design-B devices.

The above listed R scripts are given below.

(9.3_Defib_Efficacy_Pred.R)

```
##########################################################################
## 9.3_Defib_Efficacy_Pred.R                                            ##
## Make defibrillation efficacy predictions for a new design            ##
## (design-B) of an ICD device based on the data gathered for a         ##
## previous design (design-A) and a limited amount of feasibility       ##
## study data collected on the new design. A Bayesian model is          ##
## fitted with the combined data from the two studies. Based on this    ##
## model, defibrillation efficacy estimates will be computed for        ##
## a new 300-patient clinical study for design-B devices                ##
##########################################################################


# Get the required libraries
require(rjags)
require(coda)

#############################################
## SET GLOBAL VARIABLES FOR ENTIRE PROGRAM ##
#############################################

## Set-up data for JAGS Analysis for 325 patients tested (shocked) in the
## design-A devices clinical study - All shocks delivered at 60J
# S+S = 275 - Implant Success
# S+F+S+S = 11 - Implant success
# F+S+S = 12 - Implant success
# F+F+S+S = 7 - Implant success
# F+S+F+S = 3 - Implant failure
# F+F+F+S = 4 - Implant failure
# S+F+F = 2 - Implant failure
# S+F+S = 4 - Failed to induce third VF
# S = 4 - 1st shock success but Failed to induce 2nd VF
# F+F = 1 - 1st two shocks failed but 2nd VF not induced due to
clinical reasons
# F+S = 2 - 1st shock failed, 2nd shock success but 2nd VF not induced
due to
#           clinical reasons

N_ss = 275
N_sfss = 11
N_fss = 12
N_ffss = 7
N_fsfs = 3
N_fffs = 4
N_sff = 2
N_sfs = 4
N_s =4
N_ff = 1
N_fs = 2
#create a data.frame with different results in design-A clinical study
Nv <- data.frame(N_ss=N_ss,N_sfss=N_sfss,N_fss=N_fss,N_ffss=N_ffss,
N_fsfs=N_fsfs,
                   N_fffs=N_fffs,N_sff=N_sff,N_sfs=N_sfs,N_s=N_s,
N_ff=N_ff,N_fs=N_fs)

## Set-up data for JAGS Analysis for 50 patients tested (shocked) in the
```

```
## design-B devices feasibility study

# N1_s = 44; number of successes at 30J
# N1_fs = 4; number of cases where failure followed by success at 30J
# N1_ff = 2; number of cases where two shocks in a row failed at 30J

N1_s = 44
N1_fs = 4
N1_ff = 2
N1_f = N1_fs + N1_ff # First shock failures in feasibility study
N1 = N1_s + N1_fs + N1_ff  # Total number of patients tested in
design-B feasibility
                           # study
# Total number of patients tested in design-A & design-B studies
combined
N = N_ss+N_sfss+N_fss+N_ffss+N_fsfs+N_fffs+N_sff+N_sfs+N_s+N_ff+N_fs+N1
# Number of implant successes in design-A device clinical study
N_Imp_S = N_ss+N_sfss+N_fss+N_ffss
# Number of implant failures in design-A device clinical study
N_Imp_f = N_fsfs+N_fffs+N_sff
# Number of patients with 1st shock success in design-A clinical study
N_1S = N_ss+N_sfss+N_sff+N_s
# Number of patients who did not complete implant protocol in design-A
device
# clinical study
N_Not_comp = N_sfs+N_s+N_ff+N_fs

y <- t(array(c(rep(c(1,1,NA,NA),N_ss),rep(c(1,0,1,1),N_sfss),
rep(c(0,1,1,NA),N_fss),
    rep(c(0,0,1,1),N_ffss),rep(c(0,1,0,1),N_fsfs),rep(c(0,0,0,1),N_fffs),
    rep(c(1,0,0,NA),N_sff),rep(c(1,0,1,NA),N_sfs),rep(c(1,NA,NA,NA),N_s),
    rep(c(0,0,NA,NA),N_ff),rep(c(0,1,NA,NA),N_fs),rep(c(1,NA,NA,NA),N1_s),
    rep(c(0,1,NA,NA),N1_fs),rep(c(0,0,NA,NA),N1_ff)),dim=c(4,N)))
E <- rbind(array(60, dim=c(N-N1,4)), array(30, dim=c(N1,4)))
k = rowSums(!is.na(y))
Id <- c(rep(0,N-N1),rep(1,N1))

### Define JAGS model and store it in a temporary file "9.3_JAGSModel.txt"


modeltext ="
model{
  ### SET PRIORS; Some are informative priors
  G~dlnorm(1.423,13.18) # An informative prior obtained from
literature search;
                        # LogNormal(mean=4.31,std=1.21)
  delta ~ dbeta(0.5,0.5)
  kappa ~ dbeta(0.5,0.5)
  mu0 ~ dnorm(2.99,0.01) # allowed sufficient amount of
variability while matching the
                         # mean to historically obtained mean value
  tau0 ~ dgamma(0.286,0.01) # Match the mean value to a historical
average
  for (i in 1:N){
    Theta[i]  ~ dlnorm(mu0,tau0)  # dgamma(Alpha,Beta)  # <- Alpha
    log(Theta1[i]) <- pow(delta,Id[i])*log(Theta[i]) + log(1-kappa*Id[i])
```

```
    for (j in 1:k[i]) {
      y[i,j] ~ dbin(p[i,j],1)
      logit(p[i,j]) <- G*log(E[i,j]/Theta1[i])
      }
    }
}
"
writeLines(modeltext, con="9.3_JAGSModel.txt")

### SET UP INITIAL VALUES FOR THE MCMC
inits <- function(N) {list(
  list(mu0 = 3, # initial values for mu0 and tau0 were obtained to match
historical results
       tau0 = 2.9,
       G = 4.3,
       Theta = rep(40,N),
       delta = 0.5,
       kappa = 0.5,
       .RNG.name="base::Mersenne-Twister", .RNG.seed=2317 #To be able to
reproduce the results
  ))
}

# BayesianData: Data used in Bayesian analysis
BayesianData <- list(N=N,y=y,E=E,k=k,Id=Id)

# Create model
ModelObject <- jags.model(file = "9.3_JAGSModel.txt",
                         data=BayesianData, inits=inits(N),
                         n.chains = 1, n.adapt = 1000
)

# Burn-in stage
update(ModelObject, n.iter=20000)

# Run MCMC and collect posterior samples in coda format for selected
variables
codaList <- coda.samples(model=ModelObject,
             variable.names = c("mu0","tau0","G","Theta","delta","kappa"),
             n.iter = 30000, thin = 1)
#plot(codaList)
codaMatrix <- as.matrix( codaList )
#colnames(codaMatrix)
Theta <- codaMatrix[,2:376]
delta = codaMatrix[,'delta']
G <- codaMatrix[,'G']
kappa <- codaMatrix[,'kappa']
ngibbs <- dim(codaMatrix)[1] # Number of Gibbs samples
remove(codaMatrix)

# Plot the average DSE curve for Design-A device clinical study Pts.

source("9.3_Plot_DSE_Curve_For_DesA_Clin_Pts.R")
PlotDesA_Clin_DSE_Curve(nI=N,N1=N1,G=G,Theta=Theta)
```

```
  # Plot separate average DSE curves for Design-B device patients using
specified number
  # N_New) of patients in the planned study and N1 pts in the
feasibility study

  source("9.3_Plot_DSE_Curve_For_Given_Trial_Size_DesB.R")
  # Get the number of patients expected in the new deign-B device
clinical study
  N_New = 300
  PlotDesBDSECurves(nI=N,ngibbs=ngibbs,N1=N1,N_New=N_New,G=G,
Theta=Theta,delta=delta,
                    kappa=kappa)

  # Estimate design-B device patients defibrillation efficacies based
on a Trial
  # with specified size (N_New).

  source("9.3_Print_DesB_New_Study_Predictions.R")

  # Print predictions for a new design-B device study with 300 patients

  N_New = 300
  PrintDesBReslts(ngibbs=ngibbs,N1=N1,N1_s=N1_s,N_New=N_New,G=G,
Theta=Theta,delta=delta,kappa=kappa)

  # Estimate design-A device patients defibrillation efficacies from
  # the model and compare them to the observed results from the
  # clinical study (325 pts.)

  source("9.3_Comp_DesA_Obs_Predicted_Perf.R")

  # Print predictions for design-A device patients

  PrintDesAReslts(ngibbs=ngibbs,N=N,Nv=Nv,N1=N1,G=G,Theta=Theta)
```

(9.3_Plot_DSE_Curve_For_DesA_Clin_Pts.R)

```
  ################################################################################
  # The following function is used to plot the average Defibrillation
  # Efficacy (DSE) curve for design-A device clinical study patients
  ################################################################################
  PlotDesA_Clin_DSE_Curve <- function(nI,N1,G,Theta) {

  # Plot the Average DSE curve based only on 325 design-A device
clinical study
  # Patients
  E50_DesA = mean(Theta[,-c(nI-N1+1:nI)])
  dim(Theta[,-c(nI-N1+1:nI)]); q=G %*% t(rep(1,nI-N1)); dim(q);
  E_DesA = E50_DesA*seq(0.9,3.5,by=0.1)
  DSE_DesA = array(NA,length(E_DesA))
  for (iE in 1:length(E_DesA)) {
    DSE_DesA[iE]=mean(1/(1+(Theta[,-c(nI-N1+1:nI)]/E_DesA[iE])^q))
  }

  # save the plot in jpeg format
  jpeg("AvgDSE_Cruve_designA.jpeg", width = 7, height = 5, units = 'in',
res = 800)
```

```
  plot(x=E_DesA,y=100*DSE_DesA, xlim=c(20, 95), ylim=c(40,100),axes=FALSE,
       xlab="Defib Energy (Joules)", ylab="Probability of Defib (%)",
       main="Design-A Device Patients Defibrillation Efficacy Vs. Defib
Energy",
       mgp=c(2.4, 0.8, 0), type="l",lwd=2, lty=1, col="blue")
  # Get custom x and y axes
  axis(side=1, at=c(20,25,30,35,40,45,50,55,60,65,70,75,80,85,90,95),
labels=NULL,
       pos=40, lty=1, col="black", las=1)
  axis(side=2, at=c(40,45,50,55,60,65,70,75,80,85,90,95,100), labels=NULL,
pos=20,
       lty=1, col="black", las=1)
  # Get custom grid lines
  abline(h=c(45,50,55,60,65,70,75,80,85,90,95,100),lty=2,col="grey")
  abline(v=c(25,30,35,40,45,50,55,60,65,70,75,80,85,90,95),lty=2,
col="grey")
  #grid(lty=2,lwd=1)
  mtext("Based on the results of 325 design-A device clinical study
patients", side=1,
        line = 3.0, outer=FALSE, at=NA, adj=0,padj=1, cex=NA, col="blue",
font =1)
  dev.off()
  }

  ### End of plotting DSE curve for design-A device clinical study patients
```

(9.3_Plot_DSE_Curve_For_Given_Trial_Size_DesB.R)

```
  ##########################################################################
  # The following function is used to create the predicted average DSE
  # curve for design-B device patients that is based on the posterior
  # distributions of the model parameters obtained by fitting
  # the combined data from design-A clinical study (325 pts.) and design-B
  # device feasibility study (50 pts.).
  ##########################################################################

  PlotDesBDSECurves <- function(nI,ngibbs,N1,N_New,G,Theta,delta,kappa) {

    #N1 <- N1_s + N1_f

    # Estimate DSE curve for planned design-B study with N_New
number of patients
    # using the posterior distributions of the model parameters obtained by
    # fitting the scalable model to nI (325 design-B + N1 feas.) patients

    # Initialize the matrices to hold E50 values
    Theta0 <- array(0,dim=c(ngibbs,N_New))
    Theta11 <- array(0,dim=c(ngibbs,N_New))
    Id1 <- rep(1,N_New) #Indicator variable in the defib efficacy model
has value 1 for
                        #design-B device patients
    # For each row in the ngibbs Get a random sample of size N_New

    for (i in 1:ngibbs) {Theta0[i,] <- sample(Theta[i,],size=N_New,
replace=TRUE)}
    for (i in 1:ngibbs){
      Theta11[i,] = (Theta0[i,])^((delta[i])^Id1)*(1-kappa[i]*Id1)
```

```
    }
    E50_DesB_ScalMdl = mean(Theta11[,])
    dim(Theta11); q=G %*% t(rep(1,N_New)); dim(q);
    E_DesB_ScalMdl = E50_DesB_ScalMdl*seq(0.9,3.5,by=0.1)
    DSE_DesB_ScalMdl = array(NA,length(E_DesB_ScalMdl))
    for (iE in 1:length(E_DesB_ScalMdl)) {
      DSE_DesB_ScalMdl[iE]=mean(1/(1+(Theta11/E_DesB_ScalMdl[iE])^q))
    }

    # Estimate DSE curve for design-B device patients using the MCMC
samples for only
    # N1 pts in the feas. Study
    Theta12 <- array(0,dim=c(ngibbs,N1))
    Id1 <- rep(1,N1)
    for (i in 1:ngibbs){
      Theta12[i,] = (Theta[i,c((nI-N1+1):nI)])^((delta[i])^Id1)*
(1-kappa[i]*Id1)
    }
    E50_DesB_feas = mean(Theta12)
    dim(Theta12); q=G %*% t(rep(1,N1)); dim(q);
    E_DesB_feas = E50_DesB_feas*seq(0.9,3.5,by=0.1)
    DSE_DesB_feas = array(NA,length(E_DesB_feas))
    for (iE in 1:length(E_DesB_feas)) {
      DSE_DesB_feas[iE]=mean(1/(1+(Theta12/E_DesB_feas[iE])^q))
    }

    # Plot separate DSE curve for design-B device patients that are based
on the results
    # for N1 feas. patients and N_New new patients.
    # save the plot in jpeg format
    jpeg("AvgDSE_Cruves_designB.jpeg", width = 7, height = 5, units = 'in',
res = 800)
    plot(x=E_DesB_feas,y=100*DSE_DesB_feas, xlim=c(10, 70), ylim=c(40,100),
axes=FALSE,
          xlab="Defib Energy (Joules)",ylab="Probability of Defib (%)",
          main="Design-B device Avergae Defibrillation Efficacy Vs. Defib
Energy",
          mgp=c(2.4, 0.8, 0), type="l",lwd=2, lty=1, col="red")
    # Get custom x and y axes
    axis(side=1, at=c(10,15,20,25,30,35,40,45,50,55,60,65,70),
labels=NULL, pos=40,
          lty=1, col="black", las=1)
    axis(side=2, at=c(40,45,50,55,60,65,70,75,80,85,90,95,100),
labels=NULL, pos=10,
          lty=1, col="black", las=1)
    # Get custom grid lines
    abline(h=c(45,50,55,60,65,70,75,80,85,90,95,100),lty=2,col="grey")
    abline(v=c(15,20,25,30,35,40,45,50,55,60,65,70),lty=2,col="grey")
    lines(x=E_DesB_ScalMdl,y=100*DSE_DesB_ScalMdl,type = "b", lty=3,
pch=2, col="blue")
    #points(x=E_DesB_ScalMdl,y=100*DSE_DesB_ScalMdl,pch=2,col="blue")
    legend(x=30, y=70, c(paste("Based only on ",N1," feas. Pts"),
            paste("Based on ",N_New," new study Pts")), col = c("red",
"blue"),
            text.col="black", cex=0.8, lty = c(1, 3), pch = c(NA, 2),
merge=TRUE, bg="gray90")
```

```
    #grid(lty=2,lwd=1)
    #mtext(paste("feas. Pts Results: N1_s=",N1_s,"Successes out of ",N1,
" tested at 35J"), side = 1, line = 3.0, outer=FALSE, at=NA, adj=0,
    #      padj=1, cex = NA, col = "blue", font =1)
    dev.off()
     }
  ## End of plotting Average DSE curve for Design-B device patients
```

(9.3_Comp_DesA_Obs_Predicted_Perf.R)

```
  ##########################################################################
  # This program creates a function to generate and print the             ##
  # comparison results of observed and predicted defibrillation           ##
  # performance of design-A devices for the clinical study.               ##
  # Inputs: MCMC samples generated by the Bayesian analysis               ##
  # for the following model parameters:                                   ##
  # Number of patients in the design-A and design-B studies combined -N   ##
  # Number of patients in the Design-B feasibility study - N1             ##
  # Number of MCMC samples - ngibbs                                       ##
  # MCMC sample for the slope parameter - G                               ##
  # MCMC samples for the E50 parameter - Theta                            ##
  # MCMC samples for the scale and shrinkage parameters - Delta, Kappa    ##
  ##########################################################################

  # Define a function to print design-A device Defib Efficacy
estimation Results for a
  # new clinical study of a specified size

  PrintDesAReslts <- function(ngibbs,N,Nv,N1,G,Theta) {

    # Initialize the matrix Theta1 to store E50 values for design-A devices
    Npts <- N-N1 # All design-A clinical study patients
    Npts_PC <- Npts-(N_sfs+N_s+N_ff+N_fs) # Design-A patients who
completed the test
                                          # protocol

    Theta1 <- array(0,dim=c(ngibbs,Npts))
    #Get the E50 values for design-A devices from the posterior samples
    Theta1 <- Theta[,-c(Npts+1:N)]
    dim(Theta1); q=G %*% t(rep(1,Npts)); dim(q);

    ## Initialize the matrices to store estimated defib efficacy
probabilities using
    ## MCMC samples
    p60 <- matrix(0,nrow=ngibbs,ncol=Npts)
    p65 <- matrix(0,nrow=ngibbs,ncol=Npts)
    p70 <- matrix(0,nrow=ngibbs,ncol=Npts)
    # Initialize the matrices to hold binomial counts data
    ImpS_60J <- matrix(0,nrow=ngibbs,ncol=Npts)
    S70 <- matrix(0,nrow=ngibbs,ncol=Npts)
    # Initialize vectors to store various defibrillation efficacy
probabilities.
    # This stored data is used to compute 95% credible intervals
for the estimated
    # defib efficacies.
    ProbS_60J <- numeric(ngibbs)
    ProbS_65J <- numeric(ngibbs)
```

```
    ProbS_70J <- numeric(ngibbs)
    ProbSS_60J <- numeric(ngibbs)
    ProbSS_65J <- numeric(ngibbs)
    ProbSS_70J <- numeric(ngibbs)

    ProbImpS_60J <- numeric(ngibbs)
    ProbImpf_60J <- numeric(ngibbs)
    PrS70givenImpS60 <- numeric(ngibbs)

    for(i in 1:ngibbs){
      # OBTAIN PROBABILITY OF SUCCESSFUL SHOCK

      p70[i,] <- 1/(1+(Theta1[i,]/70)^G[i])
      p65[i,] <- 1/(1+(Theta1[i,]/65)^G[i])
      p60[i,] <- 1/(1+(Theta1[i,]/60)^G[i])
      # For each MCMC sample, store the average defib efficacy at each
different energy
      # level

      ProbS_60J[i] <- mean(p60[i,]) # Avg. Prob. of 1st shock success at 60J
      ProbS_65J[i] <- mean(p65[i,]) # Avg. Prob. of 1st shock success at 65J
      ProbS_70J[i] <- mean(p70[i,]) # Avg. Prob. of 1st shock success at 70J

      ProbSS_60J[i] <- mean((p60[i,])^2) # Avg. Prob. of 1st & 2nd shock
success at 60J
      ProbSS_65J[i] <- mean((p65[i,])^2) # Avg. Prob. of 1st & 2nd
shock success at 65J
      ProbSS_70J[i] <- mean((p70[i,])^2) # Avg. Prob. of 1st & 2nd shock
success at 70J
      # Compute the average probability of implant sucess (2S in a row out
of 4 shocks)
      # at 60J for each MCMC sample
      ProbImpS_60J[i] <- mean(ProbSS_60J[i] #SS
                             + p60[i,]*(1-p60[i,])*(p60[i,])^2  #SFSS
                             + (1-p60[i,])*(p60[i,])^2  #FSS
                             + (1-p60[i,])^2*(p60[i,])^2) #FFSS
      ProbImpf_60J[i] <- 1-ProbImpS_60J[i]

      # It is not possible to directly compute the conditional
probability of success at
      # 70J given implant success at 60J because the success
probabilities within
      # patients have unknown correlations. SO, we use the following
Binomial counts
      # data to compute the conditional probability

      #Generate success/fail counts for 1 - 4 shocked episodes at
different energies

      Shk1_60 <- rbinom(n=Npts,size=1,prob=p60[i,])
      Shk2_60 <- rbinom(n=Npts,size=1,prob=p60[i,])
      Shk3_60 <- rbinom(n=Npts,size=1,prob=p60[i,])
      Shk4_60 <- rbinom(n=Npts,size=1,prob=p60[i,])
      Shk1_70 <- rbinom(n=Npts,size=1,prob=p70[i,])
```

```
      # Get counts of design-A device implant sucess (2S in a row out of
4 shocks)
      # as follows
      ImpS_60J[i,] <- ifelse((((Shk1_60==1) & (Shk2_60 == 1)) |
((Shk1_60==0)
                        & (Shk2_60 == 1) & (Shk3_60 == 1)) | ((Shk1_60==1)
& (Shk2_60==0)
                        & (Shk3_60 == 1) & (Shk4_60 == 1)) | ((Shk1_60==0)
& (Shk2_60==0)
                        & (Shk3_60 == 1) & (Shk4_60 == 1))), 1,0)
      S70[i,] = ifelse(Shk1_70==1,1,0)
      # Compute observed implant success in design-A device clinical study
      ObsImpS_60J <- (Nv$N_ss+Nv$N_sfss+Nv$N_fss+Nv$N_ffss)/Npts

      # Determine conditional probability of success at 70J given that
implant success
      # at 60J
      PrS70givenImpS60[i] <- mean(S70[i,][ImpS_60J[i,]==1],na.rm=TRUE)
      }

    ## Determine average probability of 1st shock success at various
energy levels
    PrS_70 <- mean(p70)
    PrS_65 <- mean(p65)
    PrS_60 <- mean(p60)
    ObsPrs_60 <- (Nv$N_ss+Nv$N_sfss+Nv$N_sff+Nv$N_sfs+Nv$N_s)/Npts
    ## Determine average probability that both 1st and 2nd shocks are
successful at
    ## various energy levels
    PrSS_60 <- mean(p60^2); # S+S at 60J
    PrSS_65 <- mean(p65^2); # S+S at 65J
    PrSS_70 <- mean(p70^2); # S+S at 70J
    # Determine the observed probability that both 1st and 2nd shcoks
are successful
    ObsPrSS_60 <- Nv$N_ss/(Npts-Nv$N_s) #Removed the pts who had only
one shock

    # Probability calculations for additional outcomes (besides PrSS_60)
with implant
    # success at 60J
    PrSfSS_60 <- mean(p60*(1-p60)*p60*p60) #S+F+S+S at 60J
    PrfSS_60 <- mean((1-p60)*p60^2) #F+S+S at 60J
    PrffSS_60 <- mean((1-p60)^2*p60^2) #F+F+S+S at 60J

    # Compute the probability of design-A device clinical study implant
success for
    # 60J (10J safety margin with 70J max output device) shocks
    prImpSuc_60 <- PrSS_60 + PrSfSS_60 + PrfSS_60 + PrffSS_60
    # Compute observed probability of implant success based on all patients
    ObsprImpSuc_60 <- (Nv$N_ss+Nv$N_sfss+Nv$N_fss+Nv$N_ffss)/Npts
    # Compute observed probability of implant success based only on
those patients
    # who completed implant testing protocol
    ObsprImpSuc_60PC <- (Nv$N_ss+Nv$N_sfss+Nv$N_fss+Nv$N_ffss)/Npts_PC
```

```
    # Compute probabilities of design-A device clinical study implant
failure for
    # 60J shocks
    prImpfail_60 <- 1-prImpSuc_60
    ## Determine average conditional probability of Success at 70J given
    ## implant success at 60J

    PrAvgS70givenImpS60 <- mean(PrS70givenImpS60,na.rm=TRUE)

    # Determine 95% Credible Intervals for various types of
defibrillation efficacy
    # estimates
    PrS95CI_60J <- quantile(ProbS_60J, probs=c(0.025,0.975),na.rm=FALSE,
                            names=FALSE, type=7)
    PrS95CI_65J <- quantile(ProbS_65J, probs=c(0.025,0.975),na.rm=FALSE,
                            names=FALSE, type=7)
    PrS95CI_70J <- quantile(ProbS_70J, probs=c(0.025,0.975),na.rm=FALSE,
                            names=FALSE, type=7)

    PrSS95CI_60J <- quantile(ProbSS_60J, probs=c(0.025,0.975),na.rm=FALSE,
                             names=FALSE, type=7)
    PrSS95CI_65J <- quantile(ProbSS_65J, probs=c(0.025,0.975),na.rm=FALSE,
                             names=FALSE, type=7)
    PrSS95CI_70J <- quantile(ProbSS_70J, probs=c(0.025,0.975),na.rm=FALSE,
                             names=FALSE, type=7)

    PrImpS95CI_60J <- quantile(ProbImpS_60J, probs=c(0.025,0.975),
                               na.rm=FALSE, names=FALSE, type=7)

    PrImpf95CI_60J <- quantile(ProbImpf_60J, probs=c(0.025,0.975),
                               na.rm=FALSE, names=FALSE, type=7)
    Pr95CI_S70givenImpS60 <- quantile(PrS70givenImpS60, probs=c(0.025,0.975),
                             na.rm=FALSE, names=FALSE, type=7)


    cat(paste("PREDICTIONS for design-A device clinical study with",Npts,"
pts. based",
        "\non scalable model fitted to 325 design-A and",N1,"design-B
patients' data\n"))
    print(c(
      sprintf("Predicted Probability 1st shock success at 60J: %.1f
(95CI: %.1f - %.1f) -
      (model) Vs. %.1f (design-A study)", 100*PrS_60, 100*PrS95CI_60J[1],
      100*PrS95CI_60J[2],100*ObsPrs_60),
      sprintf("Predicted Probability of 1st shock success at 65J:
%.1f (95CI: %.1f -
      %.1f)", 100*PrS_65, 100*PrS95CI_65J[1], 100*PrS95CI_65J[2]),
      sprintf("Predicted Probability of 1st shock success at 70J:
%.1f (95CI: %.1f -
      %.1f)", 100*PrS_70, 100*PrS95CI_70J[1], 100*PrS95CI_70J[2]),
      sprintf("Predicted Probability of S + S (Success in 1st and
2nd shocks) at 60J: %.1f
      (95CI: %.1f - %.1f) - (model) Vs. %.1f (design-A study)",
100*PrSS_60,
      100*PrSS95CI_60J[1], 100*PrSS95CI_60J[2],100*ObsPrSS_60),
```

```
      sprintf("Predicted Probability of S + S (Success in 1st and
2nd shocks) at 65J: %.1f
      (95CI: %.1f - %.1f)",100*PrSS_65, 100*PrSS95CI_65J[1],
100*PrSS95CI_65J[2]),
      sprintf("Predicted Probability of S + S (Success in 1st and
2nd shocks) at 70J: %.1f
      (95CI: %.1f - %.1f)",100*PrSS_70, 100*PrSS95CI_70J[1],
100*PrSS95CI_70J[2]),
      sprintf("Predicted Probability of 2 of 4 implant success with
60J shocks: %.1f
      (95CI: %.1f - %.1f)- (model) Vs. %.1f (design-A study)",
100*prImpSuc_60,
      100*PrImpS95CI_60J[1], 100*PrImpS95CI_60J[2],100*ObsprImpSuc_60PC),
      sprintf("Predicted Probability of failure to pass 2 of 4 implant
criterion with 60J
      Shocks: %.1f (95CI: %.1f - %.1f)", 100*prImpfail_60,
100*PrImpf95CI_60J[1],
      100*PrImpf95CI_60J[2]),
      sprintf("Predicted Probability of success at 70J given that 2 of 4
implant success
      with 60J shocks: %.1f (95CI: %.1f - %.1f)", 100*PrAvgS70givenImpS60,
      100*Pr95CI_S70givenImpS60[1], 100*Pr95CI_S70givenImpS60[2])
      ))
  }
```

(9.3_Print_DesB_New_Study_Predictions.R)

```
  ###########################################################################
  # This program creates a function to generate and print the results     ##
  # for various defibrillation efficacy estimates for a new study with    ##
  # 300 pts for design-B devices.                                         ##
  # Inputs: MCMC samples generated by the Bayesian analysis for the       ##
  # following model parameters:                                           ##
  # Number of patients in the design-A and design-B studies combined -N  ##
  # Number of patients in the Design-B feasibility study - N1             ##
  # Number of MCMC samples - ngibbs                                       ##
  # Number of patients in the new design-B clinical study - N_New         ##
  # MCMC sample for the slope parameter - G                               ##
  # MCMC samples for the E50 parameter - Theta                            ##
  # MCMC samples for the scale and shrinkage parameters - Delta, Kappa    ##
  ###########################################################################

  # Define a function to print design-B device Defib Efficacy
estimation Results for a
  # new clinical study of a specified size

  PrintDesBReslts <- function(ngibbs,N1,N1_s,N_New,G,Theta,delta,kappa) {

    # Initialize the matrix Theta1 to store E50 values for design-B devices
    Theta0 <- array(0,dim=c(ngibbs,N_New))
    Theta1 <- array(0,dim=c(ngibbs,N_New))
    # Generate a sample of size N_New from the posterior distributions
of E50 values.
    # There are N E50 values to choose from for this sample for each
of the ngibbs MCMC
    # samples.The sampling is done with replacement
```

```
    for (i in 1:ngibbs) {Theta0[i,] <- sample(Theta[i,],size=N_New,
replace=TRUE)}
    Id1 <- rep(1,N_New)
    #dim(Theta1)
    for (i in 1:ngibbs){
      Theta1[i,] = (Theta0[i,])^((delta[i])^Id1)*(1-kappa[i]*Id1)
      }

    ## Initialize the matrices to store estimated defib efficacy
probabilities using
    ## MCMC samples
    p30 <- matrix(0,nrow=ngibbs,ncol=N_New)
    p35 <- matrix(0,nrow=ngibbs,ncol=N_New)
    p40 <- matrix(0,nrow=ngibbs,ncol=N_New)
    # Initialize the matrices to hold binomial counts data
    ImpS_30J <- matrix(0,nrow=ngibbs,ncol=N_New)
    S40 <- matrix(0,nrow=ngibbs,ncol=N_New)
    # Initialize vectors to store various defibrillation efficacy
probabilities.
    # This stored data is used to compute 95% credible intervals for the
estimated
    # defib efficacies.
    ProbS_30J <- numeric(ngibbs)
    ProbS_35J <- numeric(ngibbs)
    ProbS_40J <- numeric(ngibbs)
    ProbSS_30J <- numeric(ngibbs)
    ProbSS_35J <- numeric(ngibbs)
    ProbSS_40J <- numeric(ngibbs)

    ProbImpS_30J <- numeric(ngibbs)
    ProbImpf_30J <- numeric(ngibbs)
    PrS40givenImpS30 <- numeric(ngibbs)

    for(i in 1:ngibbs){
      # OBTAIN PROBABILITY OF SUCCESSFUL SHOCK

      p40[i,] <- 1/(1+(Theta1[i,]/40)^G[i])
      p35[i,] <- 1/(1+(Theta1[i,]/35)^G[i])
      p30[i,] <- 1/(1+(Theta1[i,]/30)^G[i])
      # For each MCMC sample, store the average (based on selected
number of pts) defib
      # efficacy at each different energy level
      ProbS_30J[i] <- mean(p30[i,]) # Avg. Prob. of 1st shock success at 30J
      ProbS_35J[i] <- mean(p35[i,]) # Avg. Prob. of 1st shock success at 35J
      ProbS_40J[i] <- mean(p40[i,]) # Avg. Prob. of 1st shock success at 40J

      ProbSS_30J[i] <- mean((p30[i,])^2) # Avg. Prob. of 1st & 2nd shock
success at 30J
      ProbSS_35J[i] <- mean((p35[i,])^2) # Avg. Prob. of 1st & 2nd
shock success at 35J
      ProbSS_40J[i] <- mean((p40[i,])^2) # Avg. Prob. of 1st & 2nd shock
success at 40J
      # Compute the average probability of implant success (2S in a row
out of 4 shocks)
      # at 30J for each MCMC sample
      ProbImpS_30J[i] <- mean(ProbSS_30J[i] #SS
```

```
                        + p30[i,]*(1-p30[i,])*(p30[i,])^2  #SFSS
                        + (1-p30[i,])*(p30[i,])^2  #FSS
                        + (1-p30[i,])^2*(p30[i,])^2) #FFSS
      ProbImpf_30J[i] <- 1-ProbImpS_30J[i]

      # It is not possible to directly compute the conditional
probability of success at
      # 40J given implant success at 30J because the success
probabilities within
      # patients have unknown correlations. So, we use the following
Binomial counts
      # data to compute the conditional probability

      #Generate success/fail counts for 1 - 4 shocked episodes at
different energies

      Shk1_30 <- rbinom(n=N_New,size=1,prob=p30[i,])
      Shk2_30 <- rbinom(n=N_New,size=1,prob=p30[i,])
      Shk3_30 <- rbinom(n=N_New,size=1,prob=p30[i,])
      Shk4_30 <- rbinom(n=N_New,size=1,prob=p30[i,])
      Shk1_40 <- rbinom(n=N_New,size=1,prob=p40[i,])

      # Get counts of design-B device implant sucess (2S in a row
out of 4 shocks)
      # as follows
      ImpS_30J[i,] <- ifelse((((Shk1_30==1) & (Shk2_30 == 1)) | ((Shk1_30==0)
                  & (Shk2_30 == 1) & (Shk3_30 == 1)) | ((Shk1_30==1)
& (Shk2_30==0)
                  & (Shk3_30 == 1) & (Shk4_30 == 1)) | ((Shk1_30==0)
& (Shk2_30==0)
                  & (Shk3_30 == 1) & (Shk4_30 == 1))), 1,0)
      S40[i,] = ifelse(Shk1_40==1,1,0)

      # Determine conditional probability of success at 40J given that
implant success
      # at 30J
      PrS40givenImpS30[i] <- mean(S40[i,][ImpS_30J[i,]==1],na.rm=TRUE)
      }

    ## Determine average probability of 1st shock success at various
energy levels
    PrS_40 <- mean(p40)
    PrS_35 <- mean(p35)
    PrS_30 <- mean(p30)
    ## Determine average probability that both 1st and 2nd shocks
are successful at
    ## various energy levels
    PrSS_30 <- mean(p30^2); # S+S at 30J
    PrSS_35 <- mean(p35^2); # S+S at 35J
    PrSS_40 <- mean(p40^2); # S+S at 40J

    # Probability calculations for additional outcomes (besides PrSS_30)
with implant
    # success at 30J
    PrSfSS_30 <- mean(p30*(1-p30)*p30*p30) #S+F+S+S at 30J
    PrfSS_30 <- mean((1-p30)*p30^2) #F+S+S at 30J
```

```
    PrffSS_30 <- mean((1-p30)^2*p30^2) #F+F+S+S at 30J

    # Compute the probability of design-B device new clinical study
implant success for
    # 30J (10J safety margin with 40J max output device) shocks
    prImpSuc_30 <- PrSS_30 + PrSfSS_30 + PrfSS_30 + PrffSS_30
    # Compute probabilities of design-B device new clinical study
implant failure for
    # 30J shocks
    prImpfail_30 <- 1-prImpSuc_30
    ## Determine average conditionla probability of Success at 40J given
    ## implant success at 30J

    PrAvgS40givenImpS30 <- mean(PrS40givenImpS30,na.rm=TRUE)

    # Determine 95% Credible Intervals for various types of
defibrillation efficacy estimates
    PrS95CI_30J <- quantile(ProbS_30J, probs=c(0.025,0.975),na.rm=FALSE,
                            names=FALSE, type=7)
    PrS95CI_35J <- quantile(ProbS_35J, probs=c(0.025,0.975),na.rm=FALSE,
                            names=FALSE, type=7)
    PrS95CI_40J <- quantile(ProbS_40J, probs=c(0.025,0.975),na.rm=FALSE,
                            names=FALSE, type=7)

    PrSS95CI_30J <- quantile(ProbSS_30J, probs=c(0.025,0.975),na.rm=FALSE,
                             names=FALSE, type=7)
    PrSS95CI_35J <- quantile(ProbSS_35J, probs=c(0.025,0.975),na.rm=FALSE,
                             names=FALSE, type=7)
    PrSS95CI_40J <- quantile(ProbSS_40J, probs=c(0.025,0.975),na.rm=FALSE,
                             names=FALSE, type=7)

    PrImpS95CI_30J <- quantile(ProbImpS_30J, probs=c(0.025,0.975),
                                na.rm=FALSE, names=FALSE, type=7)

    PrImpf95CI_30J <- quantile(ProbImpf_30J, probs=c(0.025,0.975),
                                na.rm=FALSE, names=FALSE, type=7)
    Pr95CI_S40givenImpS30 <- quantile(PrS40givenImpS30, probs=c(0.025,0.975),
                                na.rm=FALSE, names=FALSE, type=7)


    cat(paste("PREDICTIONS for design-B device clinical study with",
N_New," pts. based",
        "\non scalable model fitted to 325 design-A and",N1,"design-B
patients' data\n"))
    print(c(
      sprintf("Predicted Probability 1st shock success at 30J: %.1f
(95CI: %.1f - %.1f) -
      (model) Vs. %.1f (design-B feas.)",100*PrS_30, 100*PrS95CI_30J[1],
      100*PrS95CI_30J[2],100*(N1_s)/N1),
      sprintf("Predicted Probability of 1st shock success at 35J: %.1f
(95CI: %.1f - %.1f)",
              100*PrS_35, 100*PrS95CI_35J[1], 100*PrS95CI_35J[2]),
      sprintf("Predicted Probability of 1st shock success at 40J: %.1f
(95CI: %.1f - %.1f)",
              100*PrS_40, 100*PrS95CI_40J[1], 100*PrS95CI_40J[2]),
```

```
      sprintf("Predicted Probability of S + S (Success in 1st and 2nd
shocks) at 30J: %.1f
      (95CI: %.1f - %.1f)", 100*PrSS_30, 100*PrSS95CI_30J[1],
100*PrSS95CI_30J[2]),
      sprintf("Predicted Probability of S + S (Success in 1st and 2nd
shocks) at 35J: %.1f
      (95CI: %.1f - %.1f)",100*PrSS_35, 100*PrSS95CI_35J[1],
100*PrSS95CI_35J[2]),
      sprintf("Predicted Probability of S + S (Success in 1st and 2nd
shocks) at 40J: %.1f
      (95CI: %.1f - %.1f)",100*PrSS_40, 100*PrSS95CI_40J[1],
100*PrSS95CI_40J[2]),
      sprintf("Predicted Probability of 2 of 4 implant success with 30J
shocks: %.1f (95CI:
      %.1f - %.1f)", 100*prImpSuc_30, 100*PrImpS95CI_30J[1],
100*PrImpS95CI_30J[2]),
      sprintf("Predicted Probability of failure to pass 2 of 4 implant
criterion with 30J
      Shocks: %.1f (95CI: %.1f - %.1f)", 100*prImpfail_30,
100*PrImpf95CI_30J[1],
      100*PrImpf95CI_30J[2]),
      sprintf("Predicted Probability of success at 40J given that 2 of 4
implant success
      with 30J shocks: %.1f (95CI: %.1f - %.1f)", 100*PrAvgS40givenImpS30,
      100*Pr95CI_S40givenImpS30[1], 100*Pr95CI_S40givenImpS30[2])
      ))
  }
```

The following figures and defibrillation efficacy results were produced by running the R script 9.3_Defib_Efficacy_Pred.R. Figure 9.10 shows that on the average about 95% of the patients can be successfully defibrillated with a 70 J (maximum output) shock

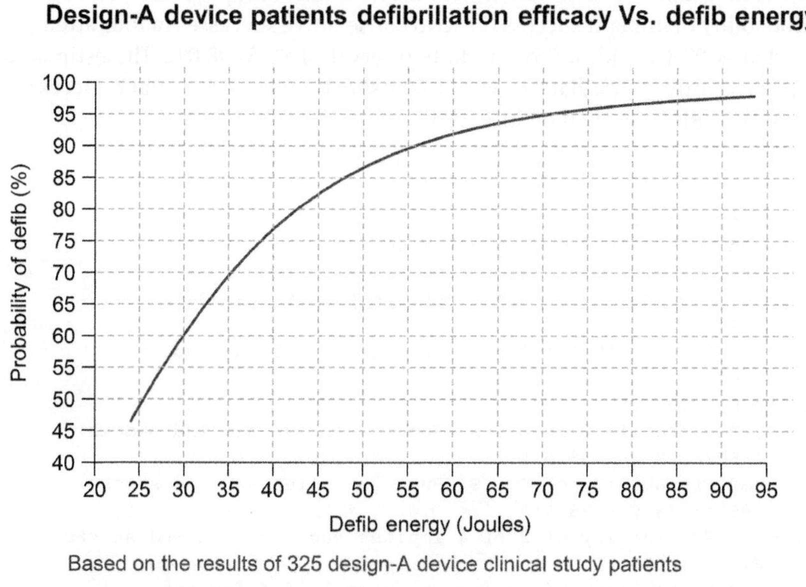

Figure 9.10 Design-*A* device patients average defibrillation efficacy versus defibrillation energy.

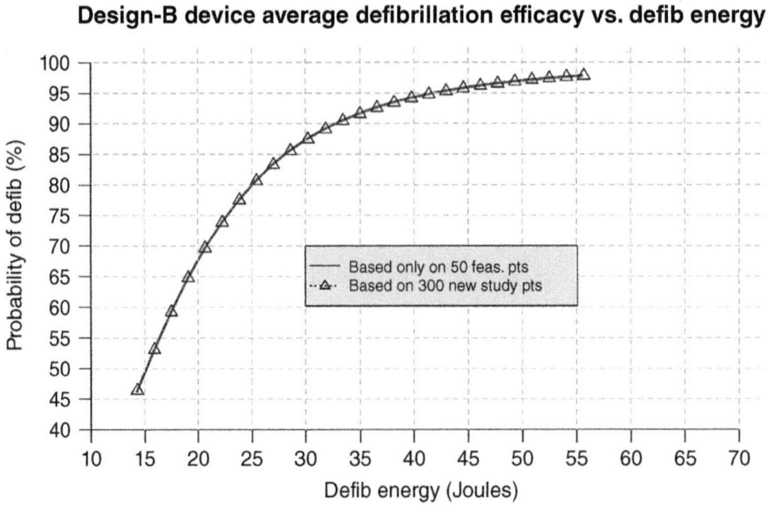

Figure 9.11 Comparing design-*B* device patients average defibrillation versus defibrillation energy estimated for 300 pts. new study and the 50 pts. feasibility study.

delivered by a design-A device. Figure 9.11 shows that the average defibrillation efficacy estimated using a 50-patient feasibility study is identical to that estimated for the planned 300-pateint clinical study.

The following results are model-based estimates for the planned 300-patient design-*B* device clinical study. All probability estimates are given as percentages without the % sign. These results show that the expected probability of successful defibrillation for a design-*B* device patient with a single shock of 30 J is 87.2% with a 95% credible interval of 77.4–94.5%. The first shock success observed in the feasibility study is 88% (=44/50). The estimated probability of implant success (i.e. two consecutive successes in four attempts with 30 J shocks) is 92.1% with a 95% credible interval of 82.5–98.0%. The estimated probability of successful defibrillation with a 40 J shock given the implant success is 95.5% with a 95% credible interval of 90.3–99.0%.

```
  [1] "Predicted Probability 1st shock success at 30J: 87.2
(95CI: 77.4 - 94.5) -
  (model) Vs. 88.0 (design-B feas.)"
  [2] "Predicted Probability of 1st shock success at 35J: 91.6
(95CI: 83.9 - 96.8)"
  [3] "Predicted Probability of 1st shock success at 40J: 94.2
(95CI: 87.9 - 98.2)"
  [4] "Predicted Probability of S + S (Success in 1st and 2nd
shocks) at 30J:
78.0 (95CI: 63.5 - 89.8)"
  [5] "Predicted Probability of S + S (Success in 1st and 2nd shocks)
at 35J: 85.0 (95CI: 72.9 - 93.9)"
  [6] "Predicted Probability of S + S (Success in 1st and 2nd shocks)
at 40J: 89.5 (95CI: 79.0 - 96.4)"
  [7] "Predicted Probability of 2 of 4 implant success with 30J shocks:
92.1 (95CI: 82.5 - 98.0)"
  [8] "Predicted Probability of failure to pass 2 of 4 implant
criterion with 30J Shocks: 7.9 (95CI: 2.0 - 17.5)"
```

```
[9] "Predicted Probability of success at 40J given that 2 of 4 implant
success with 30J shocks: 95.5 (95CI: 90.3 - 99.0)"
```

The following results are model-based estimates for a historical design-*A* device clinical study with 325 patients. All probability estimates are given as percentages without the % sign. These results show that the expected probability of successful defibrillation for a design-*A* device patient with a single shock of 60 J is 91.9% with a 95% credible interval of 89.9–93.6%. The first shock success observed in the clinical study was 91.1%. The estimated probability of implant success (i.e. two consecutive successes in four attempts with 60 J shocks) is 96.0% with a 95% credible interval of 94.4–97.4%. The observed implant success based on the patients who completed the testing protocol was 97.1% (=305/314). The observed implant success using all patients is 93.8% (=305/325). The estimated probability of successful defibrillation with a 70 J shock given the implant success is 95.6% with a 95% credible interval of 92.6–98.1%.

```
PREDICTIONS for design-A device clinical study with 325 pts. based
on scalable model fitted to 325 design-A and 50 design-B patients' data

[1] "Predicted Probability 1st shock success at 60J: 91.9
(95CI: 89.9 - 93.6) - (model) Vs. 91.1 (design-A study)"
[2] "Predicted Probability of 1st shock success at 65J: 93.6
(95CI: 91.8 - 95.2)"
[3] "Predicted Probability of 1st shock success at 70J: 94.9
(95CI: 93.0 - 96.6)"
[4] "Predicted Probability of S + S (Success in 1st and 2nd shocks)
at 60J: 85.6 (95CI: 82.0 - 88.7) - (model) Vs. 85.7 (design-A study)"
[5] "Predicted Probability of S + S (Success in 1st and 2nd shocks) at
65J: 88.4 (95CI: 85.2 - 91.2)"
[6] "Predicted Probability of S + S (Success in 1st and 2nd shocks)
at 70J: 90.6 (95CI: 87.3 - 93.6)"
[7] "Predicted Probability of 2 of 4 implant success with 60J shocks:
96.0 (95CI: 94.4 - 97.4)- (model) Vs. 97.1 (design-A study)"
[8] "Predicted Probability of failure to pass 2 of 4 implant
criterion with 60J Shocks: 4.0 (95CI: 2.6 - 5.6)"
[9] "Predicted Probability of success at 70J given that 2 of 4 implant
success with 60J shocks: 95.6 (95CI: 92.6 - 98.1)"
```

The comparison of results for design-*A* and -*B* devices suggests that the probability of successful defibrillation with a maximum energy shock given implant success is similar between the two designs.

9.4 Summary

Linear regression builds a relationship between a continuous response variable and one or more predictors. Logistic regression is used to build a relationship between a categorical response and one or more predictors. In this chapter, linear regression and binary logistic regression models using Bayesian methods were introduced with examples, including a case study.

Bayesian statistics treats unknown parameters as random variables with probability distributions assigned, rather than unknown constants. Therefore, the Bayesian solution of a coefficient in a regression model consists of a series of values. In a Bayesian framework, the uncertainty of unknown parameters can be easily propagated to the final outcome. How to make future predictions based on Bayesian regression models is demonstrated in the case study and other examples.

References

Carlin, P.C. and Louis, T.A. (2009). *Bayesian Methods for Data Analysis*. Boca Raton, FL: Taylor & Francis.

Gelman, A., Carlin, J.B., Stern, H.S. et al. (2014). *Bayesian Data Analysis*. Boca Raton, FL: Taylor & Francis.

Kruschke, J.K. (2011). *Doing Bayesian Data Analysis*. Oxford: Elsevier.

Montgomery, D.C. (2013). *Design and Analysis of Experiments*. Hoboken, NJ: Wiley.

Montgomery, D.C., Peck, E.A., and Vining, G.G. (2012). *Introduction to Linear Regression Analysis*. Hoboken, NJ: Wiley.

Myers, R.H., Montgomery, D.C., and Anderson-Cook, C.M. (2016). *Response Surface Methodology: Process and Product Optimization Using Designed Experiments*. Hoboken, NJ: Wiley.

Smits, K. and Virag, N. (2010). Estimating the parameter distributions of defibrillation shock efficacy curves in a large population. *Annals of Biomedical Engineering* 38 (4): 1314–1325.

Smits, K., Virag, N., and Swerdlow, C.D. (2013). Impact of defibrillation testing on predicted ICD shock efficacy: implications for clinical practice. *Heart Rhythm* 10 (5): 709–717.

Appendix A

Guidance for Installing R, R Studio, JAGS, and rjags

Create a folder in a writeable directory (e.g. C:\Bayesian), and install all the programs (R, RStudio, JAGS) into this folder following the steps below.

A.1 Install R

The following are the steps to install the programming language R.

1.1 Go to website https://www.r-project.org.

1.2 To download R, choose a CRAN mirror at webpage https://cran.r-project.org/mirrors.html.
e.g. click the first link under category USA, which is https://cran.cnr.berkeley.edu.
This leads you to the page to download and install R.
Follow the instructions, e.g. for Windows users, select *Download R for Windows*.
On the following page, R for Windows, select *Install R for the first time*.
On the following page, select *Download R x.x.x (version number) for Windows*.

1.3 Run the R installation program to install R. When asked to Select Destination Location, select the folder created earlier, e.g.
C:\Bayesian\R-3.3.0,
and follow the installation instructions.

A.2 Install R Studio

The following are the steps to install the R Studio editor, which provides a friendly graphical user interface for R programming.

2.1 Go to the R Studio website https://www.rstudio.com and select *Download RStudio*.

2.2 Select *Download RStudio Desktop*.

2.3 On the following page, select the appropriate program for your platform, e.g. *RStudio x.x.xxx – Windows Vista/7/8/10*.

2.4 Run the RStudio installation program to install RStudio. When asked to Choose Install Location, select the folder created earlier, e.g.
C:\Bayesian\Rstudio,
and follow the installation instructions.

Practical Applications of Bayesian Reliability, First Edition. Yan Liu and Athula I. Abeyratne.

Companion website: www.wiley.com/go/bayesian20

A.3 Install JAGS

The following are the steps to install the Bayesian sampling program JAGS.

3.1 Go to website http://mcmc-jags.sourceforge.net for instructions. To download the JAGS installation program *(JAGS –x.x.x.exe)*, visit the webpage https://sourceforge.net/projects/mcmc-jags/files.

3.2 Run the JAGS installation program to install JAGS. When asked to Choose Install Location, select the folder created earlier, e.g.
C:\Bayesian\JAGS-4.2.0,
and follow the installation instructions.

A.4 Install Package rjags

The following are the steps to install R package rjags (to let R talk to JAGS). There are two ways to do this:

1) In the RStudio menu bar at the top of the screen, select *Tools → Install Packages*, then type `rjags` under Packages
   or
2) In the R Studio console window, type `install.packages("rjags")`.

A.5 Set Working Directory

The working directory is a default location for files to be read from or written to. Use R command `getwd()` to find your current working directory. There are two ways to set the working directory in RStudio:

1) In the RStudio menu bar at the top of the screen, select *Session → Set Working Directory → Choose Directory...* (e.g. chose `C:\Bayesian` as the working directory)
   or
2) In the R Studio console window, use command `setwd`, for example `setwd ("C:/Bayesian")`.

Note: Before running the R scripts in this book, please save all the data files (.txt files), .JAGS files, and .R files in the working directory.

Appendix B

Commonly Used R Commands

B.1 How to Run R Commands

There are several ways to run R commands:

- Type a command in the R Studio console window (after the prompt symbol >), and press Enter.
- Run codes in an R script file.R in the R Studio source window (where you can read, create, and edit R scripts) once the file is open. This is recommended for readers to run the R scripts in each chapter. By clicking the Run button (located above the R script file, at the top right of the editing window), R will execute the current line (where the cursor is). You can also select (highlight) multiple lines in an R script file and click Run to execute them all at once.

B.2 General Commands

Note that the > symbol at the beginning of each R command in the R Studio console window is a prompt generated by R.

- *Get R help files for a function*
  Type "?" followed by a command name, for example:

```
> ?library
> ?matrix
```

- *Set the working directory*
  By setting the working directory in R using command `setwd`, files can be read from or written to this default location, for example

```
> setwd("C:/Bayesian")
```

- *Check the current working directory*

```
> getwd()
```

- *Install R packages*

```
> install.packages("mcmcplots")
```

- *Load packages*

```
> library(rjags)
```

Practical Applications of Bayesian Reliability, First Edition. Yan Liu and Athula I. Abeyratne.

Companion website: www.wiley.com/go/bayesian20

B.3 Generate Data

- *Assign a value to a variable*
  There are two ways to assign a value to a variable in R:

```
> x <- 3
```

or

```
> x = 3
```

Both of these commands assign the value 3 to the variable x. When you type `x`, its value will be shown.

```
> x
[1] 3
```

- *Concatenate (combine) function*

```
> c(1,3,5) # generate a vector of three values
[1] 1 3 5

> c(1,"a") # creates a vector of a numeric value and a charter sting
[1] "1" "a"
```

- *Generate repeats*

```
> rep(1, 5)
[1] 1 1 1 1 1
```

- *Colon operator*
  Generate a series of integers.

```
> 1:5
[1] 1 2 3 4 5
```

- *Generate a sequence of values*
  Generate a sequence from a starting value of 0 to an ending value of 1, with the total number of elements being 11.

```
> seq(0,1,length=11)
[1] 0.0 0.1 0.2 0.3 0.4 0.5 0.6 0.7 0.8 0.9 1.0
```

Generate a sequence from a starting value of 0 to an ending value of 1, with the increment being 0.1.

```
> seq(0,1,by=0.1)
[1] 0.0 0.1 0.2 0.3 0.4 0.5 0.6 0.7 0.8 0.9 1.0
```

Generate a sequence from a starting value of 0 to an ending value of 1, with the increment being 0.3. Note that this sequence does not exceed the ending value.

```
> seq(0,1,by=0.3)
[1] 0.0 0.3 0.6 0.9
```

Compute the Beta(1,2) density at a sequence of points between 0 and 1

```
> p <- seq(0,1,length=11)
> dbeta(p,1,2)
 [1] 2.0 1.8 1.6 1.4 1.2 1.0 0.8 0.6 0.4 0.2 0.0
```

- *Generate random numbers from a distribution*
  Generate 10 random numbers from a normal distribution with mean 10 and standard deviation 2.

```
> rnorm(10, mean = 10, sd = 2)
[1] 12.281438 7.191194 8.687122 7.469112 6.921286 9.248468 11.081916
12.590917 8.505709
[10] 9.065589
```

  Generate 10 random numbers from a standard normal distribution with mean 0 and standard deviation 1.

```
> rnorm(10)
[1] 0.6150031 0.6553519 0.5737280 1.9936318 0.7202471 0.2354205
-0.2036708 -0.4450824 -0.3848908 -0.7987034
```

  Other examples of generating random numbers from commonly used distributions are:
  Beta distribution: `rbeta(n,shape1,shape2)`
  Gamma distribution: `rgamma(n, shape, rate = 1, scale = 1/rate)`
  Lognormal distribution: `rlnorm(n, meanlog = 0, sdlog = 1)`
  Uniform distribution: `runif(n, min = 0, max = 1)`
  Binomial distribution: `rbinom(n, size, prob)`
  Weibull distribution: `rweibull(n, shape, scale = 1)`

B.4 Variable Types

- *Scalar:* A single number

```
> x <- 3
```

- *Vector:* One-dimensional list of elements of the same type
  - *Numeric vector:* A sequence of numbers

```
> x <- c(1,2,3)
```

  - *Character vector:* A sequence of characters

```
x <- c("a","b","c")
```

  - *Logical vector:* A vector of TRUE (or T), FALSE (or F), or NA

```
x <- c(T,F)
```

  Subset vectors

```
> x <- c(1,2,3)
> x[3] # get the third element in vector x
[1] 3

> x[x>1] # get all the elements in vector x that are greater than 1
[1] 2 3
```

- *Matrices:* Have two dimensions of values of the same type. Columns in a matrix must have the same length. A vector can be converted to a matrix by specifying the dimension attributes (i.e. number of rows, number of columns or both).

To create a matrix: The first argument specifies that the content of the matrix is a vector `c(1,1,2,2,3,3)`. The second argument specifies that the matrix has two rows. By default, a matrix is filled by columns. Note: the number of elements in the vector has to be a multiple of the dimension attributes.

```
> x <- matrix(c(1,1,2,2,3,3), nrow=2)

> x
 [,1] [,2] [,3]
[1,] 1 2 3
[2,] 1 2 3


Use class to check the type of the object, e.g.,
> class(x)
[1] "matrix"

To fill a matrix by row, add an argument byrow=TRUE.
> x <- matrix(c(1,1,2,2,3,3), nrow=2, byrow=TRUE)

> x
 [,1] [,2] [,3]
 [1,] 1 1  2
 [2,] 2 3  3

Use dimnames attribute to assign row and column names
> matrix(1:12, nrow=3, dimnames=list(row_names=c("R1","R2","R3"),
col_names=c("C1","C2","C3","C4")))

 col_names
row_names C1 C2 C3 C4
          R1 1 4 7 10
          R2 2 5 8 11
          R3 3 6 9 12
```

Subset matrices

```
> x <- matrix(c(1,1,2,2,3,3), nrow=2, byrow=TRUE)
```

Show elements in the first row (all columns are included):

```
> x[1,]
[1] 1 1 2
```

Show elements in the second column (all rows are included):

```
> x[,2]
[1] 1 3
```

Show the elements in the first row and the second column:

```
> x[1,2]
[1] 1
```

- *Arrays:* Similar to matrices but can have one, two or more dimensions. A matrix is a two-dimensional array.

- *Data frames:* Also have two dimensions, which is similar to matrices. The difference compared to matrices is that different columns in a data frame can be of different types (numeric, character, etc.).
  Data frame is the default data format after reading data using the command `read.table`. We will explore this later in the **Read and write data** section.
- *Lists:* A collection of several components which can be of different types.
  Create a list named `DataList` that has four elements: the first element is a numeric vector, the second element is an integer, the third element is a matrix, and the fourth element is a string named "Data."

```
> x <- c(1,3,5,3,4,7)
> DataList <-list(x=x,n=length(x), matrix=matrix(x, nrow=2),name="Data")
> DataList
$x
[1] 1 3 5 3 4 7

$n
[1] 6

$matrix
 [,1] [,2] [,3]
 [1,] 1  5 4
 [2,] 3  3 7

$name
[1] "Data"
```

  Examples of getting components in the list:

```
> DataList$n
[1] 6

> DataList$name
[1] "Data"

> DataList[3]
$matrix
 [,1] [,2] [,3]
 [1,] 1  5 4
 [2,] 3  3 7
```

B.5 Calculations and Operations

- *Arithmetic*

```
> 1 + 2*3 - 6/4
[1] 5.5
```

- *Power, logarithms, and the exponential function*

```
> 2^3 # 2 to the power of 3
 [1] 8

> log(3) # natural logarithms
```

```
 [1] 1.098612

> log10(10) # common (i.e., base 10) logarithms
 [1] 1

> exp(1) # exponential function
 [1] 2.718282
```

- *Numeric vector operations*
  The summation of two vectors generates a vector consisting of the summations of corresponding elements. Note: the two vectors have to be of the same length or the length of the longer vector has to be a multiple of the length of the shorter vector.

```
> x <- c(1,2,3)
> x+x
[1] 2 4 6
```

  The product of two vectors generates a vector consisting of the products of corresponding elements. Again, the two vectors must be same length.

```
> x*x
[1] 1 4 9
```

  If a constant is added to a vector, that constant is added to each element of the vector.

```
> x+1
[1] 2 3 4
```

B.6 Summarize Data

- *Calculate the sum of elements in a vector*

```
> v1 <- c(1,2,3)
> sum(v1)
 [1] 6

> sum(v1>1) # count the number of elements in the vector that are greater
than 1
 [1] 2
```

- *Calculate the mean of a vector*

```
> v1 <- c(1,2,3)

> mean(v1)
[1] 2
```

- *Calculate the standard deviation of a vector*

```
> sd(v1)
[1] 1
```

- *Quantiles*
  - Display 0%, 25%, 50%, 75%, and 100%

```
> quantile(vector_name)
```

- Display 5%

```
> quantile(vector_name, 0.05)
```

- Display 2.5% and 97.5%

```
> quantile(vector_name, c(0.025,0.975))
```

- *Summarize data:* Display the minimum, first quartile, median, mean, third quartile, and maximum

```
> summary(dataset_or_vector_name)
```

- *Range*

```
> range(c(1,4,8,10,-1))
[1] -1 10
```

B.7 Read and Write Data

- *Read data from a .txt file*
  Assume the following data are saved in a text file named `Mail.txt` located in the directory `C:/temp/`.

Day	Number_of_mails
1	10
2	12
3	15
4	14
5	12

The following commands can be used to read the text file `Mail.txt`.

```
        > path <- "C:/temp/"
        > data <- read.table(file=paste0(path,"Mail.txt"),header=T,sep="")
        Now we use class to check the object type and find that object data
is a data frame.
        > class(data)        # check the object type
        [1] "data.frame"

        > data
          Day NUmber_of_mails
        1   1              10
        2   2              12
        3   3              15
        4   4              14
        5   5              12

        Subset the data frame data

        > data$NUmber_of_mails
        [1] 10 12 15 14 12
```

```
List 1st 3 elements of the 2nd column of the data frame "data"

> data[1:3,2]
[1] 10 12 15
```

- *Read data from a CSV file*

```
> testData <- read.csv("C:/temp/test.csv",header=T)
```

- *Write data to a .txt file*

```
write.table( data, file="C:/temp/MailNew.txt", row.names=FALSE,
quote=FALSE )
```

- *Write data to a CSV file*

```
write.csv( data, file="C:/temp/MailNew.csv", row.names=FALSE,
quote=FALSE )
```

B.8 Plot Data

- *Histogram*

```
> hist(vector_name)

> hist(vector_name, col=rgb(0.1,0.1,0.1,0.5), main = " Histogram of
reliability", xlab="Reliability", xlim=range(0.94,1), ylim=range(0,500),
breaks=100)
> hist(rnorm(1000, mean = 10, sd = 1), main = " Histogram of a
Normal distribution (mean=10, sd=1)", xlab="x")
```

Histogram shown in Figure B.1

Combine plots

```
> par(mfrow=c(1,2)) # create a matrix of plots with 1 row and 2 columns,
filled by row
# The following two histograms will be displayed side by side
> hist(shape_samples, main="Shape histogram", breaks=100)
> hist(scale_samples, main="Scale histogram", breaks=100)
> par(mfrow=c(1,1)) # Reset par() to default state
```

- *Scatterplot*

```
> plot(rnorm(1000, mean = 10, sd = 1), ylab="Data from Normal(mean=10,
sd=1)")
```

Scatterplot shown in Figure B.2

- *x vs. y plot*

```
> plot(x_vectpr, y_vector)
```

- *Boxplot*

```
>       boxplot(rnorm(1000, mean = 10, sd = 1),ylab="Data")
```

Boxplot shown in Figure B.3

- *Density plot*

```
> p <- seq(0,1,length=101)
```

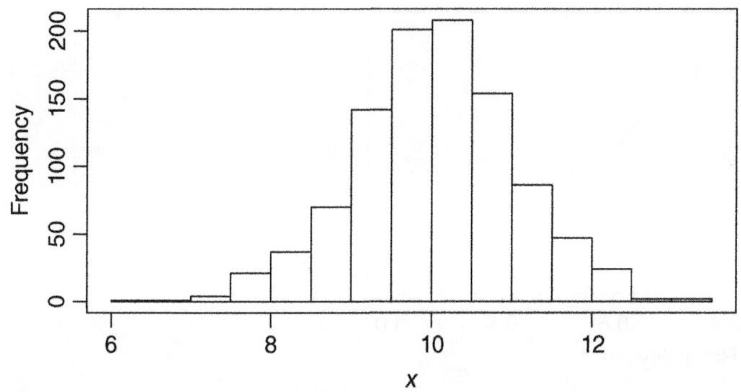

Figure B.1 Histogram of 1000 random numbers sampled from a normal distribution with a mean of 10 and a standard deviation of 1.

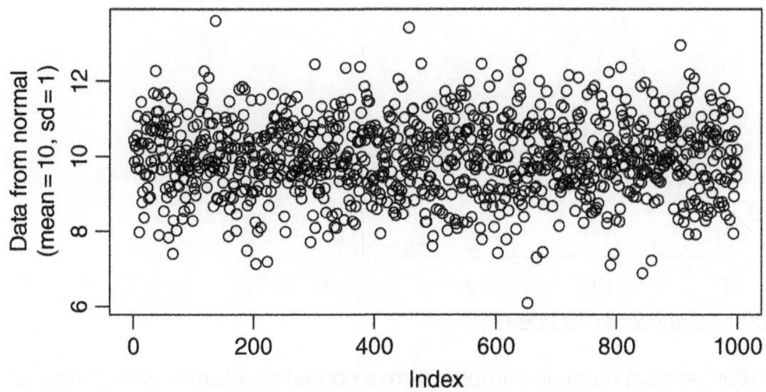

Figure B.2 Scatterplot of 1000 random numbers sampled from a normal distribution with a mean of 10 and a standard deviation of 1.

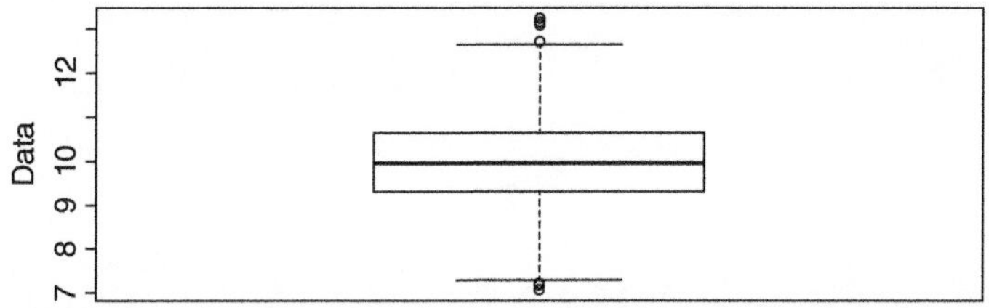

Figure B.3 Boxplot of 1000 random numbers sampled from a normal distribution with a mean of 10 and a standard deviation of 1.

```
> plot(p, dbeta(p,10,10),type="l",ylab="Density",xlab="Reliability (p)",
lty=1,lwd=2,col="red",ylim=c(0,5)) # plot likelihood
```

Density plot shown in Figure B.4

```
> x <- rnorm(1000, mean = 10, sd = 1)
> d <- density(x)
> plot(d)
```

Density plot shown in Figure B.5

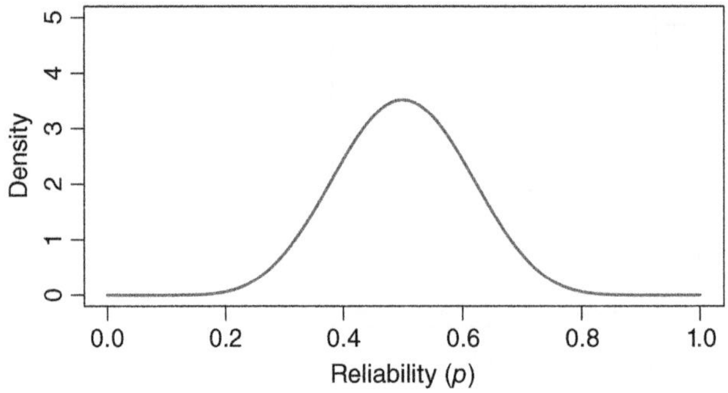

Figure B.4 Density plot of a beta distribution.

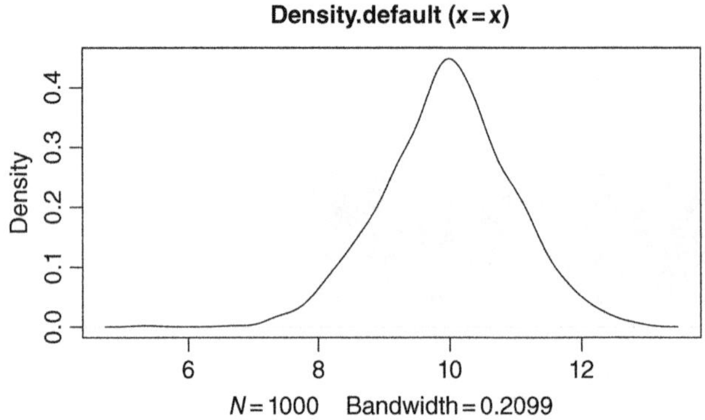

Figure B.5 Density plot of 1000 random numbers sampled from a normal distribution with a mean of 10 and a standard deviation of 1.

B.9 Loops and Conditional Statements

- *for loop*

```
for (i in 1:100) {
...

}
```

- *if statement*

```
if (...) {
 ...

}

if (...) {
 ...

} else if (...) {
 ...

} else
 ....
```

Appendix C

Probability Distributions

C.1 Discrete Distributions

C.1.1 Binomial Distribution

$$X \sim Binomial\,(n, \theta)$$

θ is the probability of getting a success in each trial, $0 \leq \theta \leq 1$,

$$x = 0,\ 1,\ 2,\ \ldots\ n$$

Probability mass function

$$f(x \mid n, \theta) = \binom{n}{x} \theta^x (1-\theta)^{n-x}$$

where

$$\binom{n}{x} = \frac{n!}{x!(n-x)!}$$

Mean and variance

$$E(X) = n\theta$$
$$Var(X) = n\theta(1-\theta)$$

C.1.2 Poisson Distribution

$$X \sim Poisson\,(\lambda)$$

$$\lambda > 0,\ x = 0,\ 1,\ 2,\ \ldots$$

Probability mass function

$$f(x \mid \lambda) = \frac{\lambda^x e^{-\lambda}}{x!}$$

Mean and variance

$$E(X) = \lambda$$

$$Var(X) = \lambda$$

Practical Applications of Bayesian Reliability, First Edition. Yan Liu and Athula I. Abeyratne.

Companion website: www.wiley.com/go/bayesian20

C.2 Continuous Distributions

C.2.1 Beta Distribution

$$X \sim Beta\,(\alpha, \beta)$$

$$0 \le X \le 1, \quad \alpha > 0, \quad \beta > 0$$

Probability density function

$$f(x \mid \alpha, \beta) = \frac{x^{\alpha-1}(1-x)^{\beta-1}}{B(\alpha, \beta)}$$

where

$$B(\alpha, \beta) = \frac{\Gamma(\alpha)\ \Gamma(\beta)}{\Gamma(\alpha+\beta)}$$

$\Gamma(x)$ is the gamma function,
$\Gamma(x) = \int_0^\infty s^{x-1}e^{-s}ds$, $x > 0$ (when $x = 1, 2, \ldots$, $\Gamma(x) = (x-1)!$)
Mean and variance

$$E(X) = \frac{\alpha}{\alpha+\beta}$$

$$Var(X) = \frac{\alpha\beta}{(\alpha+\beta)^2(\alpha+\beta+1)}$$

C.2.2 Exponential Distribution

$$X \sim Exponential\,(\lambda)$$

$x > 0$, $\lambda > 0$ (rate parameter; mean number of failures per unit time; the hazard rate)
Probability density function

$$f(x \mid \lambda) = \lambda e^{-\lambda x}$$

Mean and variance

$$E(X) = 1/\lambda$$

$$Var(X) = \frac{1}{\lambda^2}$$

C.2.3 Gamma Distribution

$$X \sim Gamma\,(a, b)$$

$X > 0$, $a > 0$ (shape parameter), $b > 0$ (rate parameter)
Probability density function

$$f(x \mid a, b) = \frac{b^a x^{a-1} e^{-bx}}{\Gamma(a)}$$

Mean and variance

$$E(X) = a/b$$

$$Var(X) = a/b^2$$

C.2.4 Inverse Gamma Distribution

$$X \sim InverseGamma\,(\alpha, \beta)$$

$$X > 0, \alpha > 0, \beta > 0$$

When $1/X \sim Gamma\,(\alpha, \beta)$, then $X \sim InverseGamma\,(\alpha, \beta)$

Probability density function

$$f(x \mid \alpha, \beta) = \frac{\beta^{\alpha} x^{-\alpha-1} e^{-\beta/x}}{\Gamma(\alpha)}$$

Mean and variance

$E(X) = \frac{\beta}{\alpha-1}$ for $\alpha > 1$

$Var(X) = \frac{\beta^2}{(\alpha-1)^2(\alpha-2)}$ for $\alpha > 2$

C.2.5 Lognormal Distribution

$$X \sim Lognormal\,(\mu, \sigma^2)$$

$$X > 0, -\infty < \mu < \infty, \sigma^2 > 0$$

If $y = \log(x)$ is normally distributed with mean μ and standard deviation σ, then the distribution of x becomes a two-parameter log-normal distribution.

Probability density function

$$f(x \mid \mu, \sigma^2) = \frac{1}{x\sqrt{2\pi\sigma^2}} e^{-\frac{(\ln x-\mu)^2}{2\sigma^2}}$$

Mean and variance

$$E(X) = e^{\mu+\frac{\sigma^2}{2}}$$

$$Var(X) = e^{2\mu+2\sigma^2} - e^{2\mu+\sigma^2}$$

C.2.6 Normal Distribution

$$X \sim Normal\,(\mu,\ \sigma^2)$$

$$-\infty < x < \infty, -\infty < \mu < \infty, \sigma^2 > 0$$

Probability density function

$$f(x \mid \mu,\ \sigma^2) = \frac{1}{\sqrt{2\pi\sigma^2}} e^{-\frac{(x-\mu)^2}{2\sigma^2}}$$

Mean and variance

$$E(X) = \mu$$

$$Var(X) = \sigma^2$$

C.2.7 Uniform Distribution

$$X \sim Uniform\,(a,\, b)$$

$$a \le x \le b$$

Probability density function

$$f(x \mid a, b) = \frac{1}{b - a}$$

Mean and variance

$$E(X) = \frac{a + b}{2}$$

$$Var(X) = \frac{(b - a)^2}{12}$$

C.2.8 Weibull Distribution

$$X \sim Weibull\,(\beta, \eta)$$

$x \ge 0$, $\beta > 0$ (shape parameter), $\eta > 0$ (scale parameter)

Probability density function

$$f(x \mid \beta, \eta) = \frac{\beta}{\eta}\left(\frac{x}{\eta}\right)^{\beta-1} e^{-(x/\eta)^{\beta}}$$

$\beta > 1$ indicates increasing hazard rate or wear-out failures. $\beta < 1$ indicates decreasing hazard rate or early failures. $\beta = 1$ indicates constant hazard rate or random failures. A Weibull distribution with $\beta = 1$ is also an exponential distribution. In the literature, η is also known as the characteristic life of the Weibull distribution, since at time η, 63.2% ($= e^{-1}$) of the units will have failed. This failure rate does not depend on the value of β.

Mean and variance

$$E(X) = \eta\Gamma\left(1 + \frac{1}{\beta}\right)$$

$$Var(X) = \eta^2\left[\Gamma\left(1 + \frac{2}{\beta}\right) - \left(\Gamma\left(1 + \frac{1}{\beta}\right)\right)^2\right]$$

Hazard function

$$h\,(x \mid \beta, \eta) = \frac{\beta}{\eta}\left(\frac{x}{\eta}\right)^{\beta-1}$$

Reliability

$$R\,(x \mid \beta, \eta) = e^{-(x/\eta)^{\beta}}$$

Appendix D

Jeffreys Prior

The Jeffreys prior is a non-informative prior distribution that is invariant under transformation (reparameterization). The Jeffreys prior is proportional to the square root of the determinant of the expected Fisher Information Matrix of the selected model

$$p(\boldsymbol{\theta}) \propto |I(\boldsymbol{\theta})|^{1/2},$$

where $I(\boldsymbol{\theta})$ is the expected Fisher Information Matrix, i.e.

$$I(\boldsymbol{\theta}) = -E_{X|\theta}\left[\frac{\partial^2}{\partial\theta^2}\log f(X \mid \boldsymbol{\theta})\right],$$

where $f(\boldsymbol{X} \mid \boldsymbol{\theta})$ is the likelihood function of the data $\boldsymbol{X}$ given the parameter vector $\boldsymbol{\theta}$.

Let $p(\theta)$ be the Jeffreys prior and $\omega = h(\theta)$ be a 1-1 transformation for a single-parameter case. This prior is invariant under reparameterization, which means that

$$[I(\omega)]^{1/2} = [I(\theta)]^{1/2}\left|\frac{d\theta}{d\omega}\right|.$$

The proof of the above equation is as follows.

According to chain rule, if $y = f(g(x))$, then $\frac{dy}{dx} = \frac{df}{dg}\cdot\frac{dg}{dx}$ and according to the product rule,

$$\frac{d^2y}{dx^2} = \frac{d^2f}{dg^2}\cdot\frac{dg}{dx}\cdot\frac{dg}{dx} + \frac{df}{dg}\cdot\frac{d^2g}{dx^2} = \frac{d^2f}{dg^2}\cdot\left(\frac{dg}{dx}\right)^2 + \frac{df}{dg}\cdot\frac{d^2g}{dx^2}.$$

$\omega = h(\theta)$, thus $\theta = h^{-1}(\omega)$. Based on the above equation ($\theta = h^{-1}(\omega)$ is equivalent to $g(x)$, and ω is equivalent to x in the above equation),

$$\frac{d^2\log p(x \mid \omega)}{d\omega^2} = \frac{d^2\log p(x \mid \omega)}{d\theta^2}\cdot\left(\frac{d\theta}{d\omega}\right)^2 + \frac{d\log p(x \mid \theta)}{d\theta}\cdot\frac{d^2\theta}{d\omega^2}.$$

So

$$\begin{aligned} I(\omega) &= -E_{x|\omega}\left[\frac{d^2}{d\omega^2}\log p(x \mid \omega)\right] \\ &= -E_{x|\omega}\left[\frac{d^2\log p(x \mid \omega)}{d\theta^2}\cdot\left(\frac{d\theta}{d\omega}\right)^2 + \frac{d\log p(x \mid \theta)}{d\theta}\cdot\frac{d^2\theta}{d\omega^2}\right] \\ &= -E_{x|\omega}\left[\frac{d^2\log p(x \mid \omega)}{d\theta^2}\right]\cdot\left(\frac{d\theta}{d\omega}\right)^2 - E_{x|\omega}\left[\frac{d\log p(x \mid \theta)}{d\theta}\right]\cdot\frac{d^2\theta}{d\omega^2}. \end{aligned}$$

Practical Applications of Bayesian Reliability, First Edition. Yan Liu and Athula I. Abeyratne.

Companion website: www.wiley.com/go/bayesian20

Since the expectation of the score statistic is zero, which is

$$E_{x|\omega}\left[\frac{d\log p(x\mid\theta)}{d\theta}\right]=0,$$

then

$$I(\omega)=-E_{x|\omega}\left[\frac{d^2}{d\omega^2}\log p(x\mid\omega)\right]=-E_{x|\theta}\left[\frac{d^2}{d\omega^2}\log p(x\mid\theta)\right]$$

$$\cdot\left(\frac{d\theta}{d\omega}\right)^2=I(\theta)\cdot\left(\frac{d\theta}{d\omega}\right)^2.$$

So

$$[I(\omega)]^{1/2}=[I(\theta)]^{1/2}\left|\frac{d\theta}{d\omega}\right|,$$

i.e. the Jeffreys prior is invariant under reparameterization.

Example D.1

When the likelihood function is a binomial distribution $x \sim Binomial\,(n,\theta)$, the Jeffreys prior of θ is $Beta\,(0.5,\ 0.5)$.

The Jeffreys prior is given by

$$p(\theta)\propto[I(\theta)]^{1/2},$$

where $I(\theta)$ is the expected Fisher Information Matrix, i.e.

$$I(\theta)=-E_{X|\theta}\left[\frac{d^2}{d\theta^2}\log f(x\mid\theta)\right].$$

A binomial likelihood function is

$$f(x\mid n,\theta)=\binom{n}{x}\theta^x(1-\theta)^{n-x},$$

where n is a constant. Then,

$$\log f(x\mid\theta)=\log\binom{n}{x}+x\log\theta+(n-x)\log(1-\theta)$$

$$\frac{d}{d\theta}\log f(x\mid\theta)=\frac{x}{\theta}-(n-x)\frac{1}{1-\theta}$$

$$\frac{d}{d\theta^2}\log f(x\mid\theta)=-\frac{x}{\theta^2}-\frac{n-x}{(1-\theta)^2}.$$

As $E(x)=n\theta$,

$$I(\theta)=-E_{x|\theta}\left[\frac{d^2}{d\theta^2}\log f(x\mid\theta)\right]=-E\left(-\frac{x}{\theta^2}-\frac{n-x}{(1-\theta)^2}\right)=\frac{n}{\theta(1-\theta)}$$

$$[I(\theta)]^{1/2}=n^{1/2}\theta^{-1/2}(1-\theta)^{-1/2}.$$

So the Jeffreys prior based on the binomial likelihood is

$$p(\theta)\propto[I(\theta)]^{1/2}\propto\theta^{-1/2}(1-\theta)^{-1/2}.$$

Since the beta distribution is

$$f(x \mid \alpha, \beta) = \frac{x^{\alpha-1}(1-x)^{\beta-1}}{B(\alpha, \beta)},$$

The Jeffreys prior based on the binomial likelihood is given by *Beta* (0.5, 0.5),

$$Beta\ \ (0.5,\ 0.5) = \frac{\theta^{-1/2}(1-\theta)^{-1/2}}{B(0.5,\ 0.5)}.$$

Index

Practical Applications of Bayesian Reliability, First Edition. Yan Liu and Athula I. Abeyratne.

Companion website: www.wiley.com/go/bayesian20